Single- and Multi-Carrier MIMO Transmission for Broadband Wireless Systems

RIVER PUBLISHERS SERIES IN COMMUNICAITONS

Volume 5

Consulting Series Editor

Marina Ruggieri
University of Roma Tor Vergata
Italy

Other books in this series:

Volume 1
4G Mobile & Wireless Communications Technologies
Sofoklis Kyriazakos, Ioannis Soldatos, George Karetsos
September 2008
ISBN: 978-87-92329-02-8

Volume 2
Advances in Broadband Communication and Networks
Johnson I. Agbinya, Oya Sevimli, Sara All, Selvakennedy Selvadurai,
Adel Al-Jumaily, Yonghui Li, Sam Reisenfeld
October 2008
ISBN: 978-87-92329-00-4

Volume 3
*Aerospace Technologies and Applications for Dual Use A New World of
Defense and Commercial in 21st Century Security*
General Pietro Finocchio, Ramjee Prasad, Marina Ruggieri
November 2008
ISBN: 978-87-92329-04-2

Volume 4
*Ultra Wideband Demystified Technologies, Applications, and System Design
Considerations*
Sunil Jogi, Manoj Choudhary
January 2009
ISBN: 978-87-92329-14-1

Single- and Multi-Carrier MIMO Transmission for Broadband Wireless Systems

Ramjee Prasad

CTIF, Aalborg University
Denmark

Muhammad Imadur Rahman

Ericsson Research, Kista
Sweden

Suvra Sekhar Das

Indian Institute of Technology, Kharagpur
India

Nicola Marchetti

CTIF, Aalborg University
Denmark

Routledge
Taylor & Francis Group

LONDON AND NEW YORK

River Publishers

Published 2009 by River Publishers
River Publishers
Alsbjergvej 10, 9260 Gistrup, Denmark
www.riverpublishers.com

Distributed exclusively by Routledge
4 Park Square, Milton Park, Abingdon, Oxon OX14 4RN
605 Third Avenue, New York, NY 10158

First published in paperback 2024

Single- And Multi-Carrier Mimo Transmission for Broadband Wireless Systems / by Ramjee Prasad, Muhammad Imadur Rahman, Suvra Sekhar Das, Nicola Marchetti.

Routledge is an imprint of the Taylor & Francis Group, an informa business

Publisher's Note
The publisher has gone to great lengths to ensure the quality of this reprint but points out that some imperfections in the original copies may be apparent.

While every effort is made to provide dependable information, the publisher, authors, and editors cannot be held responsible for any errors or omissions.

ISBN: 978-87-92329-06-6 (hbk)
ISBN: 978-87-7004-558-2 (pbk)
ISBN: 978-1-003-33953-3 (ebk)

DOI: 10.1201/9781003339533

Dedication

Ramjee Prasad would like to dedicate this book to all his students.

Muhammad Imadur Rahman would like to dedicate this book to his lovely wife, Maizura Ailin, who has always been there through all the ups and downs.

Suvra Sekhar Das would like to dedicate this book to his Parents, his sister, and his wife.

Nicola Marchetti would like to dedicate this book to his parents, Elvino and Pierina, whose hard work, honesty and morality have always been and will always be examples for him.

Acknowledgments

We would like to thank River Publishers for giving us the opportunity to write the book and for the support at all stages of the work. We would specially thank Elisabeth Jansen of River Publishers who was a great help in different stages of this project. It was indeed a challenge from both sides as for three of us this was the first book, and many new things needed to be mastered to come up with the final manuscript.

A special thank to the people that, directly or indirectly, contributed to the research work that is included in the book: Dr Ernestina Cianca, Dr Elisabeth de Carvalho, Dr Daniel V. P. Figueiredo, Yuanye Wang, Faisal Tariq, Nurul Huda Mahmood (Fuad) and Dua Idris, among others. Special thanks to a number of other colleagues at Center for TeleInFrastruktur (CTIF), Aalborg University, Denmark, who have always contributed to the discussions and knowledge sharing while performing the research work.

Muhammad Imadur Rahman, Suvra Sekhar Das, and Nicola Marchetti would like to express their heartiest thanks to Prof. Ramjee Prasad for being the supervisor, mentor, and above all a motivator for them throughout their research and academic works in Aalborg. They have done most of the work included in this book while they were his PhD students in Aalborg University. It was indeed a very pleasant association for them that they would appreciate for long time to come.

We would like to thank beforehand the readers who will provide us feedback on the technical and editorial content of the book, that will be taken into account for future versions.

Finally, we would like to thank CTIF for providing us the excellent research and academic environment where innovative and dynamic work culture has taught us a lot. We are indeed proud to be associated with such an esteemed organization.

Preface

ज्ञानं ज्ञेयं परिज्ञाता त्रिविधः कर्मचोदना ।
करणं कर्म कर्तेति त्रिविधः कर्मसंग्रहः ॥१८॥

jñānarth jñeyam parijñātā
tri-vidhā karma-codanā
karanam karma karteti
tri-vidhah karma-sangrahah

Knowledge, the object of knowledge, and the knowledgeable person are the initiators of the action. The sense powers, the action and the doer are the three tools of the action.

—The Bhagvad Gita (18.18)

Wireless communication has seen a tremendous growth in recent years. New technologies and new systems have been developed in the last two or three decades which have fundamentally changed the wireless scenario. Multi-carrier transmissions has been one of the main transmission techniques for many recently developed systems. Along with it, modified single-carrier transmission which takes advantage of some of the multi-carrier transmissions have also become very popular. In parallel, all the recent systems exploit the gains obtained by the presence of multiple antennas at one or both of sides of the transmission link.

Thus, the main focus of this book is to provide the basic understanding of the underlying techniques related to PHY–MAC design of future wireless systems which will be based on the above mentioned multi-carrier systems. It includes basic concepts related to single- and multi-carrier transmissions together with multi-antenna techniques. Discussions related to different recent standards that use single- and multi-carrier transmissions is also included in this book.

The book provides a comprehensive and holistic approach to the variety of technical solutions which have been till date dealt individually in the above mentioned books. Future system design would require these different technologies to work together, and not independently. Therefore, it is very important to analyze the effects and gains when they are put together in a unified platform. This is the prime focus of this book. Moreover, we include very recent research results which are not yet available in any existing book.

This book is of text book style. So, it can be used as a material for lectures in undergraduate and graduate level courses in universities. PhD level students will also find it useful as this book outlines the fundamental concepts and design methods for PHY and MAC layers of future wireless systems. This book can also be used as a handbook by engineers and developers in the industry as well as by researchers in academia. For professionals, system architects and managers who play a key role in the selection of a baseline system concept for future wireless standards, such as IMT-Advanced type architecture, we also include discussions, analysis and guidelines to highlight overall system level perspective.

In the introductory chapter (i.e Chapter 1), general backgrounds related to evolution of wireless systems will be given. The goal is to inform the reader about the logical development path of different wireless standards and systems from different perspective, e.g. technical, economical, user expectation, etc. Besides, an introductory chapter on the characteristics of wireless channels is included in Chapter 2. In the remaining chapters of the book, we concentrate on exploiting different properties of wireless channel, thus, this chapter will form the basic knowledge of the whole subject matter.

We present multi-carrier transmission fundamentals in Chapter 3. The benefits of using multi-carrier transmissions are described in terms of mitigation of channel frequency selectively. Impact of several PHY layer components, such as channel delay spread, guard interval length, channel coding, interleaving etc., are detailed in this chapter. Also, different design issues for multi-carrier systems are presented.

After this, several fundamental transmission techniques related to wireless communications systems are explained. The basic concepts related to access techniques are described in Chapter 4. Here, the benefits of Time Division

Multiple Access (TDMA), Code Division Multiple Access (CDMA), and Frequency Division Multiple Access (FDMA) over each other and the usability of these techniques compared to different system design requirements are presented. Based on this understanding, the access technique concepts and multi-carrier fundamentals from Chapter 3 are combined to explain multi-carrier based multiple access techniques. Several combinations of basic techniques, such as FDMA, TDMA, and CDMA, together with multi-carrier transmission are detailed in this chapter. The advantages of "inter-marriage" between different access techniques above a multi-carrier transmission in handled in this chapter.

Recent studies have shown that the main issue is not whether one goes for a multi-carrier or a single-carrier transmission, but whether one performs the equalization in time or in frequency domain. Several techniques using frequency-domain equalization, i.e. Orthogonal Frequency Division Multiplexing (OFDM), Single Carrier-Frequency Domain Equalization (SCFDE) and Single Carrier-Frequency Division Multiple Access (SC-FDMA) will be discussed and compared in Chapter 5. Synergies and potential interoperation of single-carrier and multi-carrier techniques with frequency-domain equalization will be also outlined.

The access techniques discussed in the previous part are very promising in providing robust communication system under different conditions. Among the many schemes OFDM is one of the prime concerns in this book. It is explained that OFDM can support very high spectral efficiency, however it is highly vulnerable to synchronization and channel estimation errors as well as imperfections due to nonlinear behavior of the high power amplifier.

In Chapter 6, the impact of time and frequency synchronization issues are discussed. Time synchronization includes packet detection, timing window synchronization, and sampling clock synchronization. Algorithms to compensate for the effects of time synchronization errors are presented for each kind of error. The influence of frequency synchronization errors which arise from carrier frequency offset and Doppler frequency spread are presented in this chapter. The effect of inter carrier interference is described and techniques to overcome these effects are presented.

Chapter 7 discusses the effects of channel estimation error on the performance of OFDM systems. Several channel estimation and equalization techniques are also presented in the chapter. One dimensional as well as two dimensional channel estimation techniques have been explained. These techniques can be applied to 3GPP LTE based OFDMA systems. Channel estimation using training sequence and decision directed methods have been described in details. Differential detection schemes have been included finally.

Chapter 8 presents the effect of high power amplifier nonlinearity on the performance of OFDM systems. OFDM systems are known to suffer from high peak to average power ratio. This causes severe degradation in performance. A proper understanding of the underlying phenomenon is necessary for a proper system design. Analysis of the effect of HPA nonlinear distortions includes, PAPR statistics, signal to distortion plus noise ratio, total degradation, spectral broadening. Techniques to find the optimal operating point considering several trade-offs is also presented.

After this, combination of multi-antenna benefits above any multi-carrier system is explained. Multiple antennas at any side of the communication link can be exploited to obtain diversity gain, array gain, multiplexing benefits, and interference cancelation. In Chapter 9, basic techniques to obtain the above mentioned benefits are detailed. The presence of channel information at the transmitter or at the receiver or at both locations heavily impacts the resultant system capacity, which is also explained in this chapter. Presence of channel information at the transmitter will also be explained in terms of different possibilities of signaling from multi-antenna transmitter. In this case, closed loop diversity and multiplexing schemes are discussed, and their comparative gains compared to any blind diversity and multiplexing schemes are presented.

Exploitation of multiple transmitter antennas in terms of beamforming (which obtains array gain) and transmit diversity (which obtains diversity gain) are described in Chapter 10. Issues related to the usage of Transmit Beamforming (TxBF) and transmit diversity schemes are studied in this chapter. Special attention is given to outage probability for certain outage capacity of the systems. From this point onwards, we have performed numerical evaluations to compare the usability of diversity and array gain in indoor and outdoor scenarios.

In Chapter 11, we discuss the exploitation of Cyclic Delay Diversity (CDD) technique in OFDM system. First, we have explained the CDD principle, followed by implementing CDD in OFDM receiver. A new technique, Pre-DFT Maximal Average Ratio Combining (MARC), is proposed that can be used in low delay spread environment to obtain receive diversity by utilizing Pre-DFT combining techniques. A performance comparison is presented in terms of BER with and without coding, between MRC for receiver diversity and CDD for transmitter and receiver diversity in the context of IEEE 802.11a and/or HiperLAN/2 WLAN systems. We have also described the implementation of CDD on top of a Spatial Multiplexing (SM) system, and studied the so-called Cyclic Delay Assisted Spatial Multiplexing (CDA-SM) system in this chapter.

In Chapter 12, different joint schemes where both diversity and multiplexing benefits are exploited at the same time have been introduced and studied. We call such schemes Joint Diversity and Multiplexing (JDM). We have analyzed the performance of different receiver strategies in realistic wireless conditions. JDM schemes are in general robust to spatial correlation caused by the inadequate spatial separation between antenna elements. The impact of spatial correlation caused by the LOS scenario on the proposed schemes is also studied. It is clear JDM schemes are a suitable solution when SM is used at any cellular access point, as JDM schemes in effect increase the coverage in terms of radius where SM can be supported. Thus, it can be said that JDM schemes can be effectively used to increase the system throughput.

In Chapter 13, several space–frequency coding schemes required to achieve any desired amount of the possible trade-off diversity-multiplexing in SCFDE systems are introduced and tested, comparing them with analogous space–time schemes in SCFDE and space–time and space–frequency schemes in OFDM systems. The versions for SCFDE of Alamouti and Jafarkhani codes, JDM schemes and Linear Dispersion Codes (LDC) are introduced. The impact of wireless channel impairments such as time and frequency selectivity on the proposed coding schemes is also tested. The computational complexity of the proposed schemes is discussed. Some considerations about multiple antenna techniques for SC-FDMA conclude the chapter.

We conclude the book in Chapter 14. Several current and future important research issues are spot and discussed, starting from cross layer design and optimization. Afterwards some considerations are made regarding the single-carrier vs multi-carrier and Frequency Domain Equalization (FDE) vs Time Domain Equalization (TDE) debate, multi-antenna issues, frequency overlay, radio resource management and packet scheduling, and cooperative communication. At last, one tries to identify what could be the future trends of wireless communications, i.e. software-defined and cognitive radio, spectrum sharing and flexible spectrum usage, self-organizing networks, and low power-emission/green systems.

Ramjee Prasad, CTIF, Aalborg University, Denmark
Muhammad Imadur Rahman, Ericsson Research, Kista, Sweden
Suvra Sekhar Das, Indian Institute of Technology, Kharagpur, India
Nicola Marchetti, CTIF, Aalborg University, Denmark

Contents

List of Figures   **xxiii**

List of Tables   **xxxv**

1   Introduction   **1**

1.1   History of Wireless Communications   1
1.2   Spectrum Issues   2
1.3   Drivers for Future Wireless Market   3
1.4   Forthcoming 4G Systems   3
1.5   Technical Challenges that Need to be Addressed for Future
      System Design   8

2   Wireless Channel Modeling   **15**

2.1   Wireless Channel   15
      2.1.1   Channel Parametrization   16
      2.1.2   Propagation Loss   18
      2.1.3   Shadowing   19
      2.1.4   Small Scale Fading   20
2.2   Quadrature Amplitude Modulation   26
      2.2.1   Probability of Error in QAM   28

3   Multi-Carrier Fundamentals   **31**

3.1   History and Development of OFDM   32
3.2   The Benefit of Using Multi-Carrier Transmission   33
3.3   OFDM Transceiver Systems   36
3.4   Channel Coding and Interleaving   39

3.5	Analytical Model of OFDM System	40
	3.5.1 Transmitter	40
	3.5.2 Channel	42
	3.5.3 Receiver	43
	3.5.4 Sampling	47
3.6	Single OFDM Symbol Baseband Model in Matrix Notations	47
3.7	Advantages of OFDM System	50
	3.7.1 Combating ISI and Reducing ICI	50
	3.7.2 Spectral Efficiency	53
	3.7.3 Some Other Benefits of OFDM System	54
3.8	Disadvantages of OFDM System	55
	3.8.1 Strict Synchronization Requirement	55
	3.8.2 Peak-to-Average Power Ratio (PAPR)	55
	3.8.3 Co-Channel Interference in Cellular OFDM	56
3.9	OFDM System Design Issues	57
	3.9.1 OFDM System Design Requirements	57
	3.9.2 OFDM System Design Parameters	59
4	**Multi-Carrier Based Access Techniques**	**61**
4.1	Definition of Basic Schemes	61
	4.1.1 OFDM–TDMA	61
	4.1.2 OFDMA	62
	4.1.3 OFDM–CDMA	63
	4.1.4 Relative Comparison	63
4.2	Orthogonal Frequency Division Multiple Access	65
	4.2.1 Multiple Access Model	65
	4.2.2 Static and Dynamic Sub-carrier Assignment	67
	4.2.3 Matrix Description	69
	4.2.4 Transceiver Architecture	72
	4.2.5 Specific Features	74
	4.2.6 OFDMA Based-Standards	75
	4.2.7 Performance Metric	76
	4.2.8 Error Probability Analysis in OFDMA System	77
4.3	Orthogonal Frequency Division Multiple Access-Fast Sub Carrier Hopping	78

	4.3.1	Multiple Access Model	79
	4.3.2	Benefit from Using Sub-carrier Hopping	79
	4.3.3	Main Idea	81
	4.3.4	Matrix Description	81
	4.3.5	Transceiver Architecture	83
	4.3.6	Specific Features and Further Research Topics	84
	4.3.7	Performance Metric	84
	4.3.8	OFDMA-FSCH Based Standards	85
4.4	Orthogonal Frequency Division Multiple Access-Slow Sub-Carrier Hopping		87
	4.4.1	Multi Access Model	89
	4.4.2	Matrix Description	90
	4.4.3	Transceiver Architecture	91
	4.4.4	Specific Features	91
	4.4.5	Performance Metric	92
	4.4.6	OFDMA-SSCH Based-Standards	93
4.5	Multi-Carrier Code Division Multiple Access		93
	4.5.1	Multiple Access Model	94
	4.5.2	Matrix Description	95
	4.5.3	Transceiver Architecture	98
	4.5.4	Specific Features	105
	4.5.5	Performance Metric	105
5	**Single-Carrier Transmission with Cyclic Prefix**		**109**
5.1	Single-Carrier FDE		109
	5.1.1	Single-Carrier vs Multi-Carrier, FDE vs TDE	111
	5.1.2	Analogies and Differences Between OFDM and SCFDE	112
	5.1.3	Interoperability of SCFDE and OFDM	114
5.2	Single Carrier FDMA		116
	5.2.1	Transceiver Architecture	116
	5.2.2	Sub-carrier Mapping	119
	5.2.3	Relation Between SC-FDMA, OFDMA, and DS-CDMA/FDE	121
5.3	Chapter Summary		124

6   Synchronization in Time and Frequency Domain   **127**

6.1   Introduction   128
6.2   Sensitivity to Phase Noise   129
6.3   Sensitivity to Frequency Offset   132
6.4   Sensitivity to Timing Errors   134
6.5   Synchronization Using Cyclic Extension   136
6.6   Synchronization Using Special Training Symbols   142
6.7   Optimal Training in Presence of Multi-path   147

7   Channel Estimation and Equalization   **151**

7.1   Introduction   151
7.2   Coherent Detection   151
       7.2.1   Two-Dimensional Channel Estimators   152
       7.2.2   One-Dimensional Channel Estimators   156
       7.2.3   Special Training Symbols   157
       7.2.4   Decision-Directed Channel Estimation   161
7.3   Differential Detection   162
       7.3.1   Differential Detection in the Time Domain   162
       7.3.2   Differential Detection in the Frequency Domain   167
       7.3.3   Differential Amplitude and Phase Shift Keying   171

8   High Power Amplifier and PAPR in OFDM   **173**

8.1   Introduction   173
8.2   HPA Models   173
       8.2.1   TWTA Model   174
       8.2.2   SSPA Model   174
8.3   PAPR in OFDM   177
       8.3.1   CDF of PAPR   177
8.4   Effect of HPA and BO Power   180
       8.4.1   Effect on Constellation Points   180
       8.4.2   Effect on Power Spectrum   182
       8.4.3   SDNR Plot   183
8.5   Performance of Different Modulation and Coding   187
       8.5.1   Performance in AWGN Channel   188
       8.5.2   Performance in Fading Channel   194
8.6   Conclusion   203

9 Multi-antenna Gains **205**

9.1 Gains Obtained by Exploiting the Spatial Domain 205
    9.1.1 Array Gain 206
    9.1.2 Diversity Gain 207
    9.1.3 Multiplexing Gain 207
    9.1.4 Interference Reduction 208
9.2 Multi-antenna and Diversity 209
    9.2.1 Time Diversity 209
    9.2.2 Frequency Diversity 210
    9.2.3 Space Diversity 210
9.3 Multi-antenna and Spatial Multiplexing 211
9.4 Diversity Gain vs Coding Gain 213
9.5 Capacity of MIMO Channels 216
    9.5.1 Fundamentals on Channel Capacity 216
    9.5.2 MIMO Channel Capacity: Information Theoretic
         Approach 217
    9.5.3 Limiting Capacity Results 219
9.6 Trade-Off Between Spatial Multiplexing and
    Spatial Diversity 220
9.7 Multi-antenna in OFDM 221
    9.7.1 Space Diversity 223
    9.7.2 Spatial Multiplexing 224
    9.7.3 Beam-Forming 225
    9.7.4 Usability of Multi-antenna Techniques in
         OFDM Systems 225

10 Transmit Diversity Vs Beamforming **233**

10.1 Introduction 233
10.2 A Brief Look at Diversity and Beamforming 235
    10.2.1 Beamforming 235
    10.2.2 Space–Time Block Coding (STBC) 239
    10.2.3 Receive Diversity System 241
    10.2.4 MIMO Diversity System 242
    10.2.5 SNR Statistics of Diversity and Beamforming
         Systems 242

10.3 Downlink Capacity and Error Probability Analysis   243
    10.3.1 Ergodic Capacity   243
    10.3.2 Outage Capacity   246
    10.3.3 Error Probability   248
10.4 Downlink Beamforming and Transmit Diversity in Multi-user OFDM Systems   251
    10.4.1 Issues in OFDM–TDMA   251
    10.4.2 DL-BF in OFDMA   252
    10.4.3 DL-BF in Clustered OFDMA   253
10.5 Performance Analysis and Comparison   254
    10.5.1 Channel Model   254
    10.5.2 Angular Spread and Spatial Correlation   254
    10.5.3 Simulation Parameters   256
    10.5.4 BER Results and Discussions   258
    10.5.5 Pilot Design Issue   261
10.6 Chapter Summary   262

11 Exploiting Cyclic Delay Diversity in OFDM System   265

11.1 Introduction   265
11.2 OFDM System Model   269
11.3 Post-DFT Maximum Ratio Combining   270
11.4 Benefitting from Cyclic Delay Property in OFDM System   271
    11.4.1 System Model with Cyclic Delay Diversity   273
    11.4.2 Capacity of CDD Based OFDM System   275
11.5 Pre-DFT Maximum Average Ratio Combining   277
    11.5.1 Optimum SNR for the Combined Signal   278
    11.5.2 Optimum Diversity Weights   280
    11.5.3 System Analysis with Dual Antenna Receiver   281
    11.5.4 Numerical Analysis and Discussions   285
11.6 Cyclic Delay Assisted Spatial Multiplexing   292
    11.6.1 Transmission Structure   293
    11.6.2 System Capacity of CDA-SM-OFDM   296
11.7 Chapter Summary   300

12 Joint Diversity and Multiplexing Schemes for MIMO-OFDM Systems 303

12.1 Introduction 303
12.2 System Model 307
 12.2.1 SM-OSFBC Transmission Scheme 307
 12.2.2 SM-QSFBC-OFDM Transmission Scheme 315
12.3 SNR Distribution for ZF Detection 318
12.4 Numerical Results 319
 12.4.1 System Parameters 319
 12.4.2 FER Performance 321
 12.4.3 10% Outage Spectral Efficiency 322
 12.4.4 Effect of Spatial Correlation 323
 12.4.5 Performance in Presence of LOS 326
12.5 Chapter Summary 327

13 MIMO Design in SC-FDE/SC-FDMA Systems: Diversity and Multiplexing 329

13.1 Space Diversity 330
13.2 Spatial Multiplexing 332
13.3 Space–Frequency Block Coding for SCFDE 333
 13.3.1 System Model 334
 13.3.2 Simulations and Discussions 338
 13.3.3 Computational Complexity 341
13.4 Combining Diversity and Multiplexing in SCFDE 341
 13.4.1 System Model 342
 13.4.2 Simulations and Discussions 345
 13.4.3 Conclusions 353
13.5 Linear Dispersion Codes for SCFDE 353
 13.5.1 Linear Dispersion Codes 354
 13.5.2 Space–Frequency LDC for SCFDE 355
 13.5.3 SD, SM, and JDM as Special Cases of LDC 372
 13.5.4 Conclusions 373
13.6 Multi-antenna in SC-FDMA 373
13.7 Chapter Summary 375

14 Conclusions and Perspectives   **377**

14.1 Cross-layer Design and Optimization   378
    14.1.1 Cross-layer Opportunities   381
    14.1.2 Cross-layer Design Related to System Integration   382
14.2 Technological Aspects: Future Perspective   383
    14.2.1 Single-Carrier vs Multi-Carrier and Frequency-Domain
           vs Time-Domain Equalization   384
    14.2.2 Multi-antenna Issues   385
    14.2.3 Frequency Overlay   386
    14.2.4 Radio Resource Management and Packet Scheduling   387
    14.2.5 Cooperative Communication   388
14.3 Future of Wireless Systems   389
    14.3.1 Software Defined Radio   389
    14.3.2 Cognitive Radio   390
    14.3.3 Spectrum Sharing   390
    14.3.4 Self-organizing Networks   391
    14.3.5 Low Emission — Green Systems   392

Abbreviations   **393**

References   **399**

Index   **415**

List of Figures

Figure 1.1   Time–frequency OFDM granularity that can be exploited for increasing system spectral efficiency.   10

Figure 1.2   Cellular interference scenario.   13

Figure 2.1   Characterization of fading.   16

Figure 2.2   Propagation loss.   17

Figure 2.3   Amplitude response.   20

Figure 2.4   Channel impulse response and transfer function relationship.   21

Figure 2.5   Multi-path propagation.   22

Figure 2.6   A typical power delay profile.   22

Figure 2.7   Frequency domain channel response.   24

Figure 2.8   Power spectral density vs frequency of Jakes' spectrum.   26

Figure 2.9   Power spectral density vs frequency of typical Gauss' spectrum.   26

Figure 2.10   Signal space diagram for rectangular 64-QAM.   27

Figure 2.11   Probability of symbol error for QAM.   28

Figure 3.1   Single carrier vs multi-carrier approach.   35

Figure 3.2   OFDM transceiver model.   38

Figure 3.3   Transmitter diagram for the OFDM analytical model, given by (3.1)–(3.8).   42

Figure 3.4   A diagram of the channel model given by (3.9) and (3.10). The transmitted signal passes through the channel, and noise is added.   43

Figure 3.5   Receiver diagram for the OFDM analytical model, given by (3.12)–(3.21).   46

Figure 3.6   Single OFDM symbol system model.   48

Figure 3.7   Simplified single OFDM symbol system model.   50

Figure 3.8   Role of guard intervals in combatting ISI and ICI.   51

Figure 3.9   Definition of cyclic prefix as the guard interval in
              OFDM systems.   52

Figure 3.10   Spectrum efficiency of OFDM compared to conven-
              tional FDM.   53

Figure 3.11   Design of sub-carrier spacing in OFDM systems.   57

Figure 3.12   Design of CP duration in OFDM systems.   58

Figure 4.1   Relative comparison of basic multi-carrier multiple-
              access techniques   64

Figure 4.2   Time–frequency diagram for the OFDMA scheme. For
              simplicity, one sub-carrier is assigned to each user. The
              vertical blocks on the time axis represents OFDM symbols. 66

Figure 4.3   Simplified block diagram of the OFDMA transmitter,
              receiver schemes.   66

Figure 4.4   (a) Contiguous sub-carrier assignment. (b) Interleaved
              sub-carrier assignment.   67

Figure 4.5   Examples of sub-channel conditions where users
              receive either good or bad OFDM symbols, different
              patterns are illustrated. (G = good OFDM symbol,
              B = bad OFDM symbol.)   68

Figure 4.6   Single OFDM symbol model for the OFDMA.   70

Figure 4.7   Block diagram of an OFDMA receiver architecture.   74

Figure 4.8   General block diagram for OFDMA-FSCH transceiver.   79

Figure 4.9   Time–frequency diagram for OFDMA-FSCH scheme
              at the transmitter.   81

Figure 4.10   Time–frequency diagram for OFDMA-FSCH scheme
              at the receiver.   82

Figure 4.11   OFDMA-FSCH transmitter model for DL.   83

Figure 4.12   OFDMA-FSCH receiver model at the MS_u.   83

Figure 4.13   OFDMA-SSCH general block diagram —
              Transmitter: from the BS.   88

Figure 4.14   OFDMA-SSCH general block diagram — Receiver:
              at the MS of uth user.   88

Figure 4.15   OFDMA-SSCH F-T diagram, transmitter.   89

Figure 4.16   OFDMA-SSCH F-T diagram, receiver.   90

Figure 4.17   Downlink MC-CDMA transmitter.   95

Figure 4.18   Time–frequency energy diagram for MC-CDMA.   96
Figure 4.19   MC-CDMA resource mapping.   99
Figure 4.20   An ideal MC-CDMA receiver in downlink.   100
Figure 4.21   MC-CDMA resource demapping.   103
Figure 5.1   SCFDE with linear FDE.   110
Figure 5.2   Block processing in FDE.   114
Figure 5.3   OFDM and SCFDE signal processing similarities and
              differences.   115
Figure 5.4   Potential interoperability of SCFDE and OFDM: a
              "convertible" modem.   115
Figure 5.5   Coexistence of SCFDE and OFDM: uplink/downlink
              asymmetry.   116
Figure 5.6   Transceiver structure of SC-FDMA system.   117
Figure 5.7   Transceiver structure of OFDMA system.   117
Figure 5.8   SC-FDMA transmit symbols generation.   119
Figure 5.9   Sub-carrier allocation methods for the case of 6 sub-
              carriers, 3 users and 2 sub-carriers per user.   120
Figure 5.10   Example of SC-FDMA transmit symbols in frequency
              domain for $N = 2$ sub-carriers/users, $Q = 3$ users, and
              $M = 6$ sub-carriers in the system. Both distributed and
              localized sub-carriers
              allocations are shown.   121
Figure 5.11   Example of SC-FDMA transmit symbols in time
              domain for $N = 2$ sub-carriers/users, $Q = 3$ users, and
              $M = 6$ sub-carriers in the system. Both cases corre-
              spondent to distributed and localized sub-carriers allo-
              cations are shown.   122
Figure 5.12   Detection process dissimilarities between OFDMA
              and SC-FDMA.   123
Figure 5.13   Modulated symbol durations dissimilarities between
              OFDMA and SC-FDMA.   123
Figure 5.14   Transceiver chain of DS-CDMA with FDE.   124
Figure 5.15   Conventional spreading; spreading signature of $(1,1)$
              and block size of 2.   125
Figure 5.16   Exchanged spreading; spreading signature of $(1,1)$ and
              block size of 2.   125

Figure 6.1   Phase noise power spectral density (PSD) with a single-sided -3 dB linewidth of 1 Hz and a -100 dBc/Hz density at 100 kHz offset.   130

Figure 6.2   SNR degradation in dB vs the -3 dB bandwidth of the phase noise spectrum for (a) 64-QAM $\left(\frac{E_s}{N_o} = 19 \text{ dB}\right)$, (b) 16-QAM $\left(\frac{E_s}{N_o} = 14.5 \text{ dB}\right)$, (c) QPSK $\left(\frac{E_s}{N_o} = 10.5 \text{ dB}\right)$.   131

Figure 6.3   Example of a PLL phase noise spectrum.   132

Figure 6.4   SNR degradation in dB vs the normalized frequency offset for (a) 64-QAM $\left(\frac{E_s}{N_o} = 19 \text{ dB}\right)$, (b) 16-QAM $\left(\frac{E_s}{N_o} = 14.5 \text{ dB}\right)$, (c) QPSK $\left(\frac{E_s}{N_o} = 10.5 \text{ dB}\right)$.   133

Figure 6.5   Example of an OFDM signal with three sub-carriers, showing the earliest and latest possible symbol timing instants that do not cause ISI or ICI.   135

Figure 6.6   Constellation diagram with a timing error of $T/16$ before phase correction.   135

Figure 6.7   Constellation diagram with a timing error of $T/16$ and after phase correction.   136

Figure 6.8   Synchronization using the cyclic prefix.   137

Figure 6.9   Example of correlation output amplitude for eight OFDM symbols with 192 sub-carriers and a 20% guard time.   137

Figure 6.10 Example of correlation output amplitude for eight OFDM symbols with 48 sub-carriers and a 20% guard time.   138

Figure 6.11 Vector representation of phase drift estimation.   140

Figure 6.12 Frequency estimation error normalized to the sub-carrier spacing. Solid lines are calculated; dotted lines are simulated. (a) $T_G/T_s = 1$, (b) $T_G/T_s = 0.2$, (c) $T_G/T_s = 0.1$.   141

Figure 6.13 Matched filter that is matched to a special OFDM training symbol.   143

Figure 6.14 Matched filter output vs sample number for 4 training symbols, using 48 sub-carriers and 64 samples per symbol.   144

Figure 6.15 Matched filter output vs sample number with $\{1, -1, 0\}$ values for in-phase and quadrature coefficients.   146

Figure 6.16   Raised cosine window.   147

Figure 6.17   ISI/ICI caused by multi-path signals.   148

Figure 6.18   Example of channel impulse response.   148

Figure 6.19   OFDM symbol structure.   149

Figure 7.1   Block diagram of an OFDM receiver with coherent
              detection.   152

Figure 7.2   Example of pilots (marked gray) in a block of 9 OFDM
              symbols with 16 sub-carriers.   153

Figure 7.3   Example of relative channel estimation errors vs sub-
              carrier number and symbol number.   158

Figure 7.4   Channel estimation with separable filters in frequency
              (1) and time (2) direction.   159

Figure 7.5   Example of a packet with two training symbols for
              channel estimation and two pilot sub-carriers used for
              frequency synchronization.   160

Figure 7.6   Extended training symbol for channel estimation with
              a single guard time and multiple IFFT intervals.   161

Figure 7.7   Block diagram of an OFDM receiver with differential
              detection.   162

Figure 7.8   Differential detection in the time domain. Gray sub-
              carriers are pilots that are needed as initial phase
              references.   163

Figure 7.9   Correlation between symbols vs the normalized time
              distance $f_{max}T_s$.   164

Figure 7.10   Correlation between symbols vs the normalized time
              distance $f_{max}T_s$, zoom-in of Figure 7.9.   166

Figure 7.11   Differential detection in the frequency domain. Gray
              sub-carriers are pilots.   168

Figure 7.12   Correlation between sub-carriers vs the normalized
              sub-carrier spacing τ_{rms}/T.   169

Figure 7.13 Irreducible packet error ratio vs rms delay spread, simulated for an exponentially decaying power delay profile with Rayleigh fading paths. (a) 16 Mbps with coherent QPSK and rate 1/2 coding, (b) 32 Mbps with coherent 16-QAM and rate 1/2 coding, (c) 16 Mbps with differential QPSK (in frequency domain) and rate 1/2 coding, (d) 24 Mbps with differential 8-PSK (in frequency domain) and rate 1/2 coding, (e) 32 Mbps with differential 8-PSK (in frequency domain) and rate 2/3 coding.   170

Figure 7.14 64-DAPSK constellation.   172

Figure 8.1 Transfer function and of Rapp's model with BO scenario.   175

Figure 8.2 Relation between amplifier distortion and BO power.   176

Figure 8.3 Comparison of theoretical and simulated CDF of PAPR.   178

Figure 8.4 Effect of different modulation scheme on CDF of PAPR.   179

Figure 8.5 Effect of FEC on CDF of PAPR.   180

Figure 8.6 16QAM basic constellation points.   181

Figure 8.7 Effect of BO on 16QAM constellation points.   181

Figure 8.8 Sub-carrier organization for spectrum plot.   182

Figure 8.9 Spectrum plot of OFDM signal.   182

Figure 8.10 SDNR plot for 4QAM modulation in AWGN channel.   184

Figure 8.11 SDNR plot for 16QAM modulation in AWGN channel.   185

Figure 8.12 SDNR plot for 64QAM modulation in AWGN channel.   185

Figure 8.13 SDNR plot for 4QAM modulation in fading channel.   186

Figure 8.14 SDNR plot for 16QAM modulation in fading channel.   186

Figure 8.15 SDNR plot for 64QAM modulation in fading channel.   187

Figure 8.16 BER vs SNR curve for $M = 4$, uncoded system in AWGN. 189

Figure 8.17 BER vs SNR curve for $M = 16$, uncoded system in AWGN. 189

Figure 8.18 BER vs SNR curve for $M = 64$, uncoded system in AWGN. 190

Figure 8.19 BLER vs SNR curve for $C = \frac{1}{2}$ and $M = 4$ in AWGN. 190

Figure 8.20 BLER vs SNR curve for $C = \frac{1}{2}$ and $M = 16$ in AWGN. 191

Figure 8.21 BLER vs SNR curve for $C = \frac{1}{2}$ and $M = 64$ in AWGN. 191

Figure 8.22 Spectral efficiency vs SNR curve for $C = \frac{1}{2}$ and $M = 4$ in AWGN. 192

Figure 8.23 Spectral efficiency vs SNR curve for $C = \frac{1}{2}$ and
$M = 16$ in AWGN.   193

Figure 8.24 Spectral efficiency vs SNR curve for $C = \frac{1}{2}$ and
$M = 64$ in AWGN.   193

Figure 8.25 TDe plot for $FEC = \frac{1}{2}$ with BLER threshold $= 0.1$ in
AWGN.   195

Figure 8.26 TDe plot for $FEC = \frac{1}{2}$ with BLER threshold $= 0.05$
in AWGN.   195

Figure 8.27 BER vs SNR curve for $M = 4$, uncoded system in
fading channel.   197

Figure 8.28 BER vs SNR curve for $M = 16$, uncoded system in
fading channel.   197

Figure 8.29 BER vs SNR curve for $M = 64$, uncoded system in
fading channel.   198

Figure 8.30 BLER vs SNR curve for $C = \frac{1}{2}$ and $M = 4$ in fading
channel.   198

Figure 8.31 Spectral efficiency vs SNR curve for $C = \frac{1}{2}$ and $M = 4$
in fading channel.   199

Figure 8.32 BLER vs SNR curve for $C = \frac{1}{2}$ and $M = 16$ in fading
channel.   200

Figure 8.33 Spectral efficiency vs SNR curve for $C = \frac{1}{2}$ and
$M = 16$ in fading channel.   200

Figure 8.34 BLER vs SNR curve for $C = \frac{1}{2}$ and $M = 64$ in fading
channel.   201

Figure 8.35 Spectral efficiency vs SNR curve for $C = \frac{1}{2}$ and
$M = 64$ in fading channel.   201

Figure 8.36 TDe plot for $FEC = \frac{1}{2}$ with BLER threshold $= 0.1$ in
fading channel.   202

Figure 8.37 TDe plot for $FEC = \frac{1}{2}$ with BLER threshold $= 0.05$
in fading channel.   203

Figure 9.1   Classification of multi-antenna techniques.   206
Figure 9.2   Example of space diversity.   211
Figure 9.3   An example of MIMO spatial multiplexing system.   212
Figure 9.4   Difference between diversity gain and coding gain.   214
Figure 9.5   Diversity-multiplexing trade-off.   221

Figure 9.6   Schematic of MIMO-OFDM. Each OFDM tone admits N_T inputs and N_R outputs.   222

Figure 9.7   Transmit diversity for OFDM.   223

Figure 9.8   Spatial multiplexing for OFDM.   224

Figure 9.9   OFDM system with beamforming.   226

Figure 9.10   Qualitative representation of throughput versus distance from BS for SISO, SM, and spatial diversity.   232

Figure 10.1   Beamforming and transmit diversity systems in OFDM system.   236

Figure 10.2   Time- and frequency-domain beamforming approach on an OFDMA system.   238

Figure 10.3   Probability density function (pdf) of SNR for different combinations of diversity and beamforming systems.   243

Figure 10.4   Ergodic channel capacity for transmit diversity, receive diversity, and beamforming techniques with 2, 4, and 8 antennas.   245

Figure 10.5   Outage capacity and outage probability for transmit diversity and beamforming with varying number of transmit antennas.   249

Figure 10.6   Theoretical average bit error probabilities for transmit diversity and beamforming systems.   250

Figure 10.7   Impact of antenna correlation corresponding to different angle spread on transmit diversity and beamforming.   256

Figure 10.8   BER performance for STBC/SFBC for different AS.   258

Figure 10.9   BER performance for BF for different AS.   259

Figure 10.10 BER comparison between BF and STBC for indoor channel.   260

Figure 10.11 BER comparison between BF and STBC for outdoor channel.   260

Figure 11.1   Multiple antenna receiver diversity with MRC at sub-carrier level.   267

Figure 11.2   CIR and CTF of two separate paths, their linear sum and the combined channel after applying cyclic delay principle.   272

Figure 11.3   Transmitter model for CDD based MISO-OFDM transmission scheme.   273

Figure 11.4   CDF of system capacity and 5% outage capacity for SISO and $P \times 1$ CDD systems.   277

Figure 11.5   OFDM receiver with Pre-DFT Combining CDD.   277

Figure 11.6   Diversity weight estimation method for Pre-DFT MARC with CDD receiver diversity scheme.   282

Figure 11.7   Average SNR with two-branch Pre-DFT MARC and Pre-DFT EGC with cyclic delays as a function of the delay parameter n (in samples).   284

Figure 11.8   Histogram of cyclic shifts obtained for maximum average SNR with 10000 samples.   285

Figure 11.9   Magnitudes of channel transfer functions before and after diversity combining. Pre-DFT MARC, Post-DFT MRC, and Pre-DFT EGC are shown in the figure.   286

Figure 11.10   Uncoded BER with and without application of diversity, Rayleigh fading channels with various τ_{rms}.   287

Figure 11.11   Uncoded BER with and without application of diversity, Ricean channel with $K = 4$.   288

Figure 11.12   Performance results in terms of coded BER; Rayleigh channel, $\tau_{rms} = 1$ sample; rate $\frac{1}{2}$ convolutional coding with constraint length 5.   289

Figure 11.13   Relative processing cost for Pre-DFT MARC and Post-DFT MRC in comparison to number of OFDM subcarriers and number of OFDM symbols/packet.   291

Figure 11.14   Transceiver architecture for CDA-SM-OFDM system.   293

Figure 11.15   Comparison of normalized eigenvalue spread and mean eigenvalue for 4×2 CDA-SM-OFDM system with 2×2 and 4×2 SM-OFDM system.   298

Figure 11.16   Comparison of system capacity and 5% outage capacity for 4×2 CDA-SM-OFDM system with 2×2 and 4×2 SM-OFDM system.   299

Figure 12.1   Simplified system model for SM-OSFBC/SM-QSFBC transmission scheme.   308

Figure 12.2   Simplified data flow for OSIC receiver.   313

Figure 12.3 Probability distribution functions of SNRs when different combinations of JDM schemes are used. 320

Figure 12.4 FER performance for the schemes. 321

Figure 12.5 10% outage spectral efficiency in indoor scenario. 322

Figure 12.6 FER with respect to increasing transmit correlation (i.e. decreasing antenna spacing) and fixed receive correlation at system SNR of 12 dB. 324

Figure 12.7 FER with respect to increasing receive correlation (i.e. decreasing antenna spacing) and fixed transmit correlation at system SNR of 12 dB. 324

Figure 12.8 Outage spectral efficiency with respect to decreasing transmit correlation (i.e. increasing antenna spacing) and fixed receive correlation at system SNR of 12 dB. 326

Figure 12.9 Loss in average spectral efficiency with respect to increasingly strong LOS component. 328

Figure 13.1 OFDM/SCFDE MIMO transmitter. 330

Figure 13.2 OFDM/SCFDE MIMO receiver. 330

Figure 13.3 STBC for SCFDE. 331

Figure 13.4 Spatial multiplexing for SCFDE. 332

Figure 13.5 Transmitter model of the proposed 2×1 SCFDE transmit diversity scheme. 334

Figure 13.6 Receiver model of the proposed 2×1 SCFDE transmit diversity scheme. 334

Figure 13.7 SCFDE-STBC vs the proposed scheme in low and high time selective channels. 339

Figure 13.8 Uncoded BER vs normalized coherence bandwidth at the SNR of 8 dB. 339

Figure 13.9 Uncoded BER vs normalized Doppler frequency at the SNR of 8 dB. 340

Figure 13.10 4×2 SCFDE JDM system model. 342

Figure 13.11 4×2 SCFDE QOD system model: Transmitter. 345

Figure 13.12 Uncoded average SE for low frequency selectivity ($\tau_{max} = 2.80\,\mu s$) and high time selectivity ($f_d T_s = 0.03$). 346

Figure 13.13 Uncoded average SE for moderate frequency selectivity ($\tau_{max} = 14.70\,\mu s$) and high time selectivity ($f_d T_s = 0.03$). 347

Figure 13.14 Uncoded average SE for high frequency selectivity ($\tau_{max} = 18.20\,\mu s$) and high time selectivity ($f_d T_s = 0.03$). 347

Figure 13.15 Coded average SE for high frequency selectivity ($\tau_{max} = 18.20\,\mu s$) and high time selectivity ($f_d T_s = 0.03$). 348

Figure 13.16 Uncoded average SE in presence of spatial correlation (at both transmit and receive arrays) and moderate frequency selectivity ($\tau_{max} = 14.70\,\mu s$) and high time selectivity ($f_d T_s = 0.03$). 349

Figure 13.17 Coded average SE in presence of spatial correlation (at both transmit and receive arrays) and moderate frequency selectivity ($\tau_{max} = 14.70\,\mu s$) and high time selectivity ($f_d T_s = 0.03$). 350

Figure 13.18 Uncoded average SE for low time selectivity ($f_d T_s = 0.001$) and high frequency selectivity ($\tau_{max} = 18.20\,\mu s$). 350

Figure 13.19 Uncoded average SE for moderate time selectivity ($f_d T_s = 0.015$) and high frequency selectivity ($\tau_{max} = 18.20\,\mu s$). 351

Figure 13.20 Uncoded average SE for high time selectivity ($f_d T_s = 0.03$) and high frequency selectivity ($\tau_{max} = 18.20\,\mu s$). 351

Figure 13.21 Coded average SE for high time selectivity ($f_d T_s = 0.03$) and high frequency selectivity ($\tau_{max} = 18.20\,\mu s$). 352

Figure 13.22 Space–frequency LDC transceiver. 359

Figure 13.23 Coded BER performance for 2×2 ST and SF-LDC vs ST and SF-OD, rate $= 8$, B $=$ T $= 2$, ITU vehicular A channel. 368

Figure 13.24 Coded BER performance for 4×4 ST and SF-LDC vs ST and SF-QOD, rate $= 8$, B $=$ T $= 4$, ITU vehicular A channel. 369

Figure 13.25 Impact of time-selectivity for 4×4 ST and SF-LDC vs ST and SF-QOD, rate $= 8$, B $=$ T $= 4$, SNR $= 10\,$dB. 370

Figure 13.26 Impact of frequency-selectivity for 4×4 ST and SF-LDC vs ST and SF-QOD, rate $= 8$, B $=$ T $= 4$, SNR $= 10\,$dB. 370

Figure 14.1 The cross-layer design model: performance improvement through information sharing. 379

Figure 14.2 General cross-layer opportunities. 382

List of Tables

Table 1.1   Evolution of wireless standards.   6

Table 2.1   Value of parameters for urban terrain.   19

Table 3.1   Comparison of single carrier and multi-carrier approach in terms of channel frequency selectivity.   37

Table 3.2   IEEE 802.11a OFDM PHY modulation techniques.   40

Table 4.1   A summary of multiple access scheme.   64

Table 4.2   The IEEE 802.16a sub-carrier specification. (For the Guard carriers, representing respectively the left and right number of sub-carriers.)   76

Table 4.3   The IEEE 802.16.4c four downstream modes. (For the Guard band, the two numbers are the guard interval (number of sub-carries) to the left and the right of the OFDM symbol, respectively.)   76

Table 4.4   System parameters for OFDMA-FSCH scheme under development by IEEE 802.20 working group.   86

Table 7.1   Example channel values for a block of five OFDM symbols and five sub-carriers.   155

Table 7.2   Example of covariance matrix $\mathbf{R_{pp}}$.   156

Table 7.3   Example of covariance matrix $\mathbf{R_{h\hat{p}}}$.   157

Table 7.4   Example of interpolation matrix $\mathbf{R_{h\hat{p}}} \left(\mathbf{R_{pp}} + \frac{1}{\mathrm{SNR}} \mathbf{I} \right)^{-1}$.   158

Table 7.5   Main parameters of the simulated OFDM system.   171

Table 8.1   Table for calculation of total degradation in dB.   194

Table 9.1   Array gain and diversity order for different multi-antenna configurations.   215

Table 10.1   Tarokh's $\frac{1}{2}$ rate space–time block encoding scheme; s_n denotes the transmitted symbol at nth OFDM symbol duration.   239

Table 10.2   Parameters for comparison between diversity and beamforming.   257

Table 11.1   Comparison of channel coherence time with OFDM packet duration for IEEE 802.11a WLAN Standard.   291

Table 12.1   Tags used in figures for corresponding schemes.   320

Table 12.2   Correlation for corresponding spatial separation among antennas at 3.5 GHz of carrier frequency. R and d denote spatial correlation and separation in cm across two neighboring elements respectively.   323

Table 13.1   Simulation parameters.   338

Table 13.2   Complexity in terms of complex multiplications for 2×1 case.   341

Table 13.3   Schemes characteristics.   362

Table 13.4   Simulation parameters.   368

Table 13.5   Complexity in terms of complex multiplications for 2×2 case.   371

Table 13.6   Complexity in terms of complex multiplications for 4×4 case.   372

1

Introduction

1.1 History of Wireless Communications

Modern wireless communication has its date of birth back to 1895, when Guglielmo Marconi transmitted successfully the three-dot Morse code for the letter "*S*" over a distance of three kilometers using electromagnetic waves. From this beginning, wireless communications has developed into a key element of modern society. Starting from Marconi's experiment there was a quick and impressive development of a variety of innovative techniques and related applications [1].

The history of mobile telephony can be divided into four periods: *pre-cellular*, *first generation*, *second generation*, and *third generation* cellular systems. The first or pre-cellular period is related to mobile telephones that made an exclusive use of a frequency band in a certain area, leading to problems associated with congestion and call completion.

The introduction of cellular technology lead to a big increase in the spectrum efficiency. The new idea compared to first generation was the so-called cellular concept: by breaking down a geographic area into small areas or cells, different users in different (nonadjacent) cells were able to use the same frequency for a call without interference.

The main change in Second Generation (2G) mobile systems, compared to the first generation, was the use of digital technology. The 2G had as final outcome the GSM standard, which could allow, among other things, full international roaming and relatively high quality audio. GSM is now the most widely used 2G system worldwide, in more than 130 countries.

Up to date, the latest stages in the development of mobile telephones are so-called beyond the Third Generation (3G) technologies and forthcoming Fourth Generation (4G) technologies. These systems allow for significantly increased speeds of transmission and are particularly useful for data services.

1

As applications examples, the 3G phones and forthcoming future 4G devices can more efficiently be used for e.g. e-mail services, and downloading content from the Internet, etc. An attempt to establish an international standard for 4G mobile is being moderated through the International Telecommunication Union (ITU), under its IMT-Advanced program.

1.2   Spectrum Issues

Already at first stages, it became clear that international coordination was required for wireless communication to be effective. This coordination involved two features. Locally, the inherent potential for interference in radio transmissions meant that coordination was needed to avoid the transmission of conflicting signals. Globally, coordination was necessary between countries to guarantee consistency in approach to wireless services [1].

This was the driver for government intervention to ensure the coordinated allocation of radio spectrum. The ITU is in charge of internationally coordinating the use of radio spectrum, managing interference, and setting global standards. The ITU is a specialist agency of the United Nations with over 180 members and was created by the International Telecommunications Convention in 1947. The ITU allocates certain frequencies for specific uses on either a worldwide or a regional basis. Individual countries may then further allocate frequencies within the ITU international allocation. The amount of spectrum allocated to different uses differs by country and frequency band. The ITU's procedure to deal with interference is to require member countries to follow notification and registration acts whenever they plan to assign frequency to a particular use, such as mobile telephony.

The number of different devices using wireless communications (and therefore needing radio spectrum) is rising rapidly but however, by far the most important and dramatic change in the use of wireless communications in the past two decades has been the rise of mobile telephony.

1.3   Drivers for Future Wireless Market

Over the last two decades, cellular telephony has lead to the expectation of anytime, anywhere accessibility. Consequently, with more efficient networks available in more geographical locations and the recent size reduction in mobile devices together with the surge in attractive and useful multimedia services, people can be more and more wireless (i.e. mobile) and the requirement of being connected wired for utilizing multimedia services could be removed in both the work and personal lives [2].

Though representing a good business opportunity for operators, the cost of providing mobile data service is significant; this is one of the main reasons why mobile data tariffs today are still relatively expensive compared to fixed line broadband pricing. This contributes to the evolution of mobile network technology toward more efficient networks (such as WiMAX and LTE) that address these market needs. All these recent changes have had a great impact on people's lifestyle, making them more mobile, connected and responsive.

These changes are now driving new needs that are not yet fulfilled by today's telecom services. For example, at macro level, networks are still very fragmented; today, a single user is not able to experience a seem-less connectivity, as at different times of the day he might need to interact with several different networks: e.g. home, office, and mobile networks. As another example, there is a huge potential to expand the wireless networks, by connecting objects that are nowadays stand alone and turning them into communication nodes, e.g. cars or latest devices like MP3 players or digital cameras.

The future telecommunication network will provide the users with access to any content, anytime, anywhere, on any device, at speeds such that multimedia communication will be perceived as instantaneous and perfect by the end-user, matching the Quality of Service (QoS) with the customers' expectations.

1.4 Forthcoming 4G Systems

Wireless communication has gained a momentum in the last decade of 20th century with the success of 2G of digital cellular mobile services. Worldwide successes of GSM, IS-95, PDC, IS-54/137 etc., systems have shown new way of life for the new information and communication technology era. These systems were derived from a voice legacy, thus primary services were all voice transmission. 2G systems provided better quality of services at lower cost and a better connectivity compared to previous analog cellular systems. Numerous market researches show that there is a huge demand for high-speed mobile multimedia services all over the world. With the advent of 3G wireless systems, it is promised that higher mobility with reasonable data rate (up to 2 Mbps) can be provided to meet the current user needs [3, 4]. But, 3G is not the end of the tunnel; ever increasing user demands have drawn the industry to search for better solutions to support data rates in the ranges of tens of Mbps. Naturally dealing with ever unpredictable wireless channel at high data rate communications is not an easy task. Hostile wireless channel has always been proved as a bottleneck for high speed wireless systems. This motivated the research toward finding a better solution for combating all the odds of wireless channels; thus, the idea of multi-carrier transmission has surfaced recently to be used for future generations of wireless systems.

Third Generation (3G) promises wire line quality of services via wireless channel. For wide area coverage, further evolutions of 3G systems, such as High Speed Packet Access (HSPA) systems, are already being implemented in different parts of the world. Certainly the bit rate will be much higher than 2 Mbps for such a system, up to tens of Mbps, to be exact 14.4 Mbps for High Speed Downlink Packet Access (HSDPA) and 5.76 Mbps for High Speed Uplink Packet Access (HSUPA) [5]. For local area coverage, Wireless Local Area Networks (WLANs), such as IEEE 802.11a, HiperLAN/2 or MMAC[1] standards are capable of providing data rates up to 54 Mbps. Along with these three, there are few other emerging short-range wireless applications available, such as Bluetooth, HomeRF, etc.

Wireless Local Area Networks have potentially become an established tool in different user environments, namely home, corporate, and public environment etc. They are used to connect wireless users to a fixed LAN in indoor environments. A major WLAN application is in public sectors, where WLAN can be used to connect users to the backbone network [6]. Airports, hotels, high-rise offices, city centers will be the target areas for such public WLAN usage. A popular vision of future generations of telecommunications systems suggests that they will be an amalgamation of capabilities of high data-rate wireless wide area networks (such as UMTS) and newly standardized WLANs, i.e. the future networks will provide high rate at wide area in highly mobile scenario [7]. However, systems of the near future will be required to provide data rate of greater than 100 Mbps; hence there is a need to further improve

---

[1] IEEE802.11a is an USA-standard, HiperLAN/2 is an European standard and MMAC is developed in Japan. All three of the standards are almost similar in their PHY layers.

the capacity of existing wireless systems. Although the term 4G is not yet clear to the industry, it is likely that they will enhance the 3G and beyond 3G networks in capacity, allowing greater range of applications and better universal access. Some of the visionaries term these systems as Mobile Broadband Systems (MBSs). Seamless and uninterrupted service quality for a user regardless of the system he/she is using will be one of the main goals of future systems. The expected systems will require an extensive amount of bandwidth per user.

Several technologies are considered to be candidates for future applications. With the advent of Worldwide Interoperability for Microwave Access (WiMAX) and Universal Mobile Telecommunications Systems–Long Term Evolution (UMTS–LTE) activities under IEEE 802.16 and 3GPP respectively, there is a clear direction in the possible solutions for future 4th Generation (4G) systems. The research community agrees that high data rate at wider coverage area with suitable QoS mechanisms will be required for 4G systems [8].

Though some understanding of the future 4G systems is already achieved, yet, these research efforts have so far given rise to several distinctive features of the 4G technologies. So, an unified definition, characteristics and specifications of 4G systems is still subject to ongoing discussions [7]. We consider one of the major 4G features is to be reflected in the (co)existence of heterogeneous terminals. Namely, while 2G terminals were characterized by homogenous services and platforms, it is widely believed that different types of terminals will exist in 4G. The difference will be in terms of display size, energy consumption, complexity, etc. Each type of terminal fits a specific service or a subset of supported services. This sets the requirements for variable, but high data rates systems (in the order of tens of Mbps and more). Some services should also be available with high data rate under high mobility conditions.

Since 4G systems will be deployed in a wide range of environments and over heterogeneous terminals, it is of primary importance that the air interface should be adaptive. There is a firm trend of service personalization, which imposes the tailoring of the content to the end-user device such as to enable adaptable service presentations, fitting the capabilities of the terminals in use regardless of network types. The user will have the services at disposal through soft-availability in terms of a flexible mapping of service to capabilities of the heterogeneous terminals taking the channel conditions and

Table 1.1 Evolution of wireless standards.

	1G	2G	2.5G	3G	Beyond 3G	4G
	Analog voice	Digital voice	Voice + data	Multimedia services	Broadband multimedia	Ubiquitous networks
	NMT	GSM	GPRS	WCDMA	HSPA	IMT-A ??
	AMPS	PDC	HSCSD	CDMA 2000	WiMAX	
	C-net	IS-95A	EDGE	TD-SCDMA	UMTS–LTE	
		IS-136	IS-95B		CDMA 2000 1xEV	
	FM modulation	Digital modulation	Voice + data	'Any time, any where'	High data rate	Heterogeneous networks
	Analog switching	Error control	Higher rate than 2G	Multimedia traffic	High QoS support	Adaptive air interface
	Cellular concept	Data compression		Packet based data	Broadband wide area	Guaranteed QoS
	Hard handover	High quality voice		Dynamic RRM		
	FDMA	TDMA/ CDMA/ FDMA	TDMA/ CDMA/ FDMA	WCDMA	WCDMA/ OFDMA	OFDMA
	Very low rate	9.6–28.8 kbps	57–115 kbps	0.144–2 Mbps	~10's of Mbps	~100's of Mbps

context into account. A single PHY layer technology cannot be optimal for all environments, terminal capabilities and user requirements, viz. short or long range, indoor or outdoor, low or high data rate, high or low mobility, multi-user capability, device processing power, weight and cost, service type, etc. A system capable of dynamically adapting one or more of the above parameters may offer superior performance. Hence, evaluation of different PHY technologies suited to various scenarios is necessary for selecting the optimal solution in each case. This presents the ground for implementing optimum adaptive PHY technology.

There are a number of generic PHY techniques that are expected to play major roles in the future wireless technologies:

1. Multi-carrier techniques will form the basis of PHY techniques [8], in which Orthogonal Frequency Division Multiplexing (OFDM) is the most widely recognized candidate. Multiple-access can be easily employed using Orthogonal Frequency Division Multiple Access (OFDMA) technique.

2. Multiple Input Multiple Output (MIMO) antenna techniques have been shown to offer large capacity gains under favorable environments compared to single antenna techniques. Besides, MIMO techniques can also be used to increase reliability of wireless links.

3. Advanced channel coding techniques, such as Turbo and Low Density Parity Check (LDPC) codes, through the suboptimal, but very efficient iterative decoding, have been shown to be the most efficient error-correction codes, getting closest to the Shannon limit.

4. Some forms of Code Division Multiple Access (CDMA) techniques, such as Frequency Hopping (FH) and Direct Sequence (DS), which will enable spectrum sharing in some specific multi-user and multi-cell systems.

With the emergence of mobile WiMAX and UMTS–LTE, which allows inter-operability and wide range of benefits compared to other technologies, it seems that these technologies are on tracks of their own toward the 4G goal. Thus, either working in complements or in competition, WiMAX and UMTS–LTE will be strong candidates to be the standard of next generation mobile technology [9]. These technologies propose solutions for many types of high-bandwidth and long distance applications and will enable service carriers to converge the all-IP-based network for triple-play services of data, voice, and video. With their QoS support, longer reach, and high data capacity, the envisioned 4th Generation (4G) systems are positioned for broadband access applications in both rural and urban areas; and in both Fixed Wireless Access (FWA) and MBS scenario. Among other residential applications are high speed Internet, Voice Over IP telephony and streaming video/online gaming with additional applications for enterprises such as Video conference and Video surveillance, secured Virtual Private Network (with need for high security). The future technologies will also allow covering applications with media content requesting more bandwidth [7, 8].

1.5 Technical Challenges that Need to be Addressed for Future System Design

A key requirement on a new radio-access technology is the provision of significantly enhanced services by means of higher data rates, lower latency, etc., at a substantially reduced cost, compared to current mobile-communication systems and their direct evolution. As an example, within the WINNER project, data rates in the order of 100 Mbps for wide-area coverage and up to 1 Gbps for local-area coverage are set as targets. At the same time, the delay/latency of the radio-access network should be limited to a few milliseconds.

The services that are expected to dominate the future radio-access systems are likely to evolve from the services we see in the fixed networks already today, i.e. best effort and rate-adaptive Transmission Control Protocol (TCP)-based services. In current 3G systems, services are typically divided into different QoS classes. However, in order to support TCP data rates in the range 100–1000 Mbps, the requirements on delay and packet-loss rates must be kept so stringent that for future radio-access systems it is meaningless to set different requirements for real-time, quasi real-time, and nonreal-time services.

Currently, it is not clear what spectrum a future radio-access technology is to be operated in. However, it is highly desirable that a new radio-access technology should also be deployable in other frequency bands, including existing 2G and 3G spectrum. Thus, a new radio-access technology should support a high degree of spectrum flexibility including, among other things, flexibility in terms of frequency-band-of-operation, size of available spectrum, and duplex arrangement.

Based on the requirements discussed above it is possible to give some suggestions about a future radio-access system:

- the Downlink (DL) transmission is OFDM-based with support for fast link adaptation and time- and frequency-domain scheduling;
- the Uplink (UL) modulation is Single Carrier (SC)-based and with different prefiltering operations it is possible to operate in both a pure OFDM mode and a pure SC mode;
- advanced antennas are integrated in the system design by means of LDC [10].

Some recent studies have clearly shown that the basic issue is not OFDM (OFDMA) vs SC (TDMA) but rather FDE vs Time Domain Equalization (TDE). FDE has several advantages over TDE in high mobility propagation environments (usually with long tail channel impulse responses). We can also have a SC transmission with FDE so more generally, the chosen approach should be FDE-based, keeping an eye on other possibilities.

For channels with severe delay spread, FDE is computationally simpler than corresponding TDE for the same reason OFDM [11] is simpler: because equalization is performed on a block of data at a time, and the operations on this block involve an efficient FFT operation and a simple channel inversion operation. Sari *et al.* [12, 13] pointed out that when combined with FFT processing and the use of a cyclic prefix, a SC system with FDE (SCFDE) has essentially the same performance and low complexity as an OFDM system. It should also be noticed that a frequency domain receiver processing SC modulated data shares a number of common signal processing functions with an OFDM receiver. In fact, as it is pointed out later on, SCFDE and OFDM modems can easily be configured to coexist, and significant advantages may be obtained through such coexistence.

Orthogonal Frequency Division Multiplexing (OFDM) is a special form of multi-carrier transmission where all the sub-carriers are orthogonal to each other. It promises great resilience to severe signal fading effects of the wireless channel at a reasonable level of implementation complexity [11]. Judging the recent acceptance of OFDM technology for various different wireless systems, it can be seen that there is a strong possibility that next generation wireless era belongs to OFDM technology [7]. One of the main advantages that OFDM provides is the fine granularity in the time- and frequency-domains that can be

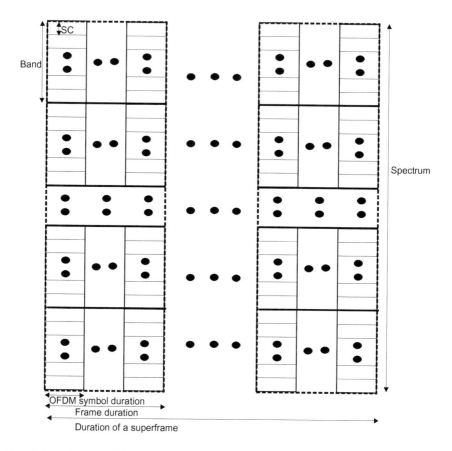

Fig. 1.1 Time–frequency OFDM granularity that can be exploited for increasing system spectral efficiency.

exploited efficiently to optimize the system performance. Figure 1.1 shows the different time and frequency units in any OFDM system. One of the main challenges in designing any multi-user wireless system is to assign the available resources efficiently. Hence, it is an important question to utilize the time-frequency resources in an OFDM-based system so that the multiple access scheme is implemented according to the user requirement and the system scenario. Several techniques are proposed for implementing the multiple access in OFDM systems, namely OFDMA [14]; Orthogonal Frequency Division Multiplexing–Time Division Multiple Access (OFDM–TDMA) [15], which are basically pure FDMA and TDMA on top of an OFDM system. There are a number of proposals where the benefits of CDMA and OFDM are combined together, such as Multi-Carrier Code Division Multiple Access (MC-CDMA) [16], and Variable Spreading Factor-Orthogonal Frequency and Code Division Multiple Access (VSF-OFCDMA) [17], etc. These techniques are suitable for some specific scenarios, as they are designed according to the requirement of particular user environments [18]. It is well-known that when Channel State Information (CSI) is known at the transmitter, then time-frequency resources as seen in Figure 1.1 can be utilized optimally [19]. For wide area systems, users experience heterogeneous channel conditions in terms of delay spread and Doppler; also the scattering environment is very different for different users. Due to these reasons, it may become prohibitive to obtain the CSI at the transmitter side for designing the proper signaling scheme.

When full or partial CSI is available, a window of opportunity appears for improving the system performance. One of the techniques is Link Adaptation (LA), which exploits the known channel conditions to adapt the link parameters accordingly, so that the fading can be avoided and appropriate amount of bits can be transmitted across certain sub-carriers (or sub-channels). Several different techniques are proposed in literature [20, 21, 22, 23, 24]. Most of these algorithms are highly complex in terms of computational requirement, though high spectral efficiency is achieved. One of the very important issues in LA related studies is the rate at which we could adapt the bit and power level at specific sub-channels. Some theoretical studies are performed in this regard, such as [25]. In these studies, it is proposed that power and bit adaptation need not be done simultaneously, as it does not provide any additional gain compared to only bit-adapted systems. It needs to be studied whether in certain scenarios, bit and power adjustments can be done simultaneously and some gains in terms of power savings can be obtained. To the best of our knowledge, these studies are not conducted for WiMAX/UMTS–LTE type scenario so far. It is also an interesting problem to study the suitable size of time-frequency domain resources of any OFDM resources (as seen in Figure 1.1) that can be utilized in LA process. So far, most of the available LA studies consider an idealized scenario, where system impairments in terms of transmitter nonlinearities, frequency de-synchronization etc., are not considered. It is expected that these impairments would require revised threshold levels for bit and power adaptation purposes.

Besides the recent advancements of OFDM techniques, MIMO algorithms will also be an important element in all future wireless systems. Numerous

MIMO diversity and multiplexing algorithms are studied in the available literature [26, 27, 28, 29, 30, 31]. In line with this research, there is a need to study advanced (and simplified) MIMO techniques for either decreasing the implementation complexity while preserving the MIMO benefits, or increasing the spectral efficiency by exploiting more spatial dimensions. In this light, we present a number of MIMO signaling techniques, where both diversity and multiplexing benefits can be obtained in the same structure. The combination is done in the space domain, in comparison to hybrid MIMO structures that combine diversity and multiplexing in the time-domain, such as [32]. We call such schemes as Joint Diversity and Multiplexing (JDM). Besides, we have also described Cyclic Delay Diversity (CDD) based receive diversity system, where a novel Pre-DFT combining receiver is developed and a trade-off between required complexity and system performance is studied. All these studies are done on a pure link level, where no interference is considered.

In a cellular system, the performance of well-known MIMO techniques can become very different than its performance in link level scenario. Due to the fact that users in neighboring cells may operate in co-channel, we may experience a scenario where the performance of any MIMO can be severely affected by Co-Channel Interference (CCI) caused by the same or other MIMO schemes. The interfering links in Figure 1.2 may use the same or any other MIMO schemes. This depends on the MIMO scheme assigned to any user based on its channel condition. In broadband systems, such as WiMAX and UMTS–LTE, aggressive frequency re-use factor will be pursued (i.e. reuse factors will be close to 1), thus, the possibility that the same or other MIMO schemes will collide in the same sub-channel will be very high. Recent studies of MIMO schemes in CCI scenario concentrate very much on theoretical capacity analysis, such as [33, 34, 35]. Besides these works, there is a number of works where some specific solutions for CCI cancelation or avoidance is studied when a MIMO link is interfered by the same MIMO link from interfering cells [36, 37]. Very little attention has been given so far about the impact on Bit Error Rate (BER) and/or system throughput when a MIMO link is interfered by the same or other MIMO schemes. It is well-known that the nature and number of interfering logical streams are important in interference cancelation schemes [38]. Besides, the number of logical streams has more impact on the desired link than the power of the interferers.

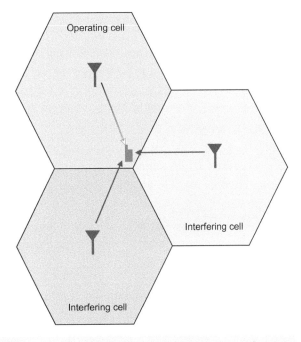

Only two interfering cells are shown for simplicity. In practice, there will be at least one full tier of interfering cells.

Fig. 1.2 Cellular interference scenario.

Since as mentioned above, for both DL and UL the candidates for future systems are FDE based, i.e. OFDM for DL and SCFDE for UL, and the multiple antenna techniques will be a key component of next generation wireless systems, the study of MIMO for Frequency Domain Equalization (FDE) systems is very important. While a lot of work has been done on the application of multiple antenna techniques to OFDM, there are much less studies for Single Carrier-Frequency Domain Equalization (SCFDE) case. The extension of MIMO techniques to SCFDE is straightforward in some cases, but it is complex for some schemes such as the ones involving coding in frequency domain. In future wireless systems, the application of diversity techniques where coding is done across antennas and sub-carriers instead of antennas and time, is important for two main reasons:

- Space–Frequency (SF) schemes can fulfill the requirement of low latency, as Space–Time (ST) schemes processing might not be suitable, since the receiver has to wait for more than one symbol period before processing the received stream. This extra-delay can be reduced when the coding is done across sub-carriers. Here the only latency considered is the one possibly due to ST coding; this delay is in relation with other processing delays (imposed by inter-leaving, channel coding, channel estimation etc.) whose study is not an objective of this book, and are left for further investigations;
- SF schemes can work efficiently over high mobility environments, where ST schemes might fail (since they have to assume the channel constant over several symbol periods).

2

Wireless Channel Modeling

2.1 Wireless Channel

In 1860s, James Clerk Maxwell developed the fundamental laws of electromagnetic theory and Heinrich Hertz proved the existence of electromagnetic waves in 1880s. In the early 1890s, Nicola Tesla demonstrated radio telegraphy and Alexander Popov build his first radio receiver in mid 1890s. Then Acharya Jagadish Chandra Bose gave his first public demonstration of electromagnetic waves (at millimeter wavelengths), by using them to ring a bell remotely and to explode some gunpowder. It was reported in *The Daily Chronicle* of England in 1896: "The inventor (J.C. Bose) has transmitted signals to a distance of nearly a mile and herein lies the first and obvious and exceedingly valuable application of this new theoretical marvel." This was followed by the first successful wireless signaling experiment by Marconi on Salisbury Plain in England in May 1897 [39]. Since then through several developmental stages we have reached an age of near ubiquitous wireless communication network.

To establish any communication system the knowledge of the channel is very important. Its characteristics drive signal design for the system. Professor Ramjee Prasad's statement is worth mentioning here that understanding of the channel is bread and butter for the communication engineer [11]. A proper understanding of the environment leads to correct parametrizations and optimization of the target solution. Characterization and modeling (for simulation) of the wireless channel is necessary in situation where the error probability computation is too complicated or might not yet have been solved. Through computer simulation we can get an idea about the performance of the system under test in the environment it is supposed to work in. The model to be used for computer simulation must be simple enough for easy and fast implementation. The correctness of the model is also very important. An erroneous characterization and modeling of the channel would lead to improper estimates of the performance of the system. So it is essential that a proper model of the wireless channel is used. In this chapter, the wireless channel models commonly used for system evaluation are presented.

2.1.1 Channel Parametrization

The systems analyzed in this book are concerned with wideband for both the indoor wireless channel and outdoor channel between 2 GHz and 6 GHz. The received signal power varies as a function of space, frequency, and time in the entire region of described environment. The variation in general is classified as either large scale or small scale fading. The different fading conditions can be largely classified as shown in Figure 2.1.

Large Scale Fading is dealt by propagation model that predicts the mean received signal strength for an arbitrary transmitter receiver separation. The large scale fading model gives such an average with measurements across 4λ to 40λ [40], where λ is the wavelength. This is useful for estimating the coverage area. Large scale fading can be broadly classified as path loss and shadowing. Path loss deals with the propagation loss due to the distance between the transmitter and the receiver while shadowing describes variation in the average signal strength due to varying environmental clutter at different locations.

Small Scale Fading deals with signal strength characteristics within small distance of the receiver location. In such region of space the average signal

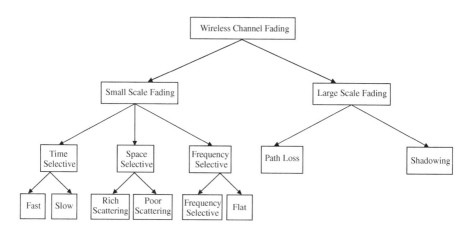

Fig. 2.1 Characterization of fading.

strength remains constant. Multi-path propagation of the electromagnetic waves is the main cause of such effects. It includes the effect of time, space and frequency selective fading characteristics. For each domain, there are broadly two kinds of conditions, one is when the variability is high and the other when the variability is very small over the observation interval.

Thus, the signal strength at a particular location depends on the large scale fading and the small scale fading. As the receiver moves, the instantaneous power of the received signal varies rapidly giving rise to small scale fading. In such a situation the received power may vary by as much as 20–40 dB over a range a few order of a fraction of a wavelength depending upon the particular environment. As the distance between the Transmitter and the Receiver increases, the local averaging of the received signal power decreases gradually, this is predicted by the large scale fading statistics. This phenomenon of a combined slow and fast fading is briefly explained in Figure 2.2.

Three main factors which influence the radio wave propagation are Reflection, Diffraction, and Scattering. Reflection is caused when the Electromagnetic Waves (EM) impinge upon surface having dimensions much larger than the wavelength of the impinging wave. Diffraction is caused due to effects of sharp edges in the path of the radio waves between the Transmitter and the Receiver. Scattering is caused when the EM waves encounter objects of dimension much smaller than the wave in the propagation medium.

Most radio propagation models use a combination of empirical and analytical methods. The empirical approach is based on fitting curves or analytical expressions that recreate a set of measured data. This has the

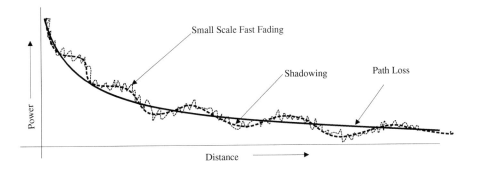

Fig. 2.2 Propagation loss.

advantage of implicitly taking into account all the propagation factors. However, the validity of an empirical model at transmitter frequencies or environments other than those used to derive the model can only be established by additional measured data in the new environment and frequencies. Propagation models and multi-path reflection models have emerged over time to enable easy simulation of the wireless channel [40].

2.1.2 Propagation Loss

There are many models for predicting the path loss such as Hata–Okumura and COST231-Hata model [40]. However, both these models are for frequency ranges up to 2000 MHz. To modify this a model is proposed in [41]. For a given close in distance,[1] d_o of 100 m, the median path loss (PL), in dB, is given by (2.1) [41].

$$\text{PL} = \text{A} + 10 n_p \log_{10} \frac{d}{d_o} + \text{s} \quad d > d_o, \tag{2.1}$$

where

$$A = 20 \log_{10}(4\pi d_o/\lambda), \tag{2.2}$$

and n_p is the path loss exponent and is given by,

$$n_p = a - b h_b + c/h_b, \quad 10\,\text{m} < h_b < 80\,\text{m}. \tag{2.3}$$

The value of a, b, and c are different for different terrain types. Values for urban area are given in Table 2.1 [42]. The shadowing factor is s and follows a log normal distribution, with a typical value around 6 dB [41]. This model is proposed for a receiver antenna height of 2 m and operating frequency of 2 GHz, and a correction factor for other frequencies and antenna heights is proposed [42]. The modified path loss in (2.1) is

$$\text{PL}_{\text{mod}} = \text{PL} + \Delta\text{PL}_f + \Delta\text{PL}_h, \tag{2.4}$$

where, PL is path loss given by (2.1), ΔPL_f is frequency correction term given by $6\log_{10}(f/2000)$, f is the frequency in MHz and ΔPL_h is receiver antenna

---

[1] The Friis free space model, which is the basis for large scale propagation models, is valid for values of d which are in the far field of the transmitting antenna, i.e. does not hold for $d = o$. Therefore, large-scale propagation models [40] use a close-distance, d_o, as known as received power reference point. The received power, $P_r(d)$, at any distance $d > d_o$ may be calculated in relation to that received at d_g The reference distance is chosen so that it lies in the distance used in the mobile communication system.

Table 2.1 Value of parameters for urban terrain.

Parameters	Urban terrain	Unit
a	4.6	
b	0.0075	m^{-1}
c	12.6	m

height correction term given by $-10.8\log_{10}(h/2)$, where h is the new receiver antenna height (m) such that $2 < h < 8$.

The propagation model used for 3GPP-LTE system can be found in [43], where different parameters have been used for different channel condition and cell orientation.

2.1.3 Shadowing

The path loss model does not capture the varying environmental clutter at different locations. However, measurements have shown that at any value d, the path loss PL(d) at a particular location is random and distributed log-normally (normal in dB) about the mean distance dependent values. Since the surrounding environmental clutter may be different at different locations the path loss will be different than the average value predicted by (2.4). This variation is mainly due to refraction and diffraction off Interfering Objects (IO) in the path of the traveling signal, and is an additive term to the path loss, with random values. This phenomenon is called shadowing. It has a log-normal distribution about the mean path loss value [40]. Therefore, the modified path loss expression is

$$Pl(d) = \overline{PL(d)} + X_\sigma \tag{2.5}$$

$$= \overline{PL(d_o)} + (10n_p)\text{Log}_{10}\left(\frac{d}{d_o}\right) + X_\sigma \tag{2.6}$$

$$Pr(d) = Pt(d) - Pl(d); \text{ antenna gains included in Pl}(d), \tag{2.7}$$

where X_σ is zero mean Gaussian distributed random variable (in dB) with standard deviation σ also in dB.

2.1.4 Small Scale Fading

In small scale fading, the signal varies rapidly over a short distance. The variation is caused by the multi-path propagation of the received signal and the Doppler frequency shift. The channel impulse response $h(t, \tau)$ is a function of two variables, time t and delay τ [44]. Due to some reflecting objects such as buildings, hills, trees, etc. some delayed versions of the transmitted signal, each with different amplitudes (A_{np}), phases (θ_n) arrive at the receiver at different delays (τ_n). The parameters (amplitude, phase, and delay) are random variables, and can be characterized by a channel impulse response. If unit impulse is transmitted and there are N_{SE} scattering elements, then the receiver would receive N_{SE} different signals. Therefore, the channel impulse response would be the sum of these N_{SE} scattered signals as given below [45].

$$h(t, \tau) = \sum_{m=1}^{N_{SE}} A_{np,m} \delta(t - \tau_m) \exp(-j\theta_m). \tag{2.8}$$

The channel impulse response is a function of time frequency and space [46]. A typical channel impulse amplitude response over a region is shown in Figure 2.3.

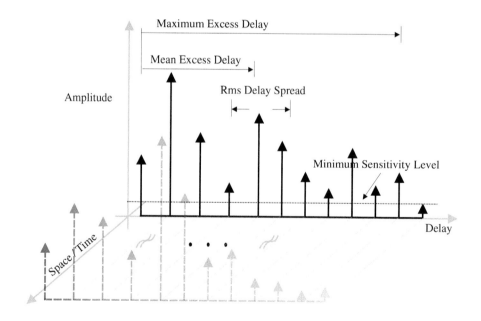

Fig. 2.3 Amplitude response.

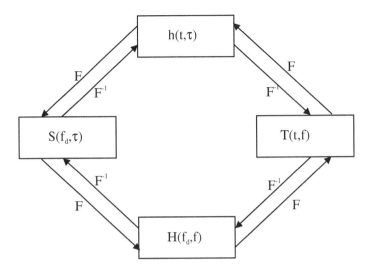

Fig. 2.4 Channel impulse response and transfer function relationship.

The relationship between the impulse response and the transfer function of the channel is shown in Figure 2.4, where f_d is Doppler frequency, τ is delay, t is time, and f is the Fourier domain representation of the delay.

Multi-path Fading Multi-path propagation as shown in Figure 2.5 gives rise to small scale fading in time and frequency domain. The multi-path properties of a given environment are usually characterized by the power delay profile. Power delay profile denotes the average power of each multipath. Figure 2.6 shows a typical power delay profile. When the first multi-path component has the highest power then it is a Ricean channel, where as when the first path does not have the highest power which usually happens in nonline of sight scenario, then it is usually a Rayleigh channel. The power delay profile of a typical Rayleigh multi-path propagation is shown to have an exponential decay profile which is a commonly used model [47]. There are several other models which consider the cluster effect, i.e. there is a double exponential decay, where each multipath is followed by a sequence of multipath during a very short interval with a steeper decay constant. Another model for the delay profile has the first few taps with same average power followed by exponential decay [48].

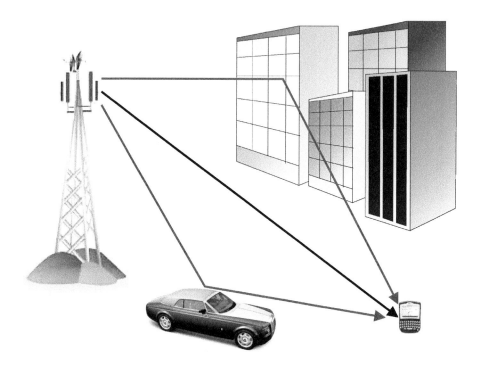

Fig. 2.5 Multi-path propagation.

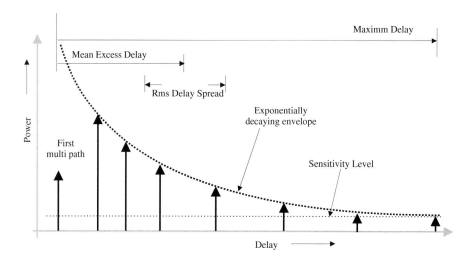

Fig. 2.6 A typical power delay profile.

The channel impulse response is an instantaneous realization of the power delay profile. A typical method of implementing it is via the Clark's methods as in [40]. Another method is the Rice method [49]. There exists other methods such as ray tracing models. The Clark's method has been mostly followed in this work. In some cases the "rayleighchan" function of Matlab® has been used. In these models, each multi-path component is generated so that they follow a desired Doppler spectrum. The Doppler spectrum can be easily integrated into the system. Though for indoor channels most of the taps in the models are supposed to have Jake's Doppler spectrum, the outdoor multi-path channel model taps can have mixed Doppler spectrum, i.e. while some of the taps are advocated to use Jakes's spectrum the one which are toward the tail of the power delay profile may have Gauss' spectrum, details which can be found in [49]. Due to the multi-path reflections a transmitted impulse gets time dispersed, i.e. spread in time domain. A measure of this time spread phenomenon is the mean excess delay, which is defined as [40]

$$\tau_m = \frac{\int_0^{\tau_{\max}} \tau E_{|h(\tau)|^2} \, d\tau}{\int_0^{\tau_{\max}} E_{|h(\tau)|^2} \, d\tau}, \tag{2.9}$$

where E denotes expectation, $\tau_{\max}$ is the maximum delay of the arriving multi paths, $h(\tau)$ is the component of the arriving multi-path at a delay of τ. The rms delay spread of the channel is defined as

$$\tau_{\text{rms}} = \sqrt{\overline{\tau^2} - \tau_m^2}, \tag{2.10}$$

where

$$\overline{\tau^2} = \frac{\int_0^{\tau_{\max}} \tau^2 E_{|h(\tau)|^2} \, d\tau}{\int_0^{\tau_{\max}} E_{|h(\tau)|^2} \, d\tau}. \tag{2.11}$$

The exponential power delay profile defines,

$$E_{|h(\tau)|^2} = \frac{e^{\frac{-\tau}{\tau_0}}}{\int_0^{\tau_{\max}} e^{\frac{-\tau}{\tau_0}}}, \quad \text{for} \quad 0 < \tau < \tau_{\max}$$

$$= 0, \qquad \text{elsewhere.} \tag{2.12}$$

In the above τ_0 is the characteristic of the power delay. The rms delay spread is the average information of a certain environment, but it is expected to have a local variation over a few hundred nano seconds [47, 50].

The small scale channel model does not generate or absorb any power, i.e.

$$\int_0^{\tau_{max}} |h(\tau)|^2 d\tau = 1. \tag{2.13}$$

This ensures that we are concerned only with short term multi-path fading scenario.

The Fourier transform of the channel impulse response is the channel transfer function, as shown in Figure 2.7. The signal experiences different levels of fading for different frequencies in the fading channel. With such characteristics, a fading channel could be either frequency or nonfrequency selective. This depends on the bandwidth of the system compared with the channel coherence bandwidth, B_c. The coherence bandwidth is defined as the frequency separation Δf such that the correlation coefficient falls below a defined real value between 0 and 1. B_c is inversely proportional to the rms

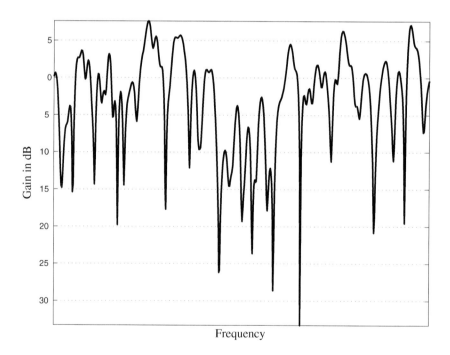

Fig. 2.7 Frequency domain channel response.

delay spread τ_{RMS} [51, 46].

$$B_c \propto \frac{1}{\tau_{RMS}}. \tag{2.14}$$

If the system bandwidth is much smaller compared to the coherence band-width, then the channel is said to be frequency nonselective. In this case, the correlation coefficient[2] of the sub-carrier channel transfer function is almost 1 for the frequencies within the system bandwidth. Physically, the frequency response within the system bandwidth is almost flat, so it is also called flat fading. On the other hand, if the system bandwidth is larger than the coherence bandwidth, then the channel is said to be frequency selective. In this case, the frequencies within the channel would experience different level of fading.

Time varying channel Similar to the channel characteristics in frequency domain, there are also signal variations in the time domain due to the time varying nature of the wireless channel. This time varying channel is mainly caused by the movement of either the transmitter, the receiver, or the reflectors, which results in the Doppler effect. With a velocity v the maximum Doppler shift is $f_m = \frac{v}{\lambda}$. The coherence time of a time varying channel is inversely pro-portional to the maximum Doppler frequency. The coherence time is defined as [40]

$$T_c = \frac{9}{16\pi f}, \tag{2.15}$$

where f is the maximum Doppler frequency. The above model is true for Jakes' spectrum. The coherence time for Gauss spectrum for the same velocity is much more than in the case of Jake's Spectrum. A Typical Jakes' spectrum is given in Figure 2.8 and a typical Gauss spectrum is given in Figure 2.9. The correlation for a tap of the multi-path channel model is given by Das and Prasad [52] as,

$$R_{h(\tau,t)h(\tau,t+\Delta t)} = \mathfrak{F}^{-1}(\tau, S(f)), \tag{2.16}$$

where $\mathfrak{F}^{-1}$ mean inverse Fourier Transform, $S(f)$ is the Doppler frequency power spectrum.

---

[2] $E_{H(f),H^*(f+\Delta f)}$, where H is the channel frequency response at frequency f.

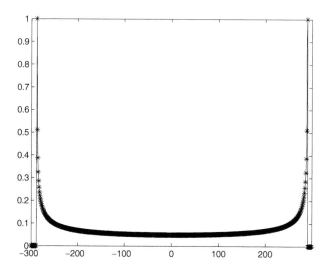

Fig. 2.8 Power spectral density vs frequency of Jakes' spectrum.

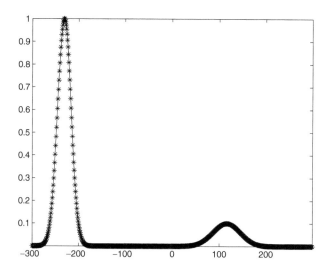

Fig. 2.9 Power spectral density vs frequency of typical Gauss' spectrum.

2.2 Quadrature Amplitude Modulation

As Quadrature Amplitude Modulation (QAM) is used in this work, the QAM is described in this section. Digital Modulation is a process that maps a digital symbol onto a signal suitable for transmission [53]. This is done in two steps: at first, the k bits of the digital signal are mapped to one symbol of the baseband signal and are represented as complex constellation points. In second step, the resultant baseband signal is then up converted to the transmitting frequency via Radio Frequency (RF) modulation. When the amplitude of the modulated signal is varied to carry the source information, the modulation is called Pulse Amplitude Modulation (PAM). On the other hand, the term Phase Shift Keying (PSK) implies the phase of the modulated waveform has the source information. QAM uses the combination of both these techniques and carries information in the phase as well as the amplitude of the modulated waveform. This can also be seen as embedding two simultaneous sequences of k bits information signal on two quadrature carriers $\cos 2\pi f_c t$ and $\sin 2\pi f_c t$. The corresponding modulated waveform can be written as [54]:

$$s_m(t) = (A_{mc} + jA_{ms})g(t)e^{j2\pi f_c t} \quad m = 1, 2, \ldots, M \qquad (2.17)$$

where A_{mc} and A_{ms} are the information-bearing signal amplitudes of the quadrature carrier and $g(t)$ is the signal pulse.

The signal space diagram of rectangular QAM for different values of M is shown in Figure 2.10 [54], where $M = 2^k$ and k is the number of information bits per modulated symbol. It is common practice to have rectangular QAM where $M = 2^{2j}$, with each symbol representing $2j$ information bits, because

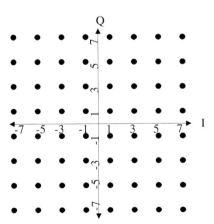

Fig. 2.10 Signal space diagram for rectangular 64-QAM.

it has the advantage of being generated as superposition of two PAM signal on quadrature carriers.

2.2.1 Probability of Error in QAM

Since rectangular M-ary QAM can be treated as the superposition of two $\sqrt{M}$-ary PAM signals, the probability of a correct decision in the case of M-ary QAM system is as given by [54],

$$P_c = (1 - P_{\sqrt{M}})^2, \qquad (2.18)$$

where $P_{\sqrt{M}}$ is the probability of error of a $\sqrt{M}$-ary PAM signal and is given by [54]:

$$P_{\sqrt{M}} = 2\left(1 - \frac{1}{\sqrt{M}}\right) Q\left(\sqrt{\frac{3}{M-1}\gamma}\right), \qquad (2.19)$$

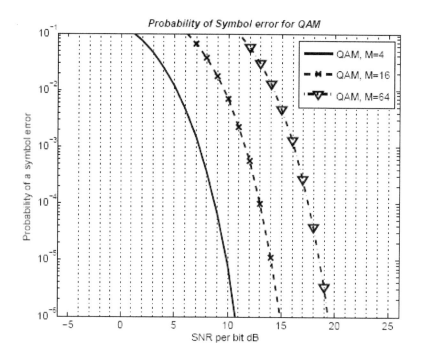

Fig. 2.11 Probability of symbol error for QAM.

where γ is the average SNR. The probability of symbol error is then given by [54]:

$$P_M = 1 - (1 - P_{\sqrt{M}})^2. \tag{2.20}$$

Probability of symbol error for QAM signal with different values of M is shown in Figure 2.11.

The channel affects the information by introducing an uncertainty in the position of the received points in constellation. As we increase the level of modulation, the phase state distance is reduced, and results in higher Bit Error Rate (BER) at a given signal-to-noise ratio. Figure 2.11 shows this scenario. Figure 2.11 also depicts the theoretical achievable limit in the performance of any system derived by Shannon. In order to increase the spectral efficiency more information bits are required and hence more states have to be allocated in the constellation diagram. However, the greater the number of states, the closer the constellation points are spaced and it makes more difficult to the detection at the receiver side.

3

Multi-Carrier Fundamentals

The nature of future wireless applications demand high data rates. Naturally dealing with ever-unpredictable wireless channel at high data rate communications is not an easy task. The idea of multi-carrier transmission has surfaced recently to be used for combating the hostility of wireless channel as high data rate communications. Orthogonal Frequency Division Multiplexing (OFDM) is a special form of multi-carrier transmission where all the sub-carriers are orthogonal to each other. OFDM promises a higher user data rate transmission capability at a reasonable complexity and precision.

At high data rates, the channel distortion to the data is very significant, and it is somewhat impossible to recover the transmitted data with a simple receiver. A very complex receiver structure is needed which makes use of computationally extensive equalization and channel estimation algorithms to correctly estimate the channel, so that the estimations can be used with the received data to recover the originally transmitted data. OFDM can drastically simplify the equalization problem by turning the frequency selective channel to a flat channel. A simple one-tap equalizer is needed to estimate the channel and recover the data.

Future telecommunication systems must be spectrally efficient to support a number of high data rate users. OFDM uses the available spectrum very efficiently which is very useful for multimedia communications. Thus, OFDM stands a good chance to become the prime technology for Forth Generation (4G). Pure OFDM or hybrid OFDM will be most likely the choice for PHY layer multiple access technique in the future generation of telecommunications systems.

3.1 History and Development of OFDM

Although OFDM has only recently been gaining interest from telecommunications industry, it had a long history of existence. It is reported that OFDM based systems were in existence during the Second World War. OFDM had been used by US military in several high frequency military systems such as KINEPLEX, ANDEFT, and KATHRYN [11]. KATHRYN used AN/GSC-10 variable rate data modem built for high frequency radio. Up to 34 parallel low rate channels using PSK modulation were generated by a frequency multiplexed set of sub-channels. Orthogonal frequency assignment was used with channel spacing of 82 Hz to provide guard time between successive signaling elements [55].

In December 1966, Robert W. Chang[1] outlined a theoretical way to transmit simultaneous data stream trough linear band limited channel without *Inter Symbol Interference* (ISI) and *Inter Carrier Interference* (ICI). Subsequently, he obtained the first US patent on OFDM in 1970 [56]. Around the same time, Saltzberg[2] performed an analysis of the performance of the OFDM system. Until this time, we needed a large number of sub-carrier oscillators to perform parallel modulations and demodulations.

A major breakthrough in the history of OFDM came in 1971 when Weinstein and Ebert[3] used *Discrete Fourier Transform* (DFT) to perform baseband modulation and demodulation focusing on efficient processing. This eliminated the need for bank of sub-carrier oscillators, thus paving the way for easier, more useful and efficient implementation of the system.

All the proposals until this time used guard spaces in frequency domain and a raised cosine windowing in time domain to combat ISI and ICI. Another milestone for OFDM history was when Peled and Ruiz[4] introduced *Cyclic Prefix* (CP) or cyclic extension in 1980. This solved the problem of maintaining

---

[1] Robert W. Chang, "Synthesis of band-limited orthogonal signals for multichannel data transmission," *The Bell Systems Technical Journal*, December 1966.

[2] B. R. Saltzberg, "Performance of an efficient parallel data transmission system," *IEEE Transactions on Communications*, Vol. COM-15, no. 6, pp. 805–811, December 1967.

[3] S. B. Weinstein and P. M. Ebert, "Data transmission of frequency division multiplexing using the discrete frequency transform," *IEEE Transactions on Communications* Vol. COM-19, no. 5, pp. 623–634, October 1971.

[4] R. Peled and A. Ruiz, "Frequency domain data transmission using reduced computational complexity algorithms," in *Proceeding of the IEEE International Conference on Acoustics, Speech, and Signal Processing, ICASSP '80*, pp. 964–967, Denver, USA, 1980.

orthogonal characteristics of the transmitted signals at severe transmission conditions. The generic idea that they placed was to use cyclic extension of OFDM symbols instead of using empty guard spaces in frequency domain. This effectively turns the channel as performing cyclic convolution, which provides orthogonality over dispersive channels when CP is longer than the channel impulse response [11]. It is obvious that introducing CP causes loss of signal energy proportional to length of CP compared to symbol length, but, on the other hand, it facilitates a zero ICI advantage which pays off.

By this time, inclusion of FFT and CP in OFDM system and substantial advancements in *Digital Signal Processing* (DSP) technology made it an important part of telecommunications landscape. In the 1990s, OFDM was exploited for wideband data communications over mobile radio FM channels, *High-bit-rate Digital Subscriber Lines* (HDSL at 1.6 Mbps), *Asymmetric Digital Subscriber Lines* (ADSL up to 6 Mbps) and *Very-high-speed Digital Subscriber Lines* (VDSL at 100 Mbps).

Digital Audio Broadcasting (DAB) was the first commercial use of OFDM technology. Development of DAB started in 1987. By 1992, DAB was proposed and the standard was formulated in 1994. DAB services came to reality in 1995 in the United Kingdom and Sweden. The development of *Digital Video Broadcasting* (DVB) was started in 1993. DVB along with *High-Definition TeleVision* (HDTV) terrestrial broadcasting standard was published in 1995. At the dawn of the 20th century, several *Wireless Local Area Network* (WLAN) standards adopted OFDM on their PHY layers. Development of European WLAN standard HiperLAN started in 1995. HiperLAN/2 was defined in June 1999 which adopts OFDM in PHY layer. Recently, IEEE 802.11a in USA has also adopted OFDM in their PHY layer.

Perhaps of even greater importance is the emergence of this technology as a competitor for future 4G wireless systems. These systems, expected to emerge by the year 2010, promise to at last deliver on the wireless Nirvana of anywhere, anytime, anything communications. Should OFDM gain prominence in this arena, and telecom giants are banking on just this scenario, then OFDM will become the technology of choice in most wireless links worldwide [57].

3.2 The Benefit of Using Multi-Carrier Transmission

Time dispersion represents a distortion of the signal that is manifested by the spreading of the modulation symbols in the time domain. It is well-known that the coherence bandwidth of the channel is always smaller than the modulation bandwidth in case of broadband multimedia communications, so, ISI is unavoidable in such wireless channels. In many instances, fading by the multipath will be frequency selective. This frequency-selectivity effect has a random pattern at any given time. This fading occurs when the channel introduces time dispersion and the delay spread is larger than the symbol period. Frequency selective fading is difficult to compensate because the fading characteristics are random and sometimes may not be easily predictable. When there is no dispersion and the delay spread is less than the symbol period, the fading will be flat, thereby affecting all frequencies in the signal equally. Practically flat fading is easily estimated and compensated with a simple equalization [44, 58].

A single carrier system suffers from ISI problem when the data rate is very high. According to previous discussions, we have seen that with a symbol duration T_{sym}, Inter-Symbol Interference (ISI) occurs when $\tau_{max} > T_{sym}$. Multi-channel transmission has surfaced to solve this problem. The idea is to increase the symbol duration and thus reduce the effect of ISI. Reducing the effect of ISI yields an easier equalization, which in turn means simpler reception techniques.

Wireless multimedia solutions require up to tens of Mbps for a reasonable Quality of Service (QoS). If we consider single carrier high speed wireless data transmission, we see that the delay spread at such high data rates will definitely be greater than the symbol duration even considering the best case outdoor scenario. Now, if we divide the high data rate channel over a number of sub-carriers, then we have larger symbol duration in the sub-carriers and the delay spread is much smaller than the symbol duration.

Figure 3.1 describes this very issue. Assuming that we have available bandwidth B of 1 MHz in a single carrier approach, we transmit the data at symbol duration of 1μs. Consider a typical outdoor scenario where the maximum delay spread can be as high as 10μs, so at the worst case scenario, at least 10 consecutive symbols will be affected by ISI due to the delay spread.

In a single carrier system, this situation is compensated by using equalization techniques. Using the estimates of channel impulse response, the equalizer multiplies complex conjugate of the estimated impulse response with the received data signal at the receiver. There are other well-known equalization algorithms available in the literature, such as adaptive equalization via Least

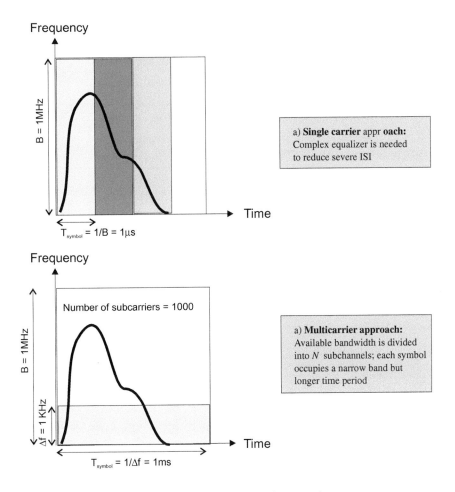

Fig. 3.1 Single carrier vs multi-carrier approach.

Mean Square (LMS), Recursive Least Squares (RLS) algorithms [59], etc. However, there are some practical computational difficulties in performing these equalization techniques at tens of Mbps with compact and low cost hardware. It is worth mentioning here that compact and low cost hardware devices do not necessarily function at very high data speed. In fact, equalization procedures take bulk of receiver resources, costing high computation power and thus overall service and hardware cost becomes high.

One way to achieve reasonable quality and solve the problems described above for broadband mobile communication is to use parallel transmission.

In a crude sense, someone can say in principle that parallel transmission is just the summation of a number of single carrier transmissions at the adjacent frequencies [11]. The difference is that the channels have lower data transmission rate than the original single carrier system and the low rate streams are orthogonal to each other. If we consider a multi-carrier approach where we have N number of sub-carriers, we can see that we can have $\frac{B}{N}$ Hz of bandwidth per sub-carrier. If $N = 1000$ and $B = 1$ MHz, then we have a sub-carrier bandwidth Δf of 1 kHz. Thus, the symbol duration in a sub-carrier will be increased to 1 ms $\left(= \frac{1}{1\,\text{kHz}} \right)$. Here each symbol occupies a narrow band but a longer time period. This clearly shows that the delay spread of 1 ms will not have any ISI effect on the received symbols in the outdoor scenario mentioned above. So, we can say that the multi-carrier approach turns the channel to a flat fading channel and thus can easily be estimated.

Theoretically increasing the number of sub-carriers should be able to give better performance in a sense that we will be able to handle larger delay spreads. But several typical implementation problems arise with a large number of sub-carriers. When we have large numbers of sub-carriers, we have to assign the sub-carriers frequencies very close to each other, if the available bandwidth is not increased. We know that the receiver needs to synchronize itself to the carrier frequency very well, otherwise a comparatively small carrier frequency offset may cause a large frequency mismatch between neighboring sub-carrier. When the sub-carrier spacing is very small, the receiver synchronization components need to be very accurate, which is still not possible with low-cost Radio Frequency (RF) hardware. Thus, a reasonable trade-off between the sub-carrier spacing and the number of sub-carriers must be achieved.

Table 3.1 describes how multi-carrier approach can convert the channel to flat fading channel from frequency selective fading channel. We have considered a multi-carrier system with respect to a single carrier system, where the system data rate requirement is 1 Mbps. When we use 128 sub-carriers for multi-carrier system, we can see that the ISI problem is clearly solved. It is obvious that if we increase the number of sub-carriers, the system will theoretically provide even better performance.

3.3 OFDM Transceiver Systems

A complete OFDM transceiver system is described in Figure 3.2. In this model, *Forward Error Control/Correction* (FEC) coding and interleaving are added in

Table 3.1 Comparison of single carrier and multi-carrier approach in terms of channel frequency selectivity.

Design parameters for outdoor channel	Required data rate	1 Mbps
	RMS delay spread, τ_{rms}	$10\,\mu s$
	Channel coherence bandwidth, $B_c = \frac{1}{5\tau_{rms}}$	20 KHz
	Frequency selectivity condition	$\sigma > \frac{T_{sym}}{10}$
Single carrier approach	Symbol duration, T_{sym}	$1\,\mu s$
	Frequency selectivity	$10\,\mu s > \frac{1\mu s}{10} \Longrightarrow$ **YES**
	ISI occurs as the channel is frequency selective	
Multi-carrier approach	Total number of sub-carriers	128
	Data rate per sub-carrier	7.8125 kbps
	Symbol duration per sub-carrier	$T_{carr} = 128\mu s$
	Frequency selectivity	$10\,\mu s > \frac{128\mu s}{10} \Longrightarrow$ **NO**
	ISI is reduced as flat fading occurs. CP completely removes the remaining ISI; and also inter-block interference is removed	

the system to obtain the robustness needed to protect against burst errors (see Section 3.4 for details). An OFDM system with addition of channel coding and interleaving is referred to as *Coded OFDM* (COFDM).

In a digital domain, binary input data is collected and FEC coded with schemes such as convolutional codes. The coded bit stream is interleaved to obtain diversity gain. Afterwards, a group of channel coded bits are gathered together (1 for BPSK, 2 for QPSK, 4 for QPSK, etc.) and mapped to corresponding constellation points. At this point, the data is represented in complex numbers and they are in serial. Known pilot symbols mapped with known mapping schemes can be inserted at this moment. A serial to parallel converter is applied and the IFFT operation is performed on the parallel complex data. The transformed data is grouped together again, as per the number of required transmission sub-carriers. Cyclic prefix is inserted in every block of data according to the system specification and the data is multiplexed to a serial fashion. At this point of time, the data is OFDM modulated and ready to be transmitted. A *Digital-to-Analog Converter* (DAC) is used to transform the time domain digital data to time domain analog data. RF modulation is performed and the signal is up-converted to transmission frequency.

After the transmission of OFDM signal from the transmitter antenna, the signals go through all the anomaly and hostility of wireless channel. After receiving the signal, the receiver down-converts the signal; and converts to digital domain using *Analog-to-Digital Converter* (ADC). At the

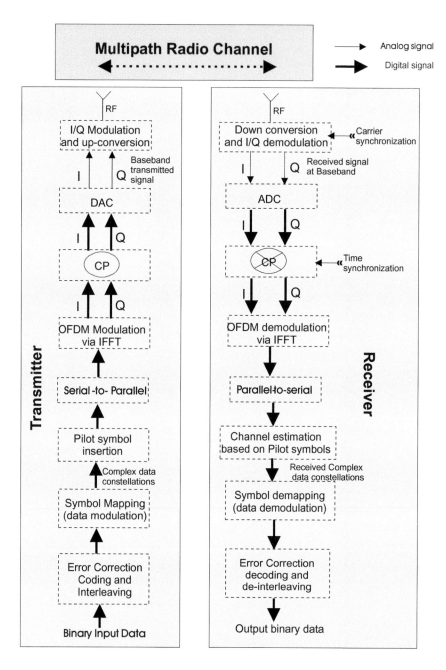

Fig. 3.2 OFDM transceiver model.

time of down-conversion of received signal, carrier frequency synchronization is performed. After ADC conversion, symbol timing synchronization is achieved. An FFT block is used to demodulate the OFDM signal. After that, channel estimation is performed using the demodulated pilots. Using the estimations, the complex received data is obtained which are demapped according to the transmission constellation diagram. At this moment, FEC decoding and de-interleaving are used to recover the originally transmitted bit stream.

3.4 Channel Coding and Interleaving

Since OFDM carriers are spread over a frequency range, there still may be some frequency selective attenuation on a time varying basis. A deep fade on a particular frequency may cause the loss of data on that frequency for that given time, thus some of the sub-carriers can be strongly attenuated and that will cause burst errors. In these situations, FEC in COFDM can fix the errors [60]. An efficient FEC coding in flat fading situations leads to a very high coding gain, especially if soft decision decoding is applied. In a single carrier modulation, if such a deep fade occurs, too many consecutive symbols may be lost and FEC may not be too effective in recovering the lost data [61].

Experiences show that basic OFDM system is not able to obtain a BER of 10^{-5} or 10^{-6} without channel coding. Thus, all OFDM systems now-a-days are converted to COFDM. The benefits of COFDM are twofold in terms of performance improvement. First, the benefit that the channel coding brings in, that is the robustness to burst error. Second, interleaving brings frequency diversity. The interleaver ensures that adjacent outputs from channel encoder are placed far apart in frequency domain. Specifically for a $\frac{1}{2}$ rate encoder, the channel encoder provides two output bits for one source bit. When they are placed far apart from each other (i.e. placed on sub-carriers that are far from each other in frequency domain), then they experience unique gain (and/or unique fade). It is very unlikely that both of the bits will face a deep fade, and thus at least one of the bits will be received intact on the receiver side, and as a result, overall BER performance will improve [57].

According to Table 3.2, IEEE 802.11a standard offers wide variety of choices for coding and modulation, this allows a chance of making trade-offs for lot of considerations. The standard enables several data rates by making use of different combinations of modulation and channel coding scheme. It

Table 3.2 IEEE 802.11a OFDM PHY modulation techniques.

Data rate (Mbps)	Modulation scheme	Coding rate	Coded bits per sub-carrier	Code bits per OFDM symbol	Data bits per OFDM symbol
6	BPSK	$\frac{1}{2}$	1	48	24
9	BPSK	$\frac{3}{4}$	1	48	36
12	QPSK	$\frac{1}{2}$	2	96	48
18	QPSK	$\frac{3}{4}$	2	96	72
24	16-QAM	$\frac{1}{2}$	4	192	96
36	16-QAM	$\frac{3}{4}$	4	192	144
48	64-QAM	$\frac{2}{3}$	6	288	192
54	64-QAM	$\frac{3}{4}$	6	288	216

is worth mentioning here that the standard demands all 802.11a complaint products to support all the data rates. Table 3.2 presents the different arrangement of modulation and coding scheme that are used to obtain the data rates [62].

3.5   Analytical Model of OFDM System

In this section, an analytical time-domain model of an OFDM transmitter and receiver, as well as a channel model, are derived.

3.5.1   Transmitter

The sth OFDM symbol is found using the sth sub-carrier block, $\mathbf{X}_s[k]$. In practice, the OFDM signal is generated using an inverse DFT. In the following model, the transmitter is assumed ideal, i.e. sampling or filtering do not affect the signal on the transmitter side. Therefore, a continuous transmitter output signal may be constructed directly using a Fourier series representation within each OFDM symbol interval.

Each OFDM symbol contains N sub-carriers, where N is an even number (frequently a power of two). The OFDM symbol duration is T_u seconds, which must be a whole number of periods for each sub-carrier. Defining the sub-carrier spacing as $\Delta\omega$, the shortest duration that meets this requirement is written as

$$T_u = \frac{2\pi}{\Delta\omega} \Leftrightarrow \Delta\omega = \frac{2\pi}{T_u} = 2\pi\,\Delta f. \tag{3.1}$$

Using this relation, the spectrum of the Fourier series for the duration of the
sth OFDM symbol is written as

$$\mathbf{X}_s(\omega) = \sum_{k=-N/2}^{N/2-1} \mathbf{X}_s[k]\delta_c(\omega - k\Delta\omega). \tag{3.2}$$

In order to provide the OFDM symbol in the time-domain, the spectrum in
(3.2) is inverse Fourier transformed and limited to a time interval of T_u. The
time-domain signal, $\tilde{x}_s(t)$, is therefore written as

$$
\begin{aligned}
\tilde{x}_s(t) &= \mathcal{F}\{\mathbf{X}_s(\omega)\}\,\Xi_{T_u}(t) \\
&= \begin{cases} \frac{1}{\sqrt{T_u}}\sum_{k=-N/2}^{N/2-1}\mathbf{X}_s[k]e^{j\Delta\omega kt} & 0 \le t < T_u \\ 0 & \text{otherwise,} \end{cases}
\end{aligned} \tag{3.3}
$$

where Ξ_{T_u} is a unity amplitude rectangular gate pulse of duration T_u. Following
the frequency- to time-domain conversion, the signal is extended, and the
cyclic prefix is added:

$$\tilde{x}'_s(t) = \begin{cases} \tilde{x}_s(t + T_u - T_g) & 0 \le t < T_g \\ \tilde{x}_s(t - T_g) & T_g < t < T_s \\ 0 & \text{otherwise,} \end{cases} \tag{3.4}$$

where T_g is the cyclic prefix duration and $T_s = T_u + T_g$ is the total OFDM
symbol duration. It should be noted, that (3.4) has the following property:

$$\tilde{x}'_s(t) = \tilde{x}'_s(t + T_u) \Leftrightarrow 0 \le t < T_g \tag{3.5}$$

that is, a periodicity property within the interval $[0, T_g]$. The transmitted complex baseband signal, $\tilde{s}(t)$, is formed by concatenating all OFDM symbols in
the time-domain:

$$\tilde{s}(t) = \sum_{s=0}^{S-1} \tilde{x}'_s(t - sT_s). \tag{3.6}$$

This signal is finally upconverted to a carrier frequency and transmitted:

$$s(t) = \Re e\left\{\tilde{s}(t)e^{j2\pi f_c t}\right\}, \tag{3.7}$$

where $s(t)$ denotes the transmitted RF signal and f_c is the RF carrier frequency.
For frequency hopping systems, the carrier frequency is changed at certain
intervals. This is written as

$$f_c[s] = f_{c,0} + f_h[s], \tag{3.8}$$

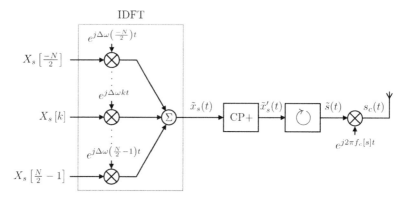

Fig. 3.3 Transmitter diagram for the OFDM analytical model, given by (3.1)–(3.8). The sub-carriers for the sth OFDM symbol each modulate a carrier, which are separated by $\Delta\omega$. The resulting waveforms are then summed, and the CP is added. The symbol $\circlearrowleft$ represents the concatenation of the OFDM symbols, given by (3.6). The resulting signal is then converted to a carrier frequency and transmitted.

where $f_c[s]$ is the carrier frequency for the sth OFDM symbol, $f_{c,0}$ is the center frequency of the band, and $f_h[s]$ is the frequency deviation from the band center when transmitting the sth OFDM symbol. The period of $f_h[s]$ is χ, where χ is the hopping sequence period measured in whole OFDM symbols.

The transmitter model described in this section is illustrated in Figure 3.3.

3.5.2 Channel

The channel is modeled as a time-domain complex-baseband transfer function, which may then be convolved with the transmitted signal to determine the signal at the receiver side. The channel baseband equivalent impulse response function for the uth user, $\tilde{h}_u(t)$ is defined as

$$\tilde{h}_u(\tau, t) = \sum_{l=0}^{L} h_{u,l}(t)\delta_c(\tau - \tau_l), \tag{3.9}$$

where $h_{u,l}(t)$ is the complex gain of the lth multi-path component for the uth user at time t. The channel is assumed to be static for the duration of one OFDM symbol, and the path gain coefficients for each path contribution are assumed to be uncorrelated. No assumption is made for the autocorrelation properties of each path, except in the case of frequency hopping systems. In such systems, the channel is assumed to be completely uncorrelated between two frequency hops, provided that the distance in frequency is sufficiently large.

As the channel is assumed to be static over each OFDM symbol, (3.9) is redefined as

$$\tilde{h}_{u,s}(t) = \sum_{l=0}^{L} h_{u,l}[s]\delta_c(t - \tau_l), \tag{3.10}$$

where

$$h_{u,l}[s] = h_{u,l}(t); \quad sT_s \le t < (s+1)T_s.$$

The corresponding frequency-domain channel transfer function, $H_{u,s}$, can then be found using Fourier transformation:

$$H_{u,s}(\omega) = \mathcal{F}\{\tilde{h}_{u,s}(t)\}$$
$$= \int_{-\infty}^{\infty} \tilde{h}_{u,s}(t)e^{j\omega t}dt. \tag{3.11}$$

The time-domain channel model is illustrated in Figure 3.4.

3.5.3 Receiver

The signal at the receiver side consists of multiple echoes of the transmitted signal, as well as thermal (white gaussian) noise and interference. The RF signal received by the uth user is written as

$$r(t) = \Re e\{(\tilde{s}(t) * \tilde{h}_{u,s}(t))e^{j2\pi f_c[s]t}\} + v(t); \quad sT_s \le t < (s+1)T_s, \tag{3.12}$$

where $v(t)$ is a real valued, passband signal combining additive noise and interference. The receiver now has to recreate the transmitted signal. Aside from noise and multi-path effects, other imperfections in the receiver may also affect this process:

Timing error: In order to demodulate the signal, the receiver must establish the correct timing. This means that the receiver must estimate which time

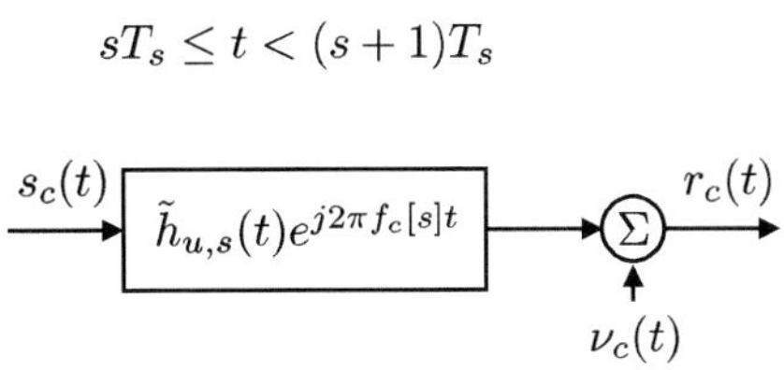

$$sT_s \le t < (s+1)T_s$$

Fig. 3.4 A diagram of the channel model given by (3.9) and (3.10). The transmitted signal passes through the channel, and noise is added.

instant corresponds to $t = 0$ in the received signal (as seen from the transmitted signal point of view). As there are different uncertainties involved, a timing error of δt is assumed.

Frequency Error: Similarly, the local oscillator of the receiver may oscillate at an angular frequency that is different from the angular frequency of the incoming signal. This difference is denoted $\delta\omega = 2\pi\,\delta f$.

The shifted timescale in the receiver is denoted by $t' = t - \delta t$. Furthermore, due to the angular frequency error $\delta\omega$, the down-converted signal spectrum is shifted in frequency. The down-converted signal is therefore written as

$$\tilde{r}(t) = (\tilde{s}(t') * \tilde{h}_{u,s}(t))e^{j\delta\omega t} + \tilde{v}(t'); \quad sT_s \le t < (s+1)T_s, \tag{3.13}$$

where $\tilde{v}(t)$ is the complex envelope of the down-converted AWGN. The signal is divided into blocks of T_s each, and the CP is removed from each of them. The sth received OFDM symbol block, $y_s'(t)$ is defined as

$$\tilde{y}_s'(t) = \tilde{r}(t' - sT_s); \quad 0 \le t < T_s. \tag{3.14}$$

The signal block corresponding to $\tilde{x}_s(t)$, $\tilde{y}_s(t)$ is found by removing the CP from each $\tilde{y}_s'(t)$:

$$\tilde{y}_s(t) = \tilde{y}_s'(t + T_g); \quad 0 \le t < T_s - T_g, \tag{3.15}$$

which can be rewritten as

$$
\begin{aligned}
\tilde{y}_s(t) &= \tilde{y}_s'(t + T_g); \quad 0 \le t \le T_u \\
&= \tilde{r}(t' + T_g - sT_s) \\
&= (\tilde{s}(t' + T_g - sT_s) * \tilde{h}_{u,s}(t))e^{j\delta\omega t} + \tilde{v}(t' + T_g - sT_s) \\
&= (\tilde{x}_s'(t' + T_g) * \tilde{h}_{u,s}(t))e^{j\delta\omega t} + \tilde{v}_s(t') \\
&= (\tilde{x}_s(t') * \tilde{h}_{u,s}(t))e^{j\delta\omega t} + \tilde{v}_s(t'),
\end{aligned}
\tag{3.16}
$$

where $\tilde{v}_s(t')$ is the noise signal block of duration T_u corresponding to the sth OFDM symbol.

In order to recreate the transmitted sub-carriers, N correlators are used, each one correlating the incoming signal with the kth sub-carrier frequency over an OFDM symbol period:

$$Y_s[k] = \frac{1}{\sqrt{T_u}} \int_0^{T_u} \tilde{y}_s(t')e^{j\Delta\omega kt}\,dt. \tag{3.17}$$

In order to determine the correlator output, (3.17) may be seen as taking the continuous Fourier transform of (3.16) multiplied by the rectangular pulse $\Xi_{T_u}(t)$ and evaluating it at the corresponding sub-carrier frequency. Assuming that the timing error is low enough to avoid ISI:

$$0 \leq \delta t < T_g - \max(\tau_l)$$

the continuous Fourier transform can be written as

$$
\begin{aligned}
Y_s(\omega) &= \mathcal{F}\{\tilde{y}_s(t)\Xi_{T_u}(t)\} \\
&= \mathcal{F}\{(\tilde{x}_s(t') * \tilde{h}_{u,s}(t))e^{j\delta\omega t} + \tilde{v}_s(t')\} * T_u e^{j\pi\frac{\omega}{\Delta\omega}} \text{sinc}\left(\frac{\omega}{\Delta\omega}\right) \\
&= \mathcal{F}\{(\tilde{x}_s(t') * \tilde{h}_{u,s}(t))e^{j\delta\omega t}\} * T_u e^{j\pi\frac{\omega}{\Delta\omega}} \text{sinc}\left(\frac{\omega}{\Delta\omega}\right) + N_s(\omega) \\
&= \mathcal{F}\{\tilde{x}_s(t') * \tilde{h}_{u,s}(t)\} * \delta_c(\omega - \delta\omega) * T_u e^{j\pi\frac{\omega}{\Delta\omega}} \text{sinc}\left(\frac{\omega}{\Delta\omega}\right) + N_s(\omega) \\
&= e^{-j\omega\delta t} \mathcal{F}\{\tilde{x}_s(t) * \tilde{h}_{u,s}(t)\} * \delta_c(\omega - \delta\omega) \\
&\quad * T_u e^{j\pi\frac{\omega}{\Delta\omega}} \text{sinc}\left(\frac{\omega}{\Delta\omega}\right) + N_s(\omega) \\
&= e^{-j\omega(\delta t + \frac{\pi}{\Delta\omega})} \sum_{k'=N/2}^{N/2-1} \mathbf{X}_s\left[k'\right] H_{u,s}(k'\Delta\omega) \\
&\quad \times \text{sinc}\left(\frac{\omega - k'\Delta\omega - \delta\omega}{\Delta\omega}\right) + N_s(\omega),
\end{aligned}
\tag{3.18}
$$

where

$$N_s(\omega) = \mathcal{F}\{\tilde{v}_s(t')\} * T_u e^{j\pi\frac{\omega}{\Delta\omega}} \text{sinc}\left(\frac{\omega}{\Delta\omega}\right) \tag{3.19}$$

is the Fourier transform of the AWGN contribution. The correlator output at the kth correlator is then found as

$$
\begin{aligned}
Y_s[k] &= Y_s(k\Delta\omega) \\
&= e^{-jk\Delta\omega(\delta t + \frac{\pi}{\Delta\omega})} \sum_{k'=N/2}^{N/2-1} \mathbf{X}_s\left[k'\right] H_{u,s}(k'\Delta\omega)\text{sinc}\left(\frac{k\Delta\omega - k'\Delta\omega - \delta\omega}{\Delta\omega}\right) \\
&\quad + N_s(k\Delta\omega).
\end{aligned}
\tag{3.20}
$$

For zero frequency error, (3.20) reduces to

$$Y_s[k] = e^{-jk\Delta\omega(\delta t + \frac{\pi}{\Delta\omega})}\mathbf{X}_s[k] H_{u,s}[k] + N_s[k]; \quad \delta\omega = 0, \tag{3.21}$$

where

$$N_s[k] = N_s(k\Delta\omega) \tag{3.22}$$

$$H_{u,s}[k] = H_{u,s}(k\Delta\omega). \tag{3.23}$$

From (3.20), it is seen that the kth correlator output, $Y_s[k]$ corresponds to the transmitted sub-carrier, $\mathbf{X}_s[k]$, with AWGN, ICI and a complex gain term (amplitude and phase shift) due to imperfect timing and channel effects. The analytical model for the receiver is illustrated in Figure 3.5.

When estimating the channel, the constant phase rotation term and the channel transfer function would be estimated jointly (as the receiver cannot discern between the two). In the following, the timing delay phase shift is omitted for clarity. Defining the equalization factor for the kth sub-carrier of the sth OFDM symbol and uth user as $Z_{u,s}[k]$, the sub-carrier estimate is written as

$$\hat{\mathbf{X}}_s[k] = Z_{u,s}[k]Y_s[k]$$

$$= Z_{u,s}[k]H_{u,s}[k]\mathbf{X}_s[k] + Z_{u,s}[k]N_s[k]. \tag{3.24}$$

Assuming a zero-forcing, frequency-domain equalizer (as well as perfect channel estimation and zero frequency error), the corresponding equalizer gain is written as

$$Z_{u,s}[k] = \frac{1}{H_{u,s}[k]}$$

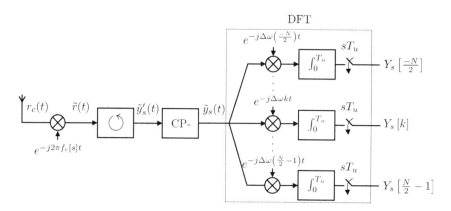

Fig. 3.5 Receiver diagram for the OFDM analytical model, given by (3.12)–(3.21). The received signal (suffering from multi-path effects and AWGN) is converted down to baseband. The symbol ↻ represents the division of the received signal into blocks, given by (3.14). The CP is removed from each block, and the signal is then correlated with each sub-carrier frequency, as shown by (3.17).

and (3.24) is rewritten as

$$\hat{\mathbf{X}}_s[k] = \mathbf{X}_s[k] + \frac{N_s[k]}{H_{u,s}[k]}.$$ (3.25)

It is observed, that although this is an unbiased estimator for $\mathbf{X}_s[k]$, the signal-to-noise ratio decreases drastically for sub-carriers in deep fades.

3.5.4 Sampling

Although the receiver may be modeled in the continuous time-domain, an OFDM receiver uses discrete signal processing to obtain the estimate of the transmitted sub-carriers.

When the received signal is modeled as a Dirac impulse train, i.e. an ideally sampled signal, (3.16) is instead written as

$$\tilde{y}_{s,d}(t) = \sum_{n=0}^{N-1} \tilde{y}_s[n]\delta_c(t - nT),$$ (3.26)

where

$$T = \frac{T_u}{N}$$ (3.27)

is the sample duration and

$$\tilde{y}_s[n] = \tilde{y}_s(nT); \quad n \in \{0, 1, \ldots, N - 1\}$$ (3.28)

is the discrete sequence corresponding to the sampled values of $\tilde{y}_s(t)$. When (3.26) is inserted into (3.17), the correlation becomes the Discrete Fourier Transform of the received signal. It can be shown, however, that (3.20)–(3.25) are still valid in the discrete-time case.

3.6 Single OFDM Symbol Baseband Model in Matrix Notations

In this section, we explain the above analytical model in matrix model, so that it becomes easier to implement in simulation programs, such as in MATLAB simulations, we have to model all the components in the transmission chain in matrix format, thus the following model will be very useful in that regard. The baseband model for a single OFDM symbol s is shown in Figure 3.6.

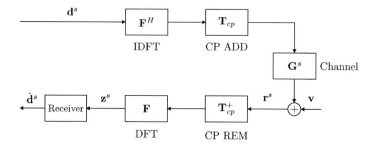

Fig. 3.6 Single OFDM symbol system model.

The single symbol received signal $\mathbf{r}^s \in \mathbb{C}^{[N_g+N+\delta_{\mathrm{g}}\times 1]}$ may be written as

$$\mathbf{r}^s = \mathbf{G}^s \mathbf{T}_{cp} \mathbf{F}^H \mathbf{d}^s + \mathbf{v} \qquad (3.29)$$

with

$$\mathbf{T}_{cp} = \begin{bmatrix}
 & \overbrace{}^{N-N_g} & \overbrace{}^{N_g} \\
 & & 1 & & \\
 & & & \ddots & \\
 & & & & 1 \\
1 & & & & \\
 & \ddots & & & \\
 & & 1 & & \\
 & & & \ddots & \\
 & & & & 1
\end{bmatrix}$$

denoting the $\left[N_g + N \times N\right]$ cyclic prefix insertion matrix, where N_g is the number of samples in the guard period, and with

$$\mathbf{G}^s = \begin{bmatrix}
g_n[0] & \mathbf{0} & \cdots & \cdots & \cdots & \cdots \\
g_{n+1}[1] & g_{n+1}[0] & \mathbf{0} & \ddots & \ddots & \ddots \\
\vdots & \vdots & \ddots & \ddots & \ddots & \ddots \\
g_{n+L-1}[L-1] & g_{n+L-1}[L-2] & \cdots & g_{n+L-1}[0] & \ddots & \ddots \\
\mathbf{0} & \ddots & \ddots & \ddots & \ddots & \ddots \\
\vdots & \ddots & g_{n+N_g+N-1}[L-1] & \cdots & \cdots & g_{n+N_g+N-1}[0]
\end{bmatrix}$$

denoting the $\left[N_g + N \times N_g + N\right]$ time-domain channel convolution matrix, where $g_n\,[l]$ represents the gain of the lth sample delayed path in respect to the first path at time n of the time-varying channel impulse response, and where L is the span of samples from the first to the last considered path. The vector $\mathbf{v} \in \mathbb{C}^{[N_g+N\times 1]}$ represents complex valued circular symmetric white Gaussian noise with variance N_0, and the matrix $\mathbf{F}^H \in \mathbb{C}^{[N\times N]}$ corresponds to the IDFT operation and is the hermitian transposed of the DFT-matrix $\mathbf{F} \in \mathbb{C}^{[N\times N]}$.

The received vector $\mathbf{z}^s \in \mathbb{C}^{[N\times 1]}$ after the DFT-operation may be expressed as

$$\mathbf{z}^s = \mathbf{FT}_{cp}^+ \mathbf{r}^s \tag{3.30}$$

$$= \mathbf{FT}_{cp}^+ \mathbf{G}^s \mathbf{T}_{cp} \mathbf{F}^H \mathbf{d}^s + \mathbf{FT}_{cp}^+ \mathbf{v} \tag{3.31}$$

with

$$\mathbf{T}_{cp}^+ = \begin{bmatrix} 1 & & & \\ & \ddots & & \\ & & & 1 \end{bmatrix}$$
$$\underbrace{}_{N_g} \underbrace{}_{N}$$

denoting the $\left[N \times N_g + N\right]$ cyclic prefix removal matrix. Assuming that $N_g \geq (L - 1)$, the linear convolution of the transmitted sequence and the channel corresponds to a circular convolution. Assuming a time-invariant channel in the discrete time interval $\left[n; n + N_g + N - 1\right]$, the received vector $\mathbf{z}$ after the DFT-operation may be expressed as

$$\mathbf{z}^s = \mathbf{H}^s \mathbf{d}^s + \mathbf{n}, \tag{3.32}$$

with $\mathbf{H}^s \in \mathbb{C}^{[N\times N]}$ denoting the frequency domain diagonal channel matrix, where the $[k, k]$th element of $\mathbf{H}^s$ corresponds to the complex-valued channel gain of the kth sub-carrier. The vector $\mathbf{n} \in \mathbb{C}^{[N\times 1]}$ represents complex valued circular symmetric white Gaussian noise with variance N_0. An estimate $\hat{\mathbf{d}}^s$ of the transmitted data symbols $\mathbf{d}^s$ may be calculated from $\mathbf{z}^s$ using various topologies as indicated in Figure 3.7.

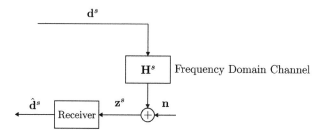

Fig. 3.7 Simplified single OFDM symbol system model.

3.7   Advantages of OFDM System

3.7.1   Combating ISI and Reducing ICI

When signal passes through a time-dispersive channel, the orthogonality of the signal can be jeopardized. CP helps to maintain orthogonality between the sub-carriers. Before CP was invented, guard interval was proposed as the solution (see Figure 3.8). Guard interval was defined by an empty space between two OFDM symbols, which serves as a buffer for the multi-path reflection. The interval must be chosen as larger than the expected maximum delay spread, such that multi-path reflection from one symbol would not interfere with another. In practice, the empty guard time introduces ICI. ICI is crosstalk between different sub-carriers, which means they are no longer orthogonal to each other [11]. A better solution was later found, that is cyclic extension of OFDM symbol or CP. CP is a copy of the last part of OFDM symbol which is appended to front the transmitted OFDM symbol [63].

Cyclic Prefix still occupies the same time interval as guard period, but it ensures that the delayed replicas of the OFDM symbols will always have a complete symbol within the FFT interval (often referred as FFT window); this makes the transmitted signal periodic. This periodicity plays a very significant role as this helps maintaining the orthogonality. The concept of being able to do this, and what it means, comes from the nature of IFFT/FFT process. When the IFFT is taken for a symbol period during OFDM modulation, the resulting time sample process is technically periodic. In a Fourier transform, all the resultant components of the original signal are orthogonal to each other. So, in short, by providing periodicity to the OFDM source signal, CP makes sure that subsequent sub-carriers are orthogonal to each other (see Figure 3.9).

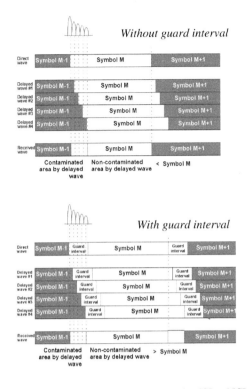

Fig. 3.8 Role of guard intervals in combatting ISI and ICI.

At the receiver side, CP is removed before any processing starts. As long as the length of CP interval is larger than maximum expected delay spread τ_{max}, all reflections of previous symbols are removed and orthogonality is restored. The orthogonality is lost when the delay spread is larger than length of CP interval. Inserting CP has its own cost, we loose a part of signal energy since it carries no information. The loss is measured as

$$\text{SNR}_{\text{loss}\_CP} = -10\log_{10}\left(1 - \frac{T_{\text{CP}}}{T_{\text{sym}}}\right) \qquad (3.33)$$

Here, T_{CP} is the interval length of CP and T_{sym} is the OFDM symbol duration. It is understood that although we loose part of signal energy, the fact that zero ICI and ISI situation pay off the loss.

To conclude, CP gives twofold advantages, first occupying the guard interval, it removes the effect of ISI and by maintaining orthogonality it completely removes the ICI. The cost in terms signal energy loss is not too significant.

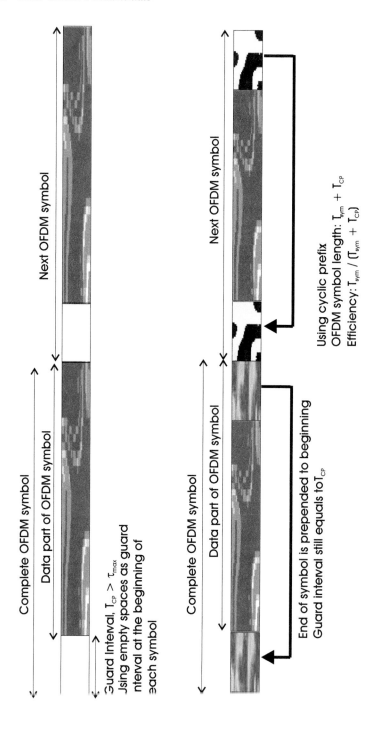

Fig. 3.9 Definition of cyclic prefix as the guard interval in OFDM systems.

3.7.2 Spectral Efficiency

Figure 3.10 illustrates the difference between conventional FDM and OFDM systems. In the case of OFDM, a better spectral efficiency is achieved by maintaining orthogonality between the sub-carriers. When orthogonality is maintained between different subchannels during transmission, then it is possible to separate the signals very easily at the receiver side. Classical FDM ensures this by inserting guard bands between subchannels. These guard bands keep the subchannels far enough so that separation of different subchannels are possible. Naturally inserting guard bands results to inefficient use of spectral resources.

Orthogonality makes it possible in OFDM to arrange the sub-carriers in such a way that the sidebands of the individual carriers overlap and still the signals are received at the receiver without being interfered by ICI. The receiver

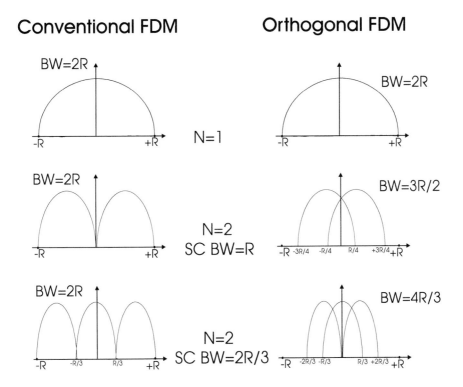

Fig. 3.10 Spectrum efficiency of OFDM compared to conventional FDM.

acts as a bank of demodulator, translating each sub-carrier down to DC, with the resulting signal integrated over a symbol period to recover raw data. If the other sub-carriers are down converted to the frequencies that, in the time domain, have a whole number of cycles in a symbol period T_{sym}, then the integration process results in zero contribution from all other carriers. Thus, the sub-carriers are linearly independent (i.e., orthogonal) if the carrier spacing is a multiple of $\frac{1}{T_{\text{sym}}}$ [64].

3.7.3 Some Other Benefits of OFDM System

1. The beauty of OFDM lies in its simplicity. One trick of the trade that makes OFDM transmitters low cost is the ability to implement the mapping of bits to unique carriers via the use of IFFT [57].

2. Unlike CDMA, OFDM receiver collects signal energy in frequency domain, thus it is able to protect energy loss at frequency domain.

3. In a relatively slow time-varying channel, it is possible to significantly enhance the capacity by adapting the data rate per sub-carrier according to SNR of that particular sub-carrier [11].

4. OFDM is more resistant to frequency selective fading than single carrier systems.

5. The OFDM transmitter simplifies the channel effect, thus a simpler receiver structure is enough for recovering transmitted data. If we use coherent modulation schemes, then very simple channel estimation (and/or equalization) is needed, on the other hand, we need no channel estimator if differential modulation schemes are used.

6. The orthogonality preservation procedures in OFDM are much simpler compared to CDMA or TDMA techniques even in very severe multi-path conditions.

7. It is possible to use maximum likelihood detection with reasonable complexity [60].

8. OFDM can be used for high-speed multimedia applications with lower service cost.

9. OFDM can support dynamic packet access.

10. Single frequency networks are possible in OFDM, which is especially attractive for broadcast applications.
11. Smart antennas can be integrated with OFDM. MIMO systems and space–time coding can be realized on OFDM and all the benefits of MIMO systems can be obtained easily. Adaptive modulation and tone/power allocation are also realizable on OFDM.

3.8 Disadvantages of OFDM System

3.8.1 Strict Synchronization Requirement

Orthogonal Frequency Division Multiplexing is highly sensitive to time and frequency synchronization errors, especially at frequency synchronization errors, everything can go wrong [13]. Demodulation of an OFDM signal with an offset in the frequency can lead to a high bit error rate.

The source of synchronization errors are two; first one being the difference between local oscillator frequencies in transmitter and receiver, second relative motion between the transmitter and the receiver that gives Doppler spread. Local oscillator frequencies at both points must match as closely as they can. For higher number of subchannels, the matching should be even more perfect. Motion of transmitter and receiver causes the other frequency error. So, OFDM may show significant performance degradation at high-speed moving vehicles [56].

To optimize the performance of an OFDM link, accurate synchronization is a prime importance. Synchronization needs to be done in three factors: symbol, carrier frequency, and sampling frequency synchronization. A good description of synchronization procedures is given in [65]. We have discussed the synchronization issues in detail in Chapter 6.

3.8.2 Peak-to-Average Power Ratio (PAPR)

Peak to Average Power Ratio (PAPR) is proportional to the number of sub-carriers used for OFDM systems. An OFDM system with large number of sub-carriers will thus have a very large PAPR when the sub-carriers add up coherently. Large PAPR of a system makes the implementation of DAC and ADC to be extremely difficult. The design of RF amplifier also becomes increasingly difficult as the PAPR increases.

There are basically three techniques that are used at present to reduce PAPR, they are *Signal Distortion Techniques*, *Coding Techniques*, and finally the *Scrambling Technique*. Since OFDM is characterized by

$$x(t) = \frac{1}{\sqrt{N}} \sum_{n=1}^{N} a_n e^{j w_n t}. \tag{3.34}$$

Here a_n is the modulating signal. For Large number of a_n both the real and imaginary parts tend to be Gaussian distributed, thus the amplitude of the OFDM symbol has a Rayleigh distribution, while the power distribution is central chi squared.

The clipping and windowing technique reduces PAPR by nonlinear distortion of the OFDM signal. It thus introduces self interference as the maximum amplitude level is limited to a fixed level. It also increases the out of band radiation, but this is the simplest method to reduce the PAPR. To reduce the error rate, additional Forward error correcting codes can be used in conjunction with the the clipping and windowing method.

Another technique called *Linear Peak Cancelation* can also be used to reduce the PAPR. In this method, time shifted and scaled reference function is subtracted from the signal, such that each subtracted reference function reduces the peak power of at least one signal sample. By selecting an appropriate reference function with approximately the same bandwidth as the transmitted function, it can be assured that the peak power reduction does not cause out of band interference. One example of a suitable reference function is a *raised cosine window*. Detailed discussion about coding methods to reduce PAPR can be found in [11].

3.8.3   Co-Channel Interference in Cellular OFDM

In cellular communications systems, CCI is combated by combining adaptive antenna techniques, such as sectorization, directive antenna, antenna arrays, etc. Using OFDM in cellular systems will give rise to CCI. Similarly with the traditional techniques, with the aid of beam steering, it is possible to focus the receiver's antenna beam on the served user, while attenuating the co-channel interferers. This is significant since OFDM is sensitive to CCI.

3.9 OFDM System Design Issues

System design always needs a complete and comprehensive understanding and consideration of critical parameters. OFDM system design is of no exception, it deals with some critical, and often conflicting parameters. Basic OFDM philosophy is to decrease data rate at the sub-carriers, so that the symbol duration increases, thus the multi-paths are effectively removed (see Figure 3.11). This poses a challenging problem, as higher value for CP interval will give better result, but it will increase the loss of energy due to insertion of CP. Thus, a trade-off between these two must be obtained for a reasonable design (see Figure 3.12).

3.9.1 OFDM System Design Requirements

OFDM systems depend on four system requirement:

- *Available bandwidth*: Bandwidth is always the scarce resource, so the mother of the system design should be available for bandwidth

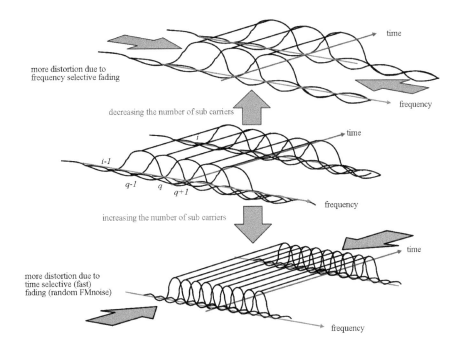

Fig. 3.11 Design of sub-carrier spacing in OFDM systems.

more energy loss

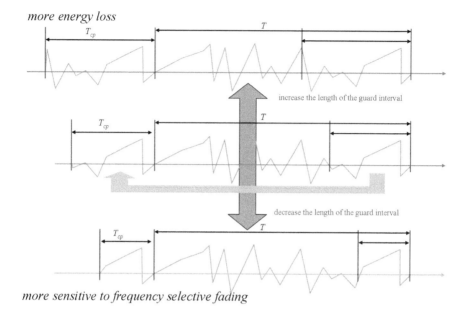

more sensitive to frequency selective fading

Fig. 3.12 Design of CP duration in OFDM systems.

for operation. The amount of bandwidth will play a significant role in determining number of sub-carriers, because with a large bandwidth, we can easily fit in large number of sub-carriers with reasonable guard space.

- *Required bit rate*: The overall system should be able to support the data rate required by the users. For example, to support broadband wireless multimedia communication, the system should operate at more than 10 Mbps at least.

- *Tolerable delay spread*: Tolerable delay spread will depend on the user environment. Measurements show that indoor environment experiences maximum delay spread of few hundreds of $n$sec at most, whereas outdoor environment can experience up to 10μs. So the length of CP should be determined according to the tolerable delay spread.

- *Doppler values*: Users on a high speed vehicle will experience higher Doppler shift whereas pedestrians will experience smaller Doppler shift. These considerations must be taken into account.

3.9.2 OFDM System Design Parameters

The design parameters are derived according to the system requirements. The requirement of the system design must be fulfilled by the system parameters. Following are the design parameters for an OFDM system [11]:

- *Number of sub-carriers*: Increasing number of sub-carriers will reduce the data rate via each sub-carrier, which will make sure that the relative amount of dispersion in time caused by multi-path delay will be decreased. But when there are large numbers of sub-carriers, the synchronization at the receiver side will be extremely difficult.

- *Guard time (CP interval) and symbol duration*: A good ratio between the CP interval and symbol duration should be found, so that all multi-paths are resolved and not significant amount of energy is lost due to CP. As a thumb rule, the CP interval must be two to four times larger than the *Root-Mean-Square* (RMS) delay spread. Symbol duration should be much larger than the guard time to minimize the loss of SNR, but within reasonable amount. It cannot be arbitrarily large, because larger symbol time means that more sub-carriers can fit within the symbol time. More sub-carriers increase the signal processing load at both the transmitter and the receiver, increasing the cost and complexity of the resulting device [66].

- *Sub-carrier spacing*: Sub-carrier spacing must be kept at a level so that synchronization is achievable. This parameter will largely depend on available bandwidth and the required number of sub-channels.

- *Modulation type per sub-carrier*: This is trivial, because different modulation scheme will give different performance. Adaptive modulation and bit loading may be needed depending on the performance requirement. It is interesting to note that the performance of OFDM systems with differential modulation compares quite well with systems using nondifferential and coherent demodulation [67]. Furthermore, the computation complexity in the demodulation process is quite low for differential modulations.

- *FEC coding*: Choice of FEC code will play a vital role also. A suitable FEC coding will make sure that the channel is robust to all the random errors.

4

Multi-Carrier Based Access Techniques

In this chapter, we present some of the main multiple access techniques that can be defined based on the original Orthogonal Frequency Division Multiplexing (OFDM)-type multi-carrier techniques.

4.1 Definition of Basic Schemes

It is imperative to understand the basic properties of three fundamental multi-carrier based multiple access techniques, namely Orthogonal Frequency Division Multiple Access (OFDMA), Orthogonal Frequency Division Multiplexing–Time Division Multiple Access (OFDM–TDMA), and Orthogonal Frequency Division Multiplexing–Code Division Multiple Access (OFDM–CDMA) before embarking on studies related to probable access technique for 4th Generation (4G) wireless communication systems, thus, we here briefly summarize the basic properties of these three access schemes.

4.1.1 OFDM–TDMA

In OFDM–TDMA, a particular user is given all the sub-carriers of the system for any specific OFDM symbol duration. Thus, the users are separated via time slots. All symbols allocated to all users are combined to form a OFDM–TDMA frame. The number of OFDM symbols per frame can be varied based on each users requirement. Frequently, an error correcting code is applied to the data to compensate for the channel nulls experienced by several random bits. This scheme allows Mobile Station (MS) to reduce its power consumption, as the MS shall process only OFDM symbols which are dedicated to it. On the other hand, the data is sent to each user in bursts, thus degrading performance for delay constrained systems delay constrained systems [68].

Different OFDM symbols can be allocated to different users based on certain allocation conditions. Since the OFDM–TDMA concept allocates the whole bandwidth to a single user, a reaction to different sub-carrier attenuations could consist of leaving out highly distorted sub-carriers [15]. The number of OFDM symbols per user in each frame can be adapted accordingly to support heterogeneous data rate requirements. An efficient multiple access scheme should grant a high flexibility when it comes to the allocation of time-bandwidth resources. On the one hand, the behavior of the frequency-selective radio channel should be taken into account, while on the other hand the user requirements for different and/or changing data rates have to be met [69]. For example, both for OFDMA and OFDM–TDMA, the usage of Adaptive Modulation and Coding (AMC) on different sub-carriers, as proposed in [70], may increase overall system throughput and help in further exploiting Channel State Information (CSI).

4.1.2   OFDMA

In OFDMA, available sub-carriers are distributed among all the users for transmission at any time instant. The sub-carrier assignment is made for the user lifetime, or at least for a considerable time frame. The scheme was first proposed for CATV systems [71], and later adopted for wireless communication systems.

Orthogonal Frequency Division Multiple Access (OFDMA) can support a number of identical downstreams, or different user data rates, [e.g. assigning a different number of sub-carriers to each user]. Based on the sub-channel condition, different baseband modulation schemes can be used for the individual sub-channels, e.g. QPSK, 16-QAM, and 64-QAM etc. This is investigated in numerous papers and referred to as adaptive sub-carrier, bit, and power allocation or Quality of Service (QoS) allocation [70, 72, 19, 73].

In OFDMA, frequency hopping, one form of spread spectrum, can be employed to provide security and resilience to inter-cell interference.

In OFDMA, the granularity of resource allocation is higher than that of OFDM–TDMA, i.e. the flexibility can be accomplished by suitably choosing the sub-carriers associated with each user. Here, the fact that each user experiences a different radio channel can be exploited by allocating only "good" sub-carriers with high Signal-to-Noise Ratio (SNR) to each user. Furthermore, the number of sub-channels for a specific user can be varied, according to the required data rate. Thus, multi-rate system can be achieved without increasing system complexity very much.

4.1.3 OFDM–CDMA

In OFDM–CDMA, user data is spread over several sub-carriers and/or OFDM symbols using spreading codes, and combined with signal from other users [16]. The idea of OFDM–CDMA can be attributed to several researchers working independently at almost the same time on hybrid access schemes combining the benefits of OFDM and CDMA. OFDM provides a simple method to overcome the Inter-Symbol Interference (ISI) effect of the multi-path frequency selective wireless channel, while Code Division Multiple Access (CDMA) provides the frequency diversity and the multi-user access scheme. Different types of spreading codes have been investigated. Orthogonal codes are preferred in case of Downlink (DL), since loss of orthogonality is not as severe in DL as it is in Uplink (UL).

Several users transmit over the same sub-carrier. In essence this implies frequency domain spreading, rather than time domain spreading, as it is conceived in a Direct Sequence Code Division Multiple Access (DS-CDMA) system. The channel equalization can be highly simplified in DL, because of the one tap channel equalization benefit offered by OFDM.

In OFDM–CDMA, the flexibility lies in the allocation of all available codes to the users, depending on the required data rates. As OFDM–CDMA is applied using coherent modulation, the necessary channel estimation provides information about the sub-carrier attenuations; this information can be used when performing an equalization in the receiver [74].

4.1.4 Relative Comparison

As shown in the previous discussions, we can consider OFDM–TDMA as the most basic multiple-access scheme, while OFDMA scheme is an extension of OFDM–TDMA, and in turn, OFDM–CDMA scheme as an extension of OFDMA (see Figure 4.1). Going from OFDM–TDMA to OFDM–CDMA, we have increased the level of flexibility in multiple-access of the system, but at the same time, increased the complexity. The OFDM–CDMA shall observe

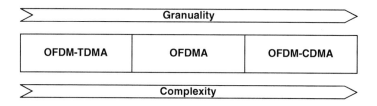

Fig. 4.1 Relative comparison of basic multi-carrier multiple-access techniques

Table 4.1 A summary of multiple access scheme.

	Advantages	Disadvantages
OFDM–TDMA	Power savings (only receives own symbols). Simple resource allocation. Easiest to implement.	Relatively high latency. Frequency-reuse factor ≥ 3. Lowest flexibility.
OFDMA	Simple implementation flexibility.	Frequency-reuse factor ≥ 3.
OFDM–CDMA	Spectral efficiency. Frequency diversity Multiple Access Interference (MAI) and inter-cell interference resistance. Frequency-reuse factor $= 1$ Soft handover capability Highest flexibility.	Requirement of power control. Implementation complexity.

all requirements from OFDMA, plus its owned requirements. And similarly, OFDMA must fulfill all requirements of OFDM–TDMA.

Table 4.1 summarizes advantages and disadvantages of three basic multi-carrier multiple-access schemes:

OFDM–TDMA: OFDM–TDMA is simple to implement, but it may lack in highly delay constraint system. For DL, basic OFDM–TDMA may not perform very well compared to other two schemes, but in UL, it may be very worthy. In UL, user time and frequency offset can cause real havoc in the system, and both OFDMA and OFDM–CDMA have to implement substantial procedure to combat the offsets, while OFDM–TDMA may be able to handle them quite easily. This is based on the fact that the entire bandwidth is allocated to a single user for several OFDM symbols, thus practically avoiding Multiple Access Interference (MAI).

OFDMA: The OFDMA scheme is distinguished by its simplicity, where the multi-access is obtained by allocating a fraction of sub-carriers to different users. The benefit is that the receiver can be implemented in a relatively simple manner.

Orthogonal Frequency Division Multiple Access (OFDMA) is already in use in some standards, e.g. IEEE 802.16a, and can be used for both DL and UL. For the UL case, issues like synchronization are a big issue, and are studied in several papers. Regarding the assignment of sub-carriers, the literature provides no definitive answer whether it should be static/dynamic or contiguous/interleaved.

OFDM–CDMA: This scheme is potentially a very good scheme in DL due to its ability to exploit available frequency diversity, even coded-OFDMA can only make use of limited frequency diversity. It has been pointed that this scheme is vulnerable to near–far effect as a normal CDMA system is. Hence this scheme suits best mainly in an indoor DL scenario [3]. Now one is easily led to argue that in an indoor situation the coherence bandwidth is very large. In 5 GHz band, it ranges from 6 to 20 MHz. Thus to make use of its special advantage of providing frequency diversity, the system has to use a very wide band. Otherwise, even with a 20 MHz channel it will get as much frequency diversity as a coded interleaved OFDM system. In outdoor scenario, the loss of orthogonality due to severe channel coding may diminish the frequency diversity effect and introduce MAI to reduce the Bit Error Rate (BER) performance.

4.2 Orthogonal Frequency Division Multiple Access

This section describes the OFDMA scheme, also referred to OFDM–FDMA. OFDMA is already in use in some standards [75].

4.2.1 Multiple Access Model

In OFDMA, a fraction of OFDM sub-carriers are assigned to a sub-channel. Multiple access is realized by allocating different sub-channels to users. The scheme was first proposed for CATV systems [71], and later adopted for wireless communication systems.

In principle, sub-carriers can be assigned arbitrarily among the sub-channels. This study assumes that they are equally distributed among the sub-channels. The sub-carrier assignment is made for the user lifetime, or at

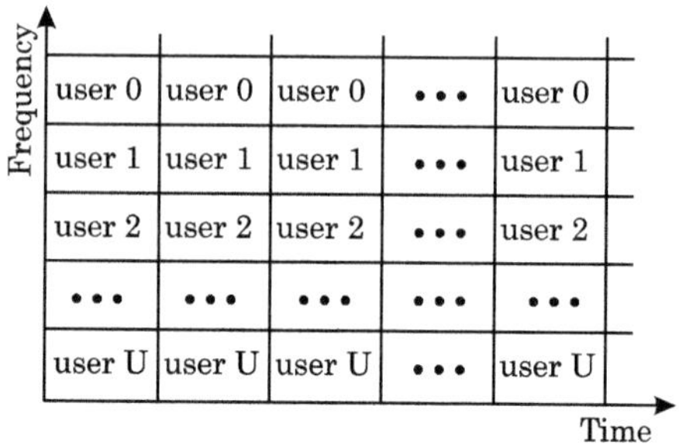

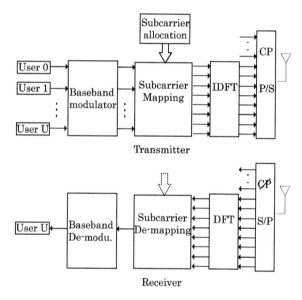

Fig. 4.2 Time–frequency diagram for the OFDMA scheme. For simplicity, one sub-carrier is assigned to each user. The vertical blocks on the time axis represents OFDM symbols.

Fig. 4.3 Simplified block diagram of the OFDMA transmitter, receiver schemes.

least for a considerable time frame, illustrated in Figure 4.2. In later sections, sub-carrier assignment schemes and assignment lifetime are examined in more detail.

In Figure 4.3, a simplified block diagram of the OFDMA transmitter scheme is illustrated. User data is modulated using a baseband modulation scheme, e.g. 16-QAM, such that the user symbols match the number of allocated sub-carriers. The symbols are assigned to sub-carriers, using the assignment map defined by the sub-carrier assignment scheme. When all users are mapped into the OFDM symbol, the symbol is transmitted, incorporating all traditional features of OFDM, with the addition of multi-user capabilities. At the receiver side, data of the uth user can be received by knowledge of the sub-carrier assignment.

Orthogonal Frequency Division Multiple Access (OFDMA) can support a number of identical downstreams, or different user data rates, [e.g. assigning a different number of sub-carriers to each user]. Based on the sub-channel condition, different baseband modulation schemes can be used for the individual sub-channels, e.g. QPSK, 16-QAM, and 64-QAM. This is investigated in numerous papers and referred to as adaptive sub-carrier, bit, and power allocation or QoS allocation [19, 70, 72, 73]. These issues will not be studied further here, as it is assumed that each user has identical data rate, using an equal number of sub-carriers in the sub-channels.

4.2.2 Static and Dynamic Sub-carrier Assignment

There are basically two approaches for the sub-carrier assignment; static and dynamic. The sub-carriers are assigned into sub-channels in a contiguous or interleaved manner [76].

For contiguous assignment, the sub-carriers are divided into contiguous sub-channels, where one benefit is the simplicity, but a disadvantage is the possible throughput degradation due to channel fading. In the interleaved case, the sub-carriers are successively assigned to the different users and then interleaved over the total number of sub-carriers. One important advantage is that it has potential to reap more channel diversity gain. The disadvantage is that issues like synchronization for the entire OFDM symbol become crucial, as the user sub-channel is scattered over the OFDM symbol. In Figure 4.4, examples of contiguous and interleaved assignment are illustrated.

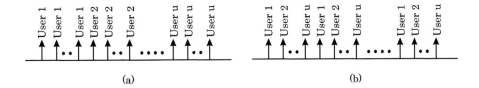

Fig. 4.4 (a) Continuous sub-carrier assignment. (b) Interleaved sub-carrier assignment.

In the dynamic sub-carrier assignment case, channel information is used to assign the sub-carriers best suited for each user. This is advantageous in the sense that users at different locations have different channel conditions, and most likely different optimal sub-carriers. The benefit is that high throughput rate can be obtained, the disadvantages are that; channel information is needed, sub-carriers must be reassigned whenever conditions change, leading to additional signaling overhead whenever sub-carriers are reassigned [70].

Assignment of Frequency: The frequency with which the sub-carriers are reassigned generally corresponds to the lifetime of the user, or at least for a significant time frame. In Figure 4.5, different sub-channel conditions are illustrated, assuming contiguous sub-carrier assignment. For user 0 and 1 the OFDM symbols are permanently good and bad, respectively. User 2 has alternating good and bad OFDM symbols, and user 3 only has few bad OFDM symbols. In the following, it is discussed for which condition static and dynamic assignment is the most profitable choice.

In the best (worst) case, the sub-channel is respectively good (bad), for static channel conditions. Static assignment is made, without information of the channel, meaning that the sub-channel is either good (bad) for the assignment duration. For dynamic assignment, perfect channel information is assumed, and the user will always be assigned a good sub-channel.

In cases with alternating good and bad sub-channels, assuming channel coherence time corresponding to a considerable number of OFDM symbols, the situation is somewhat different. In the static assignment, the sub-channel will be good for some time intervals and bad for the rest. Static sub-carrier

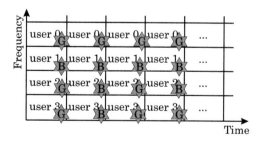

Fig. 4.5 Examples of sub-channel conditions where users receive either good or bad OFDM symbols, different patterns are illustrated. (G = good OFDM symbol, B = bad OFDM symbol.)

assignment is in this case a poor solution, as the user sub-channel will be bad for a number of OFDM symbols. For dynamic assignment, it is favorable to make the assignment within a time period corresponding to the channel coherence time, thereby always assigning good sub-channels. For dynamic assignment, perfect channel information is normally assumed [73].

For a time-variant channel, meaning that the channel coherence time is within the same range as the OFDM symbol time, the static assignment scheme could be a good solution. As the channel is time-variant, errors will have a burst-like pattern. Static assignment could be a solution, as some of the same properties from sub-carrier hopping are obtained; in sub-carrier hopping channel fading is combated by averaging the fading. The same condition is realized by having static assigned sub-carriers, as the time-variant channel is constant changing; the difference from sub-carrier hopping is that the channel is "hopping." For dynamic assignment, the reassignment must be made at a rate corresponding to the channel coherence time, e.g. every OFDM symbol, introducing large overheads for signaling.

4.2.3 Matrix Description

Using the defined matrix-vector notation the OFDMA access scheme is described. Only considering one OFDM symbol, the OFDMA scheme is illustrated in Figure 4.6. Assuming that user data is partitioned into OFDM symbols and baseband modulated, the $\mathbf{d}^s$ vector has the form:

$$
\mathbf{d}^s = \begin{bmatrix} d_0[0] \\ \vdots \\ d_0[\eta-1] \\ \vdots \\ d_{U-1}[0] \\ \vdots \\ d_{U-1}[\eta-1] \end{bmatrix}_{[N \times 1]}, \tag{4.1}
$$

where d_u is the uth user data symbols, and η is the number of sub-carriers assigned to the user. The sub-carrier mapping matrix $\mathbf{M}^s$ maps each user data

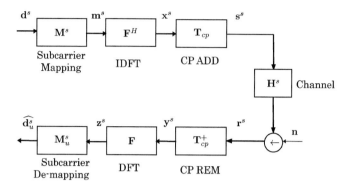

Fig. 4.6 Single OFDM symbol model for the OFDMA.

symbols into the OFDM sub-carriers by,

$$\mathbf{m}^s_{[N \times 1]} = \mathbf{M}^s_{[N \times N]} \mathbf{d}^s_{[N \times 1]}. \tag{4.2}$$

An OFDM symbol can be divided into a number of sub-channels. For each sub-channel a number of sub-carriers are assigned as

$$\Delta_u := \{I \in [0, N]\}_\eta, \tag{4.3}$$

where Δ_u are the sub-carriers belonging to uth user, such that

$$\Delta_0 \cup \Delta_1 \cup \Delta_{U-1} \le N, \text{ and } \Delta_i \cap \Delta_j = 0$$

meaning that the number of allocated sub-carriers can be less but not exceed the total sub-carriers, and that allocated sub-carriers are unique to a given user. The sub-carrier mapping matrix is defined as,

$$\left[\mathbf{M}^s\right]_{[\Delta_u]_i, \eta_{u+i}} \quad \text{for } 0 \le i \le \eta \tag{4.4}$$

such that each row and column only contain one 1 and the rest zeros, and the dimensions are $[N \times N]$. This is most readily shown by an example, considering three users, each having two data symbols. The data vector is as follows:

$$\mathbf{d}^s = \begin{bmatrix} d_0[0] \\ d_0[1] \\ d_1[0] \\ d_1[1] \\ d_2[0] \\ d_2[1] \end{bmatrix}.$$

Using a contiguous sub-carrier assignment scheme, it is decided that $d_0[x]$ goes to sub-carriers $\mathbf{m}_{5:6}$, $d_1[x]$ to $\mathbf{m}_{3:4}$, and $d_2[x]$ to $\mathbf{m}_{1:2}$. Using this definition and inserting into (4.2) the $\mathbf{M}^s$ becomes,

$$
\begin{bmatrix}
\mathbf{m}_2[0] \\
\mathbf{m}_2[1] \\
\mathbf{m}_1[0] \\
\mathbf{m}_1[1] \\
\mathbf{m}_0[0] \\
\mathbf{m}_0[1]
\end{bmatrix}
=
\begin{bmatrix}
 & & & & & 1 \\
 & & & & 1 & \\
 & & 1 & & & \\
 & & & 1 & & \\
1 & & & & & \\
 & 1 & & & &
\end{bmatrix}
\begin{bmatrix}
d_0[0] \\
d_0[1] \\
d_1[0] \\
d_1[1] \\
d_2[0] \\
d_2[1]
\end{bmatrix}.
$$

In the $\mathbf{M}^s$ matrix it is observed that the columns represent the users placement, e.g. for the first user there are 1 elements in the first two columns, second user the next two columns, and so on.

Using the $\mathbf{M}^s$ it is easy to define other allocations schemes. In Figure 4.4, an interleaved sub-carrier assignment is illustrated, and using the interleaved scheme the equation in (4.2) becomes, assuming the previous user setup,

$$
\begin{bmatrix}
\mathbf{m}_0[0] \\
\mathbf{m}_1[0] \\
\mathbf{m}_2[0] \\
\mathbf{m}_0[1] \\
\mathbf{m}_1[1] \\
\mathbf{m}_2[1]
\end{bmatrix}
=
\begin{bmatrix}
1 & & & & & \\
 & & 1 & & & \\
 & & & & 1 & \\
 & 1 & & & & \\
 & & & 1 & & \\
 & & & & & 1
\end{bmatrix}
\begin{bmatrix}
d_0[0] \\
d_0[1] \\
d_1[0] \\
d_1[1] \\
d_2[0] \\
d_2[1]
\end{bmatrix}.
$$

At the receiver side only the uth user data symbols must be retrieved, and assuming perfect transition and recovery the transmitted symbol, the received symbol becomes:

$$\mathbf{z}^s \equiv \mathbf{m}^s.$$

In $\mathbf{z}^s$ all user's data symbols are allocated, the purpose for the de-mapping matrix $\mathbf{M}_u^s$ is to select the uth user symbols and reject the rest

$$\widehat{d}^s_{u\,[\eta \times 1]} = \mathbf{M}^s_{u\,[\eta \times N]} \mathbf{z}^s_{[N \times 1]}. \tag{4.5}$$

The operation for finding $\mathbf{M}_u^s$ is to select the sub-matrix from $\mathbf{M}^s$ that refer to the user and then transpose this into an $\eta \times N$ matrix. Following the previous example, for the contiguous case and for first user,

(4.5) becomes:

$$
\begin{bmatrix} \hat{d}_0[0] \\ \hat{d}_0[1] \end{bmatrix} = \begin{bmatrix} & 1 & \\ & & 1 \end{bmatrix} \begin{bmatrix} Y_2[0] \\ Y_2[1] \\ Y_1[0] \\ Y_1[1] \\ Y_0[0] \\ Y_0[1] \end{bmatrix}.
$$

For multiple symbols there are basically two cases for OFDMA, as discussed in Section 4.2.2. For the static case, the $\mathbf{M}$ matrix will contain elements of identical $\mathbf{M}^s$ elements. The reason is that users are assigned to the same sub-carriers for the complete transmission.

$$
\mathbf{M} = \begin{bmatrix} \mathbf{M}^s & & & \\ & \mathbf{M}^s & & \\ & & \ddots & \\ & & & \mathbf{M}^s \end{bmatrix}
$$

For the dynamic assignment we assume channel knowledge and therefore the assignment is made every OFDM symbol, or after a number of symbols. In the case of new assignments, the sub-carrier mapping matrix becomes:

$$
\mathbf{M} = \begin{bmatrix} \mathbf{M}^0 & & & \\ & \mathbf{M}^1 & & \\ & & \ddots & \\ & & & \mathbf{M}^{S-1} \end{bmatrix}
$$

In either case, the transmitter and the receiver is identical to the traditional OFDM scheme. The effects of channels and Cyclic Prefix (CP) will have the same benefits and drawbacks.

4.2.4　Transceiver Architecture

Considering a traditional OFDM scheme the difference is that the data input is a combination of different user data. Meaning that the user's data is partitioned into length that combined with error coding and baseband modulation fits the assigned number of sub-carriers.

Transmitter:　In Chapter 3, the traditional OFDM transmitter structure is studied. The difference from traditional OFDM comparing to OFDMA is that only a fraction of the available sub-carriers is allocated to a user.

For the sth OFDM symbol, and the uth user, each user is provided with $|\Delta_u|$ sub-carriers, where the sub-carrier index is found using the elements of Δ_u. Defining the indexing function for the uth user, $\mathcal{K}_u\{v\}$, as a function that bijectively maps a sequential index from 0 to $|\Delta_u| - 1$ onto all sub-carrier indices in Δ_u:

$$\mathcal{K}_u\{v\} \in \Delta_u; \quad \forall m \in \{0, 1, ..., |\Delta_u| - 1\}$$
$$\mathcal{K}_u\{v\} \neq \mathcal{K}_u\{v'\}; \quad v \neq v'$$

which means that there is a one-to-one correspondence between the sequential index and the sub-carrier index. The inverse indexing function, i.e. the function that takes the sub-carrier index and maps it onto a sequential index valid for the uth user is defined such that:

$$\mathcal{K}_u^{-1}\{k\} \in \{0, 1, ..., |\Delta_u| - 1\} \Leftrightarrow k \in \Delta_u;$$
$$\mathcal{K}_u\{\mathcal{K}_u^{-1}\{k\}\} = k$$
$$\mathcal{K}_u^{-1}\{\mathcal{K}_u\{v\}\} = v.$$

The kth sub-carrier for the uth user and s^{th} OFDM symbol, $\mathbf{X}_{u,s}[k]$, is found by

$$\mathbf{X}_{u,s}[k] = \begin{cases} d_u\left[sV_u^s + \mathcal{K}_u^{-1}\{k\}\right] & k \in \Delta_u \\ 0 & \text{otherwise} \end{cases} ; \quad k \in \{0, 1, ..., N - 1\},$$

(4.6)

where $V_u^s = |\Delta_u|$ is the number of source symbols transmitted in every OFDM symbol for the uth user. The transmitted OFDM symbol for the uth user can then be written as

$$\tilde{x}_{u,s}(t) = \frac{1}{\sqrt{T_u}} \sum_{k \in \Delta_u} \mathbf{X}_{u,s}[k] e^{j\Delta\omega k t}; \quad 0 \leq t < T_u.$$

(4.7)

In OFDMA, the users are combined into a OFDM symbol, where the sth OFDM symbol is

$$\tilde{x}_s(t) = \sum_{u=0}^{U-1} \tilde{x}_{u,s}(t)$$

$$= \frac{1}{\sqrt{T_u}} \sum_{u=0}^{U-1} \sum_{k \in \Delta_u} \mathbf{X}_{u,s}[k] e^{j\Delta\omega k t}; \quad 0 \leq t < T_u.$$

(4.8)

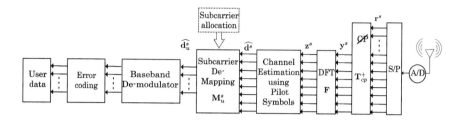

Fig. 4.7 Block diagram of an OFDMA receiver architecture.

From this point the OFDMA is identical to the traditional OFDM, described by (3.5–3.26).

Receiver: In Figure 4.7 the receiver architecture is illustrated.

In the OFDMA receiver, the complete transmitted signal will be received, having the same properties as the OFDM, see Section 3.5. Also meaning that the DFT, channel estimation is made on the complete OFDM symbol. In (3.26), the received estimated signal for the kth sub-carrier in the sth symbol, $\hat{X}_s[k]$ is written as

$$\hat{X}_s[k] = X_s[k] + \frac{N_s[k]}{H_{u,s}[k]}. \tag{4.9}$$

From $\hat{X}_s[k]$, only a fraction of the sub-carriers belong to the uth user. Using the uth user sub-carrier mapping information, the estimated data symbol sequence $\hat{d}_u$ is found using:

$$\hat{d}_u\left[sV_u^s + v\right] = \hat{X}_s[\mathcal{K}_u\{v\}]. \tag{4.10}$$

4.2.5   Specific Features

Compared to other OFDM based systems, OFDMA is simple in nature. In the literature, both DL and UL scenarios have been studied. In the DL case, OFDMA has all the advantages and disadvantages of basic OFDM, with the addition of multi-access capability. If it is assumed that the Base Station (BS) knows the instantaneous channel, sub-carriers can be assigned to sub-channels, such that maximized throughput is achieved.

Although this study has focused on the DL case, OFDMA can also be used for UL. In UL, issues such as synchronization become important, best illustrated by an example. Assume an OFDMA system with two users, where one user is much closer to the BS ("near") than the other ("far"). For the far user, the travelling time of the signal is much longer that of the near user. As the multiple-access method here is performed using the superposition of all transmitted signals, a small delay between signals will result in a loss of orthogonality within the OFDM symbol. In the literature, this problem is solved using time and frequency synchronization or estimators capable of compensating for these effects [77, 78]. Others propose an edge sidelobe scheme where guard bands are inserted between sub-channels [79, 80].

When considering mobility for OFDMA, the question becomes the relation between the channel coherence time and the OFDM symbol duration. As previously discussed, this can be overcome using a suitable sub-carrier assignment.

4.2.6 OFDMA Based-Standards

In 1998, the IEEE 802.16 Working Group started the preparation of various proposals for Wireless Metropolitan Area Networks (WirelessMAN), also known as the "last mile" networks. The initial interest was the 10–66 GHz band, desired for LOS communication. Later extended to the 2–11 GHz band, as it was realized that NLOS communication schemes are preferable. In 2003, IEEE published the 802.16a standard consisting of a common MAC layer and four different PHY layers. One of these is the WirelessMAN-OFDMA PHY layer, obviously using the OFDMA access technology.

The OFDMA symbol contain 2048 sub-carriers, containing data and pilot carriers. In DL the pilots are first allocated, which contain fixed and variable pilots. The carrier indices for the fixed pilots never change, whereas the variable pilots are rotated with a four OFDM symbol duration. To allocation the data sub-channels, the remaining carriers are partitioned into groups of contiguous carriers, where the individual data sub-channels are assigned one carrier in each group. Meaning that data sub-channel is interleaved over the remaining usable carriers. In UL the procedure is different, first the carriers are partitioned into sub-channels (interleaved over the carriers, as described for the DL), containing 53 carriers with 48 data carriers, one fixed pilot carrier and 4 variable pilot carriers.

In Table 4.2 a summary of the carrier specification can be seen.

Table 4.2 The IEEE 802.16a sub-carrier specification. (For the Guard carriers, representing respectively the left and right number of sub-carriers.)

Parameters	Downlink	Uplink
Number of FFT points	2048	2048
Usable carriers	1702	1696
Number of sub-channels	32	32
Number carriers per sub-channel	48	53 (5 pilots)
Data carriers	1536	1536
Number of pilots	166	160
Guard carriers	173, 172	176, 175

Table 4.3 The IEEE 802.16.4c four downstream modes. (For the Guard band, the two numbers are the guard interval (number of sub-carries) to the left and the right of the OFDM symbol, respectively.)

Parameters	Mode 2k	Mode 1k	Mode 256	Mode 64
Number of FFT points	2048	1024	256	64
Usable carriers	1696	848	212	53
Number of sub-channels	32	16	4	1
Number carriers per sub-channel	53	53	53	53
Guard band	176, 175	88, 87	22, 21	6, 5
Symbol duration	102.4 μs	51.2 μs	12.8 μs	3.2 μs

As mentioned previously, OFDMA is used in the IEEE 802.16 standards, Under the same umbrella the IEEE 802.16.4c proposal, is another example of a OFDMA standard [81]. For the downstream, the standard has the configuration listed in Table 4.3. The standard has four modes (defined by the FFT length). For sub-carrier assignment, the standard uses a permutation code that is defined for all four modes. The standard supports data rates of up to 65.4 Mbps, depending on the modulation and code rate used.

4.2.7   Performance Metric

BER performance	The BER of an OFDMA system is given by the modulation scheme used and averaged along channel distribution. For the same channel conditions, the overall BER of an OFDMA system does not differ from the one found for OFDM systems.
Spectral Efficiency	The spectral efficiency is equal to that of OFDM for a single user case. The spectral efficiency depends on the baseband modulation scheme, and the code rate used.
Delay Spread Tolerance	OFDMA is tolerant if the CP time T_g is larger than the maximum delay spread.

(Continued)

(*Continued*)

Mobility	Mobility can simply be evaluated on the ratio between OFDM symbol duration and channel coherence time, where $T_s < T_c$. If compared to the IEEE 802.14.4c downstream mode from Table 4.2, it is obvious that for the 2k mode, extreme mobility cannot be handled, whereas low mobility is not a problem.
Latency	In OFDMA the processing is made on one OFDM symbol, giving that the latency is equal to T_s.
Inter-Cell Interference	In a multi-cell environment there will be interference, assuming that the same carrier frequency is used. The reason is that sub-carriers are assigned to sub-channels, and when a neighbor cell uses the same sub-carriers there will be interference.
Intra-Cell Interference	For OFDMA there will be no Intra-Cell interference, since all users are using a unique sub-set of the available sub-carriers.
Frequency-Reuse Factor	In [17] it was shown that the reuse factor for OFDMA in multi-cell environments is 3. This is shown by simulation.
Implementation Complexity	In the case of statically assigned sub-carriers, the complexity of OFDMA is identical to a traditional OFDM, only adding the mapping and de-mapping of sub-channels. For the dynamic sub-carrier assignment it is assumed that channel information is obtained. Resulting in that continuous channel estimation and thereby sub-carrier assignment must be made. The effect is that significant signaling is needed for both channel information and sub-carrier location.

4.2.8 Error Probability Analysis in OFDMA System

The effective SNR per bit in an OFDM system, γ_b, is given by

$$\gamma_b' = \left(1 - \frac{T_g}{T_s}\right)\gamma_b \tag{4.11}$$

due to the loss of performance introduced by the CP.

In a AWGN channel, the bit error probability, P_b, of a coherent PSK modulation in OFDM will be the same as in a single carrier system [44]. Hence, when BPSK or QPSK is applied to all sub-carriers the P_b will be given by (4.12) [82].

$$P_b = \frac{1}{2}\text{erfc}\left(\sqrt{\gamma_b}\right). \tag{4.12}$$

For a Rayleigh fading channel, some of the sub-carriers will be affected by fading due to frequency selectivity. The performance of the system will depend on the sub-carrier recovery method employed [82]. However, a lower bound to the P_b can be found when assuming a perfect sub-carrier recovery.

The Rayleigh pdf for effective γ_b' yields:

$$p(\gamma_b') = \frac{1}{\overline{\gamma_b'}} e^{-\frac{\gamma_b'}{\overline{\gamma_b'}}}. \tag{4.13}$$

Consequently, a lower bound for P_b in a Rayleigh fading channel can be obtained out of averaging (4.12), resulting in (4.14). This fact results in a general lower bound regardless of frequency and time selectivity [82].

$$P_b = \int_0^\infty \frac{1}{2} \text{erfc}\left(\sqrt{\gamma_b'}\right) p\left(\gamma_b'\right) d\gamma_b' = \frac{1}{2} \left(1 - \sqrt{\frac{\overline{\gamma_b'}}{1 + \overline{\gamma_b'}}} \right). \tag{4.14}$$

For an OFDMA system, the overall P_b will be similar to the one already mentioned for OFDM systems, provided that all the sub-carriers are in use.

Summary: The OFDMA scheme is distinguished by its simplicity, where the multi-access is obtained by allocating a fraction of sub-carriers to different users. The befit is that the receiver can be implemented in a relatively simple manner.

Orthogonal Frequency Division Multiple Access (OFDMA) is already in use in some standards, e.g. IEEE 802.16a, and can be used for both DL and UL. For the UL case issues like synchronization are a big issue, and are studied in several papers. Regarding the assignment of sub-carriers, the literature provides no definitive answer whether it should be static/dynamic or contiguous/interleaved.

4.3   Orthogonal Frequency Division Multiple Access-Fast Sub-Carrier Hopping

This section describes the OFDMA-Fast Sub-Carrier Hopping (OFDMA-FSCH) access scheme. Frequency hopping in OFDMA systems can be obtained in two ways:

- Contiguous sub-band hopping.
- Sub-carrier hopping.

In this deliverable, access schemes which include both of these techniques are presented. In order to differentiate sub-carrier hopping from contiguous sub-band hopping, OFDMA-FSCH terminology is used rather than frequency hopping terminology.

OFDMA-FSCH system combines all the capabilities of an OFDMA system with sub-carrier hopping spread-spectrum techniques.

In the following, a multiple access model of an OFDMA-FSCH transceiver structure is introduced and a matrix description of transmitted and received symbols is provided. Finally, a detailed transceiver model is presented including matrix representation of transmitted and received data symbols.

4.3.1 Multiple Access Model

In an OFDMA-FSCH system, once sub-carriers are allocated to users, the index of the assigned sub-carrier is changed, i.e. hopped at every OFDMA symbol. Users are separated by nonoverlapping sub-carrier hopping patterns [83]. These patterns constitute the hop-set available at the BS. The arrangement of the hop-set available at neighboring cells depends on the cellular structure of the system.

A basic model for an OFDMA-FSCH transceiver is presented in Figure 4.8. The only difference from an OFDMA system is the insertion of fast sub-carrier hopping. In the following subsection, the hopping structure is described in more detail.

4.3.2 Benefit from Using Sub-carrier Hopping

In an OFDMA system, each user is assigned a defined number of sub-carriers constituting the subchannel for that user. Arising from the fact that the channel

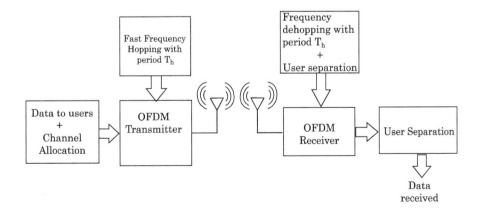

Fig. 4.8 General block diagram for OFDMA-FSCH transceiver.

suffers from frequency selective fading, some of the sub-carriers will be faded. It may happen that a few contiguous sub-carriers are affected by fading. In a simple user sub-carrier assignment, the user sub-carriers are placed next to each other. When a burst error occurs due to contiguous sub-carriers in fade, the user performance is highly affected. The channel coding will recover part of those sub-carriers in fade. However, if the number of sub-carriers in fade from one user surpasses the limit of recovering from channel codes, the data will be lost. Therefore, when a burst error occurs, it may happen that one user gets all its sub-carriers faded, hence losing its data, while the others keep their sub-carriers in good state. On the other hand, when faded sub-carriers are spread among users, the channel coding is effective in recovering the data from the few faded sub-carriers each user has.

The benefit of sub-carrier hopping becomes available with channel coding impact. Using these sub-carrier recovery methods, data may be recovered fully even if one (or more) sub-carriers are faded. However, there is a limit to the amount of faded sub-carriers channel coding is able to handle. When this limit is exceeded, performance suffers.

sub-carrier hopping reduces the probability of burst errors without *a priori* knowledge of the channel. By dividing faded sub-carriers among all users, sub-carrier hopping increases the effectiveness of channel coding and increases the overall performance of an OFDMA system.

In order to quantify the increase of system performance in terms of P_b, the type of channel coding used, as well as its recovery limitation, needs to be taken into account.

To summarize, OFDMA-FSCH introduces the following advantages to a simple OFDMA system:

- With fast sub-carrier hopping, inter-cell interference is minimized [84].
- Fast sub-carrier hopping minimizes the impact of fading. Users do not see bursty errors due to severe fading effects. Thus, a considerable improvement in the average BER per user is expected at the price of increased signaling. This improvement over a conventional OFDMA system is presented in [85].
- With fast frequency hopping channel state information is not needed for sub-channel allocation to users.

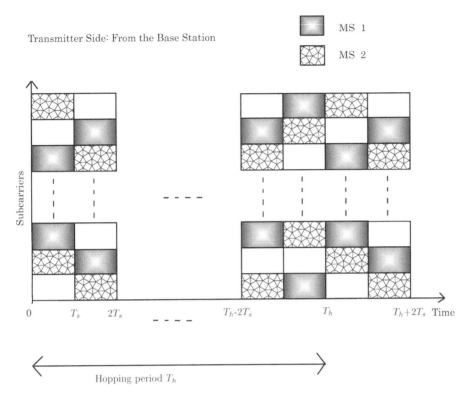

Fig. 4.9 Time–frequency diagram for OFDMA-FSCH scheme at the transmitter.

4.3.3 Main Idea

As shown in Figures 4.9 and 4.10, the sub-carrier hop-set in an OFDMA-FSCH system has a period of T_h seconds, or in other words T_h/T_s OFDM symbols. According to sub-carrier allocation algorithm used, each MS is assigned a finite number of sub-carriers denoted by α_u. For the sake of simplicity in visualization, α_u has been set to one in Figures 4.9 and 4.10. Once α_u is granted by the BS, the MS transmits over α_u sub-carriers but the allocated sub-carrier index changes at each OFDM symbol with a period of T_h/T_s OFDM symbols.

4.3.4 Matrix Description

The general matrix model presented in Section 4.2.3 can also be used to describe an OFDMA-FSCH system. For the OFDMA-FSCH system, $\mathbf{M}^s$ and

Receiver Side: @ the Mobile Station MS$_u$ (e.g. u=2)

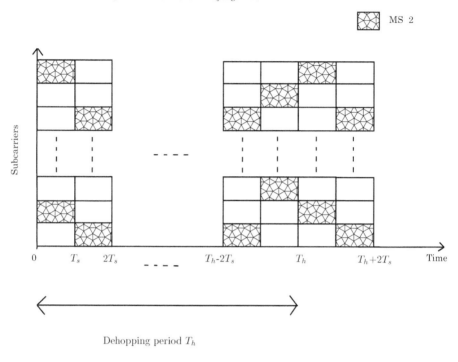

Fig. 4.10 Time–frequency diagram for OFDMA-FSCH scheme at the receiver.

$\mathbf{M}_u^s$ is changed for every OFDM symbol index, i.e. every s and they are defined in Section 4.2.3. Since the sub-carrier hop-set has a period of T_h, $\mathbf{M}^s$ is equal to $\mathbf{M}^{s+T_h/T_s}$ and $\mathbf{M}_u^s$ is equal to $\mathbf{M}_u^{s+T_h/T_s}$. The matrix of the form:

$$\mathbf{M} = \begin{bmatrix} \mathbf{M}^0 & & & & \\ & \ddots & & & \\ & & \mathbf{M}^s & & \\ & & & \ddots & \\ & & & & \mathbf{M}^{S-1} \end{bmatrix}$$

denotes the sub-carrier hopping matrix for S OFDM symbols, where the $\mathbf{M}_u^s$ corresponds to the indexes of the assigned sub-carriers to user u.

4.3.5 Transceiver Architecture

In this section, detailed OFDMA-FSCH transmitter and receiver architectures are presented. The matrix representations of transmitted and received signals in discrete-time domain is the same as in Sections 3.2.1 and 3.2.2. The only difference is the way $\mathbf{M}^s$ is changed. For the OFDMA-FSCH system, $\mathbf{M}^s$ is changed for every OFDM symbol index, i.e. every s. Figure 4.11 shows a detailed diagram for the transmitter at the BS for the DL. In Figure 4.12 the detailed diagram of the receiver side at the MS_u is shown.

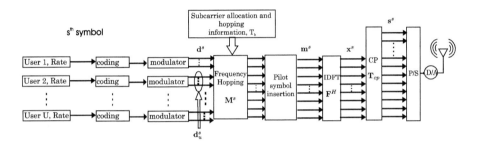

Fig. 4.11 OFDMA-FSCH transmitter model for DL.

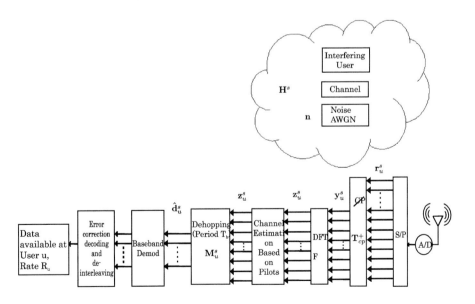

Fig. 4.12 OFDMA-FSCH receiver model at the MS_u.

4.3.6 Specific Features and Further Research Topics

When orthogonal hopping sequences are used among users, users within each cell are completely orthogonal to each other and there is no intra-cell interference due to the inherent property of OFDM. However, the intra-cell interference increases as the number of users increases in a conventional CDMA system. To prevent inter-cell interference, each BS has an orthogonal hop-set compared to neighboring cells. The design of hop-sets to reduce inter-cell interference and increase the user capacity of the overall cellular system in further research area. Since the total number of sub-carriers is limited, it is not possible to find completely orthogonal hop-set for each BS. Thus, collisions from neighboring cells are inevitable but last only for one OFDMA symbol duration. However, inter-cell interference can be eliminated deterministically by designing semi-orthogonal hop-sets for neighboring base stations [86]. Thus, users see an average inter-cell interference within each cell. Since the sub-carriers are orthogonal to each other and perfect synchronization is assumed, there is no inter-carrier interference (ICI) experienced within each cell. If CP is assumed to be larger than the maximum excess delay, no ISI is experienced.

Usually, T_s is chosen to be smaller than T_c in designing an OFDMA system. This means that, in an OFDMA-FSCH system, the indices of the sub-carriers allocated to each user are changed faster than the changes in the channel, although this may seem unnecessary. However, by these means, the bad sub-carriers are distributed to users evenly whereby decreasing the error rate of each user.

4.3.7 Performance Metric

BER performance	BER averaged overall users is reduced dramatically compared to a conventional OFDMA system for a specific channel realization [85]. With fast sub-carrier hopping, symbol recovery is possible via suitable coding with the aid of increased frequency diversity. However, this is not possible for a conventional OFDMA system in the case of bursty errors.
Spectral Efficiency	OFDMA-FSCH does not introduce spectral efficiency improvement over a conventional OFDMA system.

(Continued)

(*Continued*)

Delay Spread Tolerance	As in the case of OFDMA system, an OFDMA-FSCH system must be designed such that $T_g > \tau_{max}$. Therefore, the guard interval duration must be designed according to the environment which is determining the maximum delay spread, τ_{max}.
Mobility	Mobility of a user determines the coherence time of the channel being observed. As mobility increases the coherence time decreases. To have a static channel over one OFDMA symbol, T_s must be designed to be smaller than T_c. flash-OFDM is designed with $T_u = 80\,\mu s$. For a speed of 50 km/h and a carrier frequency below 3.5 GHz, which is the case for flash-OFDM, the coherence time is larger than 2.61 ms. This is also larger than the symbol duration which is slightly larger than T_u by an amount equal to guard interval duration. The method for these calculations is given in Chapter 3.
Latency	Latency of an OFDMA-FSCH system is the same as in a conventional OFDMA system. The latency is T_s.
Inter-Cell Interference	Fast sub-carrier hopping reduces the number of collisions from neighboring cells whereby decreases the inter-cell interference. If completely orthogonal hop-sets are used in neighboring cells, then inter-cell interference is zero with perfect synchronization. However, the usage of completely orthogonal hop-sets in neighboring cells decreases the total number of users that can be supported in each cell. If a certain amount of inter-cell interference is tolerated, than the user capacity can be increased. This is completely a cellular system design issue.
Intra-Cell Interference	Since users are using orthogonal sub-channels, there is no intra-cell interference if perfect synchronization is assumed. However, for UL, there can be a nonzero intra-cell interference due to collisions on the access channel.
Frequency-Reuse Factor	To avoid severe inter-cell interference, the frequency reuse factor of an OFDMA and also OFDMA-FSCH system must be more than 3 [17].
Implementation Complexity	Since the system requires signaling overhead to tell users their allocated hop-sets, system overhead is increased compared to a conventional OFDMA system. However, since CSI is not required, it is much more simple to assign sub-channels to users. This comes with the price of increased need to pilot sub-carriers in both UL and DL.

4.3.8 OFDMA-FSCH Based Standards

An OFDMA-FSCH based standard is called "Fast Low-Latency Access with Seamless Handoff (flash) OFDM" [84]. flash-OFDM operates below the 3.5 GHz range and designed for Frequency Division Duplex (FDD) operation. It has a bandwidth of 1.25 or 5 MHz. Sub-carriers are separated by 12.5 kHz. This corresponds to a total of 400 sub-carriers for 5 MHz bandwidth and 100 for 1.25 MHz bandwidth. The sub-carriers are assigned to individual users when they have data to send. Each sub-carrier is adaptively modulated [84].

Orthogonal Frequency Division Multiple Access-Fast Sub-Carrier Hopping (OFDMA-FSCH) scheme has been proposed to IEEE 802.20 working group [86]. This scheme is still under development by IEEE 802.20 working group. Similar to flash-OFDM, channel bandwidth is 1.25 MHz. The system parameters are listed in Table 4.4 [86].

The access intervals are separated from data intervals in the UL. Access intervals have a duration of 0.8 ms, i.e. 8 symbols [86]. Like in flash-OFDM, FDD is used. Adaptive coding and modulation is supported by allowing users to report their channel conditions to the BS. For more details on the frame structures, the reader can refer to [86].

One disadvantage of OFDMA based access schemes is their susceptibility to frequency offset errors. The maximum doppler frequency shift is less than 162 Hz for a vehicle speed of 50 km/h and for a carrier frequency below 3.5 GHz. At the maximum doppler frequency shift, the sub-carrier separation of 12.5 kHz as standardized in flash-OFDM corresponds to a relative frequency offset error which is less than 1.3% for a carrier frequency less than 3.5 GHz. This corresponds to an SNR degradation less than 0.2 dB for QPSK modulation scheme and approximately 1 dB for 64-QAM modulation scheme [87]. The largest constellation in IEEE.802.11, i.e. 64-QAM, cannot tolerate beyond 2% of frequency error [87]. The problem of frequency offset errors can be solved by inserting pilot symbols. However, as the sub-carrier separation bandwidth decreases, more pilot symbols are needed to cope up with doppler frequency shift effects.

Table 4.4 System parameters for OFDMA-FSCH scheme under development by IEEE 802.20 working group.

Basic system description parameter	Value
Carrier frequency	$\leq 3.5\,\text{GHz}$
Channel bandwidth	1.25 MHz
Sub-carrier separation	11.25 kHz
Available sub-carriers	113
Symbol duration	0.1 ms
Cyclic prefix duration	$11.1\,\mu s$ (16 symbols)
Modulation	QPSK or 16 QAM
Peak rates	DL > 4 Mbps, UL > 800 Kbps
Slow sub-carrier hopping in UL	Tones hop every 7 OFDM symbols
Coding (Low-Density Parity-Check Codes) rates	1/6 to 5/6

Summary: The basic advantage of OFDMA-FSCH scheme is that it does not require CSI in the process of subchannel assignment. Therefore, it is less complex compared to OFDMA scheme with dynamic sub-carrier assignment. It decreases the average BER that each user encounters. Once the frequency hopping sequence is agreed for each user and once it is signaled to users, the signaling overhead of an OFDMA-FSCH and an OFDMA system with static sub-carrier assignment is the same. Due to fast sub-carrier hopping, total number of pilot symbols needed for an OFDMA-FSCH scheme is larger than that of a simple OFDMA system with static sub-carrier assignment where sub-carriers are assigned into sub-channels in a contiguous manner. In the latter case, channel estimation is done in discrete points in frequency and time. The channel at other frequencies and time instants is estimated by a suitable interpolation technique. In order for dynamic sub-carrier assignment to be done in an OFDMA system, CSI is to be measurable for each and every sub-carrier. Thus overhead of pilots of an OFDMA system with dynamic sub-carrier assignment scheme and OFDMA-FSCH scheme becomes comparable.

Orthogonal Frequency Division Multiple Access-Fast Sub-Carrier Hopping (OFDMA-FSCH) system can operate both in UL and DL and it is a good solution for indoor environments where channel does not change very frequently. However, due to frequency hopping at every OFDM symbol, synchronization of individual users becomes much more critical than that of an OFDMA system.

4.4 Orthogonal Frequency Division Multiple Access-Slow Sub-Carrier Hopping

The aim of the present section is to describe OFDMA systems joint with Slow Sub-Carrier Hopping (SSCH) (OFDMA-SSCH). Sub-carrier Hopping is advantageous in the absence of channel knowledge, as the sub-carriers are randomly assigned in order to average the fading experienced by each user. The main differences between SSCH and FSCH are

- in SSCH more than one symbol is transmitted within one hop, in FSCH there is a hop at each symbol;
- in SSCH error control coding schemes are required, in FSCH systems are already robust against multi-path fading (diversity gain).

In SSCH coherent and noncoherent detection are both possible. In the UL case for FSCH some problems can arise for the coherent detection, while for the DL case the complexity is not a problem if a sufficient number of pilots is dedicated to the channel estimation. In fact the pilots set at disposition of each MS (UL) is smaller than the total number at disposition of the BS (DL) and it becomes in general more difficult to do a coherent detection, especially when FSCH is present. For sure a trade-off between the goodness of the channel estimation and the expense of BW due to the insertion of pilots has to be taken into account.

The general block diagram of OFDMA-SSCH is given in Figure 4.13 for the transmitter and in Figure 4.14 for the receiver.

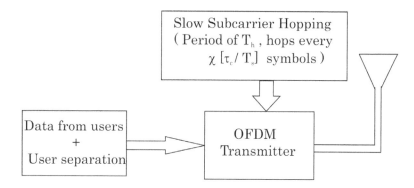

Fig. 4.13 OFDMA-SSCH general block diagram — Transmitter: from the BS.

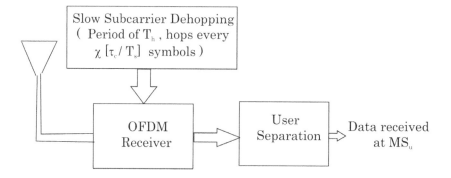

Fig. 4.14 OFDMA-SSCH general block diagram — Receiver: at the MS of uth user.

The combination of OFDMA with SSCH is a possible solution for a wide-band radio multiple-access system [88]. The OFDM component combats the multi-path interference while the SSCH provides multiaccess capability. Multiple users are allowed to transmit simultaneously using different hopping sequences, and MUI can be avoided if no frequencies are employed by several users at the same time. To make the system particularly robust to narrowband interference, it is desirable to select uniformly distributed carrier frequencies [14].

4.4.1 Multi Access Model

Figures 4.15 and 4.16 show how the different sub-carriers per user are allocated in frequency and time, respectively for transmitter and receiver, both for DL. One points out that χ is a design parameter, i.e. a coefficient which multiplies $\left\lfloor \frac{T_c}{T_s} \right\rfloor$, where T_c is the coherence time and T_s is the total OFDM symbol duration. The χ value indicates the frequency of the hops, the higher the χ the slower the hopping and vice-versa. It has sub-carriers hops after one OFDM packet (here it is defined as $\chi \left\lfloor \frac{T_c}{T_s} \right\rfloor$ OFDM symbols). After a certain

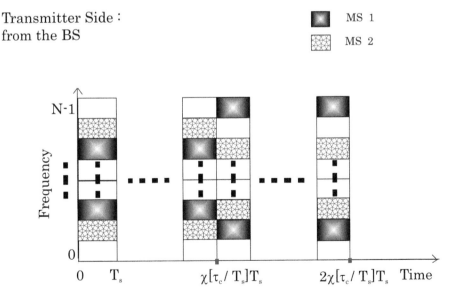

Fig. 4.15 OFDMA-SSCH F-T diagram, transmitter.

Receiver Side: at the MS_u (e.g. u=2)

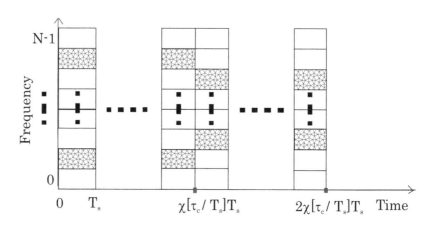

Fig. 4.16 OFDMA-SSCH F-T diagram, receiver.

number of OFDM packets (i.e. the hopping period T_h), the hopping sequence starts again from the beginning.

4.4.2 Matrix Description

Considering the whole OFDM symbol, it can be written that the following $\mathbf{M} \in \mathbb{C}^{US \times NS}$:

$$\mathbf{M} = \begin{bmatrix} \mathbf{M}^0 & & & & \\ & \ddots & & & \\ & & \mathbf{M}^s & & \\ & & & \ddots & \\ & & & & \mathbf{M}^{S-1} \end{bmatrix} \tag{4.15}$$

and in the case of OFDMA-SSCH for $s = 0, \ldots, \chi\left[\frac{T_c}{T_s}\right] - 1$ (see Subsection 4.4.1) the value of these $\mathbf{M}^s$ are equal, afterwards for $\chi\left[\frac{T_c}{T_s}\right], \ldots, 2\chi\left[\frac{T_c}{T_s}\right] - 1$ the same, and so on. Since the sub-carrier hopset period is T_h, $\mathbf{M}^s = \mathbf{M}^{s+T_h/T_s}$.

4.4.3 Transceiver Architecture

In the OFDMA-SSCH the transceiver architecture does not change respect to the OFDMA-FSCH, the only difference is that the hopping period T_h is longer than in the fast hopping case, in fact the hops happen less frequently, i.e. not each OFDM symbol, but each several OFDM symbols. Thus, for the transmitter architecture one refers to Figure 4.11 and for the receiver architecture to Figure 4.12.

4.4.4 Specific Features

Since in an outdoor environment the coherence time is smaller than in an indoor one, SSCH tends to coincide with FSCH. Therefore, the distinction between Fast and Slow Hopping loses importance.

The present study is focused on the DL, but this scheme is also applicable to the UL. Of course, one of the main problems is that the receiver has to hop synchronously with the transmitter to recover the transmitted information; as in the DL transmissions experience a single common channel to reach a particular user u, while in the UL there are several channel from each user to the BS, it is easier to synchronize in the DL [85]. In fact, the BS is unique and can decide when to send information to the mobiles, and usually the BS is a huge terminal with high computational capacity. Instead the problem with the UL is that all mobiles experience different pathloss and fading and can decide to transmit at different time instances to the BS, so it is difficult for the BS to detect which data comes from which users. Few works have been published on an UL scenario using SCH and OFDMA, so this problem is still an open issue.

Concerning the signaling topic, an advantage of the Slow Hopping, compared with the fast one, is that since the hops occur less frequently, the complexity of signaling for synchronization between the BS and the MS decreases. For example if the MS knows the hopping sequence, the MS needs to synchronize with the BS only at the first OFDM symbol, as the subsequent hops are known *a priori*.

Another interesting future development concerning random SCH-OFDMA is the possible benefits of using Space Frequency Coding (SFC). Random hopping implies occasional collisions whose detection and recovery is mandatory for efficient transmission of information. In [89] it is proved that SFC can avoid some collisions in OFDMA, thereby reducing the MUI and increasing the signal quality (it is worth noting that this feature is typical of SFC and cannot be attainable with coding in space and time only). Of course considering the transmit diversity together with the SSCH-OFDMA will provide further advantages.

Concerning modulation, in [90] using MFSK, the number of frequency slots available to the users is increased, for a fixed spread spectrum bandwidth, if compared to conventional noncoherent scheme used in SCH systems. Therefore the effect of MAI is reduced and the required data bandwidth is minimized.

The use of SSCH can take advantage of the inherent benefits of the SSCH access: the *near far* resistance and the possibility to employ side information (about the reliability of the received symbols) for error and erasure decoding. This side information is governed by the occurrence of hits (events in which more users occupy the same frequency at the same time). The use of error and erasures decoding, exploiting the side information, can outperform errors-only decoding based only on hard decisions [88].

4.4.5   Performance Metric

BER performance	It decreases compared to an OFDMA system, caused by the averaging of the fading, due to SSCH.
Spectral Efficiency	In an OFDMA system the same spectral efficiency is achieved as in the case of single user OFDM, and obviously this consideration is still valid when SSCH is present.
Delay Spread Tolerance	As in the case of OFDMA and OFDMA-FSCH systems, the system must be designed such that $T_g > \tau_{\max}$.
Mobility	The same consideration done for OFDMA and OFDMA-FSCH are still valid: as mobility increases, T_c decreases. Furthermore, to have a static channel over one OFDM symbol, T_S must be designed such that $T_S < T_c$.
Latency	It is same as in a conventional OFDMA system and its value is T_S.
Inter-Cell Interference	It is managed putting orthogonal hopping sequences for adjacent cells, to reduce collisions from neighboring cells.
Intra-Cell Interference	There are possible collisions among users when they access the channel for the first time, after that there are no collisions because all users use orthogonal sub-channels.
Frequency-Reuse Factor	As mentioned in the previous chapters, it must be more than 3.
Implementation Complexity	Due to signaling issues, the complexity of OFDMA-SSCH is higher than in OFDMA, but lower than in OFDMA-FSCH.

4.4.6 OFDMA-SSCH Based-Standards

For the OFDMA-SSCH standard, it refers to *flash* technology [84] (see Chapter 4, in particular Table 4.4). In fact, considering the UL for that technology, sub-carriers hop once for every seven OFDM symbols, so this kind of sub-carrier hopping can be classified as *slow*. The seven symbols mentioned before are named *dwell time* (among them, one is a reference symbol and the other six are data symbols). Since there are no shared UL pilots, data is modulated in each dwell time with the reference symbol.

About UL access, in each *superslot* (11.3 ms long) there is an *access interval* of eight symbols (that is 0.8 ms). The access intervals are separated from *data intervals* and they are used by access mobiles (which are not yet UL-synchronized) and by existing mobiles (for periodic timing tracking). Concerning signaling (a very important issue whenever it refers about sub-carrier hopping), an access signal is a multi-tone signal which provides diversity and timing resolution. The access signals can be detected with low processing complexity. Only this kind of signals are contention-based, after access all signaling is contention-free.

Summary: In the absence of channel knowledge, a good solution is to use SSCH or FSCH, whereby the fading is averaged. The main advantage is that the BER is smaller than the one in an OFDMA system.

Because of the necessity of signaling between transmitter and receivers after hops, the complexity is higher than in OFDMA, but lower than FSCH-OFDMA, in fact in that case the hops happen more frequently.

4.5 Multi-Carrier Code Division Multiple Access

The idea of MC-CDMA can be attributed to several researchers working simultaneously at almost the same time on hybrid access schemes combining the benefits of OFDM and CDMA. The OFDM provides a simple method to overcome the ISI effect of the multi-path frequency selective wireless channel, while CDMA provides the frequency diversity and the multi-user access scheme. Different types of spreading codes have been investigated. Orthogonal codes are preferred in case of downlink since orthogonality loss is not very severe in downlink as it is in uplink. The implementation has a particular user's data symbol (modulated data symbol) spread, then converted from serial to parallel, interleaved and then Inverse Discrete Fourier Transformed. Several users transmit over the same sub-carrier. In essence this implies frequency domain spreading, rather than time domain spreading as is conceived in a direct sequence CDMA system. The channel equalization can be highly simplified in downlink, because of the one tap channel equalization benefit offered by OFDM. The complexity depends upon the detection strategy used. Equal Gain Combining (EGC), Maximum Ratio Combining (MRC), Controlled Equalization (CE) are some of the primary simple methods of detection. Several other complex detection schemes exists such as Soft Interference Cancellation (SIF) and Maximum Likelihood Symbol by Symbol Estimation (MLSSE). The performance of the scheme is increased with the complexity of the detection scheme. It is a purely implementation trade-off that needs to be decided upon for a particular situation. This scheme has been so far hailed to be a very efficient in downlink transmission. In uplink, asynchronism has a severe effect on the orthogonal codes used. The downlink transmitter model can be represented as given in Figure 4.17.

4.5.1   Multiple Access Model

The main idea of sharing time frequency and energy between different users can be depicted in the so-called information cube as given in Figure 4.18. Each plane of the figure symbolizes a user data, which has been spread in the frequency domain. It can be noted from the figure that one user symbol does not completely occupy the entire frequency band. This is of course a design criteria. The number of symbols to be transmitted in one OFDM symbol is an implementation issue. In the figure, it has been shown that the spread chips of a symbol occupy contiguous sub-carriers. But that is actually a logical representation. In practice the chips of different symbols are interleaved to achieve frequency diversity. The gain in frequency diversity is dependent on the coherence bandwidth of the system. The chips of each data symbols needs to be so spread that each chip of a symbol are separated at least by a coherence bandwidth. As is depicted in Figure 4.18, all users transmit at the same time at all frequencies.

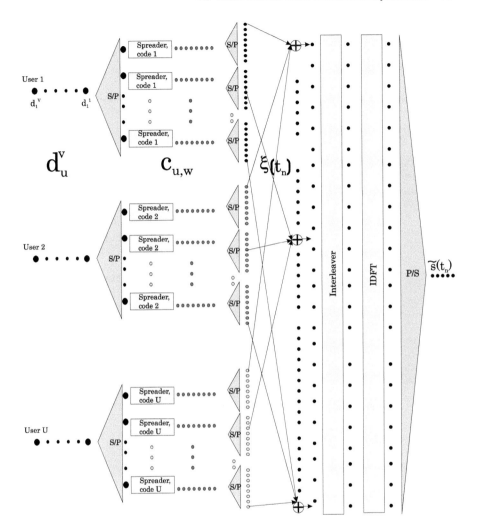

Fig. 4.17 Downlink MC-CDMA transmitter.

4.5.2 Matrix Description

MC-CDMA also known as OFDM–CDMA can be described with the common multi-user, multi-carrier, matrix model as described in Chapter 3 on system model.

The specific resource mapping and code matrices are described here.

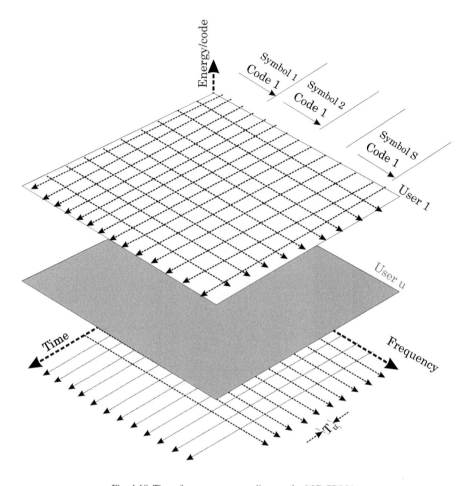

Fig. 4.18 Time–frequency energy diagram for MC-CDMA.

Frequency Domain: The data symbol vector for user u, $\mathbf{d}_u$, for each user consists of V subsequent data symbols:

$$\mathbf{d}_u = \begin{bmatrix} d_u[0] \\ d_u[1] \\ \vdots \\ d_u[V-1] \end{bmatrix}_{[V \times 1]}, \tag{4.16}$$

that is, the first serial-to-parallel conversion is $1:V$. The data vector for all users is written as

$$
\mathbf{d} = \begin{bmatrix} \mathbf{d}_0 \\ \mathbf{d}_1 \\ \vdots \\ \mathbf{d}_{U-1} \end{bmatrix}_{[VU \times 1]}. \tag{4.17}
$$

The sub-carrier mapping matrix, $\mathbf{M}$ for OFDM–CDMA can then be written as

$$
\mathbf{M} = \begin{bmatrix} \mathbf{C}_{0,0} & \mathbf{C}_{0,1} & \cdots & \mathbf{C}_{0,U-1} \\ \mathbf{C}_{1,0} & \mathbf{C}_{1,1} & \cdots & \mathbf{C}_{2,U-1} \\ \vdots & \vdots & \ddots & \vdots \\ \mathbf{C}_{V-1,0} & \mathbf{C}_{V-1,0} & \cdots & \mathbf{C}_{V-1,U-1} \end{bmatrix}_{[VW \times VU]}, \tag{4.18}
$$

where

$$
\mathbf{C}_{v,u} = \begin{bmatrix} \vdots & \vdots & \vdots \\ \mathbf{0}_{[W \times v]} & \mathbf{c}_u & \mathbf{0}_{[W \times (V-1-v)]} \\ \vdots & \vdots & \vdots \end{bmatrix}_{[W \times V]}, \tag{4.19}
$$

that is, a matrix where the vth column vector equals the code vector for the uth user, $\mathbf{c}_u$, and the remainder of the matrix equals zero. Note, that this is an extension of (4.18) as all data symbols for all users are combined into a single vector.

The sub-carrier vector, $\mathbf{m}$ is found using:

$$
\mathbf{m} = \mathbf{Md}. \tag{4.20}
$$

As this model operates in the frequency domain, the IFFT and FFT operations are skipped, as well as the guard interval addition and removal. The received sub-carrier vector may therefore be found directly using:

$$
\mathbf{z} = \mathbf{Hm} = \mathbf{HMd} + \mathbf{n}, \tag{4.21}
$$

where $\mathbf{n}$ is the frequency-transformed additive noise vector.

At the receiver the de-mapped data symbols can be written, for uth user $\hat{\mathbf{d}}$, as (assuming a simple zero forcing solution and neglecting noise variance):

$$
\hat{\mathbf{d}} = \mathbf{M}^H \mathbf{H}^{-1} \mathbf{z} + \mathbf{M}^H \mathbf{H}^{-1} \mathbf{n}. \tag{4.22}
$$

Time Domain: The extension to time domain model is as given in Chapter 3 on system model.

4.5.3   Transceiver Architecture

Transmitter: The analytical model will be described here. Following the data flow models (spreading of source data symbols and serial to parallel conversion in blocks of N spread symbols (chips) as in Figure 4.17 the spread data signal can be written as

$$\xi[n] = \sum_{u=0}^{U-1} d_u \left[\left\lfloor \frac{n}{W} \right\rfloor \right] c_u \left[(n)_W \right], \tag{4.23}$$

where

$$\left\lfloor \frac{n}{W} \right\rfloor = \text{floor}\left(\frac{n}{W} \right) \tag{4.24}$$

and where $(n)_W$ denotes the number n modulo W, i.e.

$$(n)_W = \min(n - cW) \geq 0; \qquad n, c \in \mathbb{N}_0, \, W \in \mathbb{N}$$

$$n \in \{-\infty, \ldots, 0, \ldots, \infty\} \tag{4.25}$$

Also,

$$NS = VW$$

$$\frac{N}{W} = \frac{V}{S} = V_u^s, \tag{4.26}$$

where V_u^s indicates the number of user source data symbols per OFDM symbol. The following are mathematical notations used to structure the transmitted signal. The sequence of transmitted spread signal is converted into blocks of N chips for transmission over OFDM modem. Accordingly, the following can be defined

$$n = sN + \varpi,$$

where

$$\varpi \in \{0, 1, \ldots, N - 1\}. \tag{4.27}$$

The spread sequence block for the sth OFDM symbol can therefore be written as

$$\xi_s[\varpi] = \xi[sN + \varpi] \tag{4.28}$$

and the chips (spread symbols) in one OFDM symbol can be written as

$$\xi_s[\varpi] = \sum_{u=0}^{U-1} d_u \left[V^s s + \left\lfloor \frac{\varpi}{W} \right\rfloor \right] c_u[(\varpi)w]. \tag{4.29}$$

When these blocks are mapped to the sub-carriers, there is a translation of index that is performed in order to keep in accordance to IDFT procedure. Hence, the sub-carrier index is related to the samples in a block as follows:

$$k = \varpi - \frac{N}{2}; \qquad k \in \left\{ -\frac{N}{2}, \ldots, \frac{N}{2} - 1 \right\}. \tag{4.30}$$

When mapping the sequence to the sub-carriers, the sth sub-carrier block and the sth spread sequence block are related by:

$$\mathbf{X}_s[k] = \xi_s[\varpi]. \tag{4.31}$$

The sub-carrier blocks are then OFDM modulated and transmitted over the channel. The OFDM section of the transmitter is described by (3.2–3.8). The resource mapping described in this section is shown in Figure 4.19.

Receiver: It has already been mentioned that $H_u[k]$ is the channel response of the kth sub-carrier of the uth user. It has been well explained and known that if

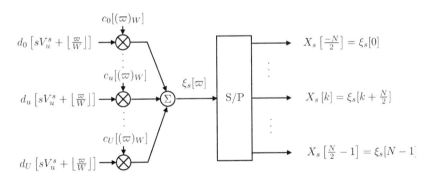

Fig. 4.19 MC-CDMA resource mapping. The data symbols for the different users are spread and combined, resulting in the N point block sequence $\xi_s[m]$. For each OFDM symbol, $V_u^s = \frac{N}{W}$ symbols are transmitted. The transmitted sub-carriers for the sth OFDM symbol are found by serial-to-parallel converting the spread sequence $\xi_s[m]$. This implies that MC-CDMA employs frequency-domain spreading.

the guard interval is sufficiently larger than the channel impulse response and if the sub-carrier spacing is such that, it is less than the coherence bandwidth, then each sub-carrier faces a flat fading condition. And thus instead of writing the convolution equation one can use the equivalent frequency domain expression for the channel effect.

The MC-CDMA receiver architecture is diagrammatically explained in Figure 4.20.

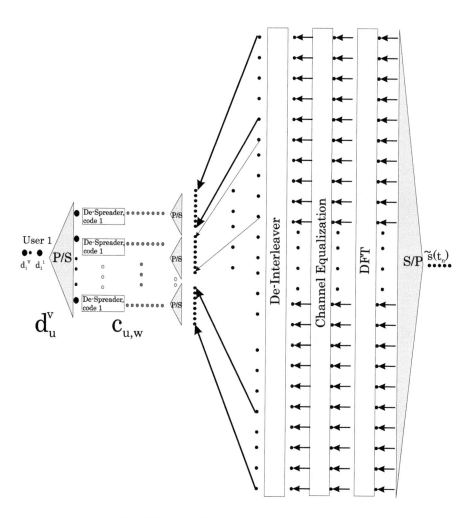

Fig. 4.20 An ideal MC-CDMA receiver in downlink.

After analogue to digital conversion and Fourier Transform, i.e. after OFDM de-multiplexing (described by (3.13–3.25)), the sth OFDM symbol can be written as (assuming perfect synchronization; scaling factor removed)

$$\hat{X}_s[k] = \left(\sum_{u=0}^{U-1} H_{s,u}[k]d_u\left[\left\lfloor\frac{k+\frac{N}{2}}{W}\right\rfloor\right]c_u\left[\left(k+\frac{N}{2}\right)_W\right] + N_f[k]\right)Z_s[k].$$

(4.32)

The receivers can be broadly categorized by the type of combining used. It can be Orthogonality Restoring Combining (ORC), EGC, MRC, CE, MMSE etc. Different combining schemes use different $Z_{s,u}[k]$. Thus the recovered source data for a particular user, (in this case the reference user is taken as 0, and removing the scaling factor) can be expressed as

$$\hat{d}_0[V^s s + v]$$

$$= \sum_{w=vW}^{(v+1)W-1}\hat{X}_s\left[w-\frac{N}{2}\right]Z_{s,0}\left[w-\frac{N}{2}\right]c_0[(w)_W]; \qquad v \in \{0, 1, ..., V_u^s - 1\}$$

$$= \sum_{w=vW}^{(v+1)W-1}\left\{\sum_{u=0}^{U-1}H_{s,u}\left[w-\frac{N}{2}\right]d_u\left[V^s s + \left\lfloor\frac{w}{W}\right\rfloor\right]c_u[(w)_W]\right.$$

$$+ N_f\left[w-\frac{N}{2}\right]\right\}Z_{s,0}\left[w-\frac{N}{2}\right]c_0[(w)_W]$$

$$= \sum_{w=vW}^{(v+1)W-1}\left(\sum_{u=0}^{U-1}H_{s,u}\left[w-\frac{N}{2}\right]d_u\left[V^s s + \left\lfloor\frac{w}{W}\right\rfloor\right]c_u[(w)_W]\right)$$

$$\times Z_{s,0}\left[w-\frac{N}{2}\right]c_0[(w)_W] + \sum_{w=vW}^{(v+1)W-1}N_f\left[w-\frac{N}{2}\right]$$

$$\times Z_{s,0}\left[w-\frac{N}{2}\right]c_0[(w)_W]$$

$$= \sum_{w=vW}^{(v+1)W-1}\left(H_{s,0}\left[w-\frac{N}{2}\right]d_0\left[V^s s + \left\lfloor\frac{w}{W}\right\rfloor\right]c_0[(w)_W]\right)$$

$$\times Z_{s,0}\left[w-\frac{N}{2}\right]c_0[(w)_W]$$

$$+ \sum_{w=vW}^{(v+1)W-1} \left(\sum_{u=1}^{U-1} H_{s,u}\left[w - \frac{N}{2}\right] d_u \left[V^s s + \left\lfloor \frac{w}{W} \right\rfloor\right] c_u\left[(w)_W\right] \right)$$
$$\times Z_{s,0}\left[w - \frac{N}{2}\right] c_0\left[(w)_W\right]$$
$$\underbrace{\qquad\qquad\qquad\qquad\qquad\qquad}_{=\beta[sV^s+v]}$$

$$+ \sum_{w=vW}^{(v+1)W-1} N_f\left[w - \frac{N}{2}\right] Z_{s,0}\left[w - \frac{N}{2}\right] c_0\left[(w)_W\right]. \qquad (4.33)$$
$$\underbrace{\qquad\qquad\qquad\qquad\qquad\qquad}_{=\vartheta[sV^s+v]}$$

To be noted in downlink, at the particular receiver, the channel experienced by the data of different users is the same. The receiver resource demapping described in this section is shown in Figure 4.21. There are several receiver architectures that are possible, these are based on the type of combining technique (the equalizer gain $Z_{s,u}[k]$) used. In the present study the following are considered.

- Equal Gain Combining (EGC).
- Maximum Ratio Combining (MRC).
- Controlled Equalization (CE).

In the current analysis, it is assumed that almost ideal phase compensation of the channel can be performed while the amplitude compensation may not be perfect. Further, for the present analytical results for BPSK modulation and Rayleigh channel are presented.

It has been assumed for now that each sub-carrier experiences independent fading. The local mean power of the kth sub-carrier of the uth user can be expressed as

$$\bar{P}_u[k] = E[H_{s,u}[k]^2] = \frac{1}{2}E_{|H_{s,u}[k]|^2}, \qquad (4.34)$$

where iid assumptions imply all sub-carriers have the same local mean power.

It must be noted that in downlink transmission, when it is assumed that the 0th user is being considered,

$$II_{s,u}[k] = H_{s,0}[k] \quad \forall u. \qquad (4.35)$$

$$\hat{\xi}_s[\varpi] = \hat{X}_s\left[\varpi - \tfrac{N}{2}\right]$$

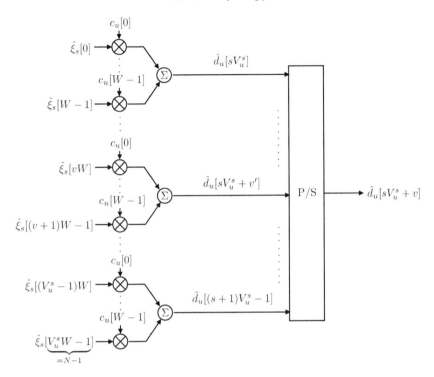

Fig. 4.21 MC-CDMA resource demapping. The received sub-carriers are divided into blocks of W and multiplied by successive elements of the spreading code for the uth user, c_u [0, 1, ..., $W-1$]. By summing the results, the estimated data symbol corresponding to that particular block is found. When all V_u^s blocks have been processed in that way, the source symbol sequence estimate, $\hat{d}_u$ is found by parallel-to-serial conversion of the different summations.

The following are also defined

$$\sum_{w=0}^{W-1} c_{u1}[w]\,c_{u2}[w] = \begin{cases} 0 & \text{for } u1 \neq u \\ 1 & \text{for } u1 = u2. \end{cases} \qquad (4.36)$$

Further $c_{u1}[w]\,c_{u2}[w] = \{-1, 1\}$ implies that a set $\{c_{u1}[w]\,c_{u2}[w]\}_{w=0}^{W-1}$ contains $\frac{W}{2}$ elements which are $a_i + 1$ and an equal number of $b_i - 1$. Where $i \in \left\{0, 1, \ldots, \frac{W}{2}\right\}$. The multi-user interference can be written as

$$\sigma_\beta^2 = 2(u - 1)\left[1 - \frac{\pi}{4}\right]\bar{P}_u[k]. \qquad (4.37)$$

$$EGC: \quad Z_{s,u}[k] = \frac{H_{s,u}[k]^*}{|H_{s,u}[k]|}.$$

In case of EGC, the interference noise can be considered Gaussian [16] with the following properties

$$E_\beta = 0$$

and

$$\sigma_\beta^2 = (U - 1)\bar{P}_u \qquad (4.38)$$

The probability of error can be expressed by using Law of Large numbers as

$$P_v(\text{error}|\bar{P}_u) \cong \frac{1}{2}\text{erfc}\left(\sqrt{\frac{\pi}{4} \frac{\bar{P}_u T_u}{2\frac{U-1}{W}\left[1 - \frac{\pi}{4}\right]\bar{P}_u T_u + N_0}}\right). \qquad (4.39)$$

MRC: $Z_{s,u}[k] = H_{s,u}[k]^*$.

The symbol error probability can be written by using Law of Large numbers as [16]

$$P_v(\text{error}|\bar{P}_u) \cong \frac{1}{2}\text{erfc}\left(\sqrt{\frac{\bar{P}_u T_u}{2\frac{U-1}{W}\bar{P}_u T_u + N_0}}\right). \qquad (4.40)$$

CE: In this not all sub-carriers (chips) of the spread symbol are utilized. Those which are above a certain threshold are kept and others are simply ignored. This is done to reduce the effect of noise and interference enhancement. If a threshold is defined as $|H_{s,u}[k]|_{\text{thres}}$ then [16]

$$\begin{aligned} Z_{s,u}[k] &= \frac{H_{s,u}[k]^*}{|H_{s,u}[k]|^?} & \text{if} & \quad |H_{s,u}[k]| \geq |H_{s,u}[k]|_{\text{thres}} \\ &= 0 & \text{if} & \quad |H_{s,u}[k]| < |H_{s,u}[k]|_{\text{thres}}. \qquad (4.41) \end{aligned}$$

The expression for symbol error rate is

$$P_v \cong p_{n_o}(n_o) \sum_\beta p_{\beta|n_o}(\beta|n_o)\frac{1}{2}\text{erfc}\left(\frac{n_o - \beta}{\sqrt{2}\sigma_{\vartheta|n_o}}\right), \qquad (4.42)$$

where the distribution of number of sub-carriers above the threshold, n_o is described by

$$p_{n_o}(n_o) = \binom{N}{n_o}\{pr_{on}\}^{n_o}\{1 - pr_{n_o}\}^{N-n_o}, \qquad (4.43)$$

where that the probability that a sub-carrier is above a threshold for Rayleigh fading is

$$pr_{on} = e^{-\frac{\bar{P}_u}{2\bar{P}_{0,k}}} \tag{4.44}$$

And

$$p_{\beta|n_o}(\beta|n_o) = \frac{\left(\frac{\frac{N}{2}}{\frac{n_o+\beta}{2}}\right)\left(\frac{\frac{N}{2}}{\frac{n_o-\beta}{2}}\right)}{\left(\frac{N}{n_o}\right)}. \tag{4.45}$$

4.5.4 Specific Features

MC-CDMA is one of the candidates to be used as one the future downlink access schemes. With its high frequency diversity, low complex equalization and high spectral efficiency, it is one of the more probable choices for downlink in next generation systems. It uses the benefits of CDMA while avoiding the complexity of a rake receiver. Another variation of this scheme namely OFDMA-CDM, where one user is given a set of sub-carriers and all the data symbols it has to transmit, is spread and transmitted just as in MC-CDMA. There is need of such variation of the original MC-CDMA scheme to make it even more usable. The aforementioned scheme has the advantage that it does not introduce MAI as MC-CDMA does. But, it does introduce self-interference. This make it a viable scheme for uplink in small indoor scenarios. If used with other variants or hybrid technologies such as Frequency Hopping to avoid the near far effect or asynchronous transmission, it might be used for uplink. It can be said that although this is a good scheme for downlink especially in indoor conditions, it does not qualify to be a good scheme in uplink scenario.

4.5.5 Performance Metric

BER performance	Dependent on combining scheme, MAI is present. Using schemes such as soft interference cancellation or MLSSE can improve the performance. Otherwise the performance depends a lot on the channel estimation and equalization.

(Continued)

(*Continued*)

Spectral Efficiency	It is highly spectral efficient in downlink, as long as orthogonality of the codes can be kept. In nonideal conditions it is not possible to operate under full load.
Average Throughput	High throughput under ideal conditions. Several factors affect the throughput. Due to the available frequency diversity, BER performance improves, which in turn also improves the throughput.
Delay Spread Tolerance	Delay tolerance is dependent on the guard interval length design as in normal OFDM based systems. This is an implementation issue. The guard interval as usual in OFDM schemes needs to be longer than the maximum delay spread of the channel.
Mobility	Maximum doppler spread tolerable is dependent on the sub-carrier spacing. As such there is MAI due to nonideal channel compensation under synchronous environment as well. Doppler spread leading to frequency spread will add to it since inter sub-carrier interference will come into play, further worsening the situation. But if the sub-carrier spacing is large enough, which is again a system design issue as in an OFDM system, such that the doppler frequency is very small compared to the sub-carrier spacing, i.e. normalized doppler frequency is small the ICI will be less. Otherwise the system will become interference limited. When compared against un-coded OFDM system, the doppler tolerance is better in MC-CDMA.
Latency	Latency is as in a normal OFDM system.
Inter-Cell interference	With proper choice and allocation of codes to users based on location and co-ordination between base station this can be kept to a quite low value, but in general, the system is susceptible to inter-cell interference.
Intra-Cell interference	It has already been pointed out that the system needs some additional changes to be implemented in uplink. In downlink the multi-user interference is already explained, and has been noted that channel estimation error highly influence the performance of the system.
Frequency-Reuse Factor	Does not have unit frequency re-use factor. But using PN-sequence might provide this benefit.
Implementation Complexity	Simpler compared to CDMA rake receiver, frequency domain combining.

Summary: This scheme is potentially a very good scheme in downlink to exploit the frequency diversity. This is where it scores over OFDMA. Even coded OFDMA can use limited frequency diversity. One very important observation can be made about the system. It has been pointed that this scheme is vulnerable to near–far effect as a normal CDMA system is. Hence this scheme suits bests mainly in an indoor downlink scenario. Now one is easily led to argue that in an indoor situation the coherence bandwidth is very large. In 5 GHz band it ranges from 6 to 20 MHz. Thus to make use of its special advantage of providing frequency diversity, the system has to use a very wide band. Otherwise, even with a 20 MHz channel it will get as much frequency diversity as a coded interleaved OFDM system. This brings out the fact that, no one scheme can be attributed to be the best scheme.

5

Single-Carrier Transmission with Cyclic Prefix

This chapter will clarify as the key issue is not whether one adopts a multi-carrier or a single-carrier transmission, but instead the domain where equalization is done, i.e. Time Domain Equalization (TDE) or Frequency Domain Equalization (FDE). Several techniques using FDE, i.e. OFDM, SCFDE, and SC-FDMA are discussed and compared, spotting potential synergies and interoperation.

5.1 Single-Carrier FDE

A conventional anti-multipath approach, which was pioneered in voiceband telephone modems and has been applied in many other digital communications systems, is to transmit a single carrier, modulated by data using, for example Quadrature Amplitude Modulation (QAM), and to use an adaptive equalizer at the receiver to compensate for Inter-Symbol Interference (ISI) [91]. Adaptive equalizion in the time domain to compensate for ISI [91] in Single Carrier (SC) systems was pioneered in voiceband telephone modems and has been applied in many other digital communications systems. Its main components are one or more transversal filters for which the number of adaptive tap coefficients is on the order of the number of data symbols spanned by the multipath. For tens of Megasymbols per second and more than about 30–50 symbols ISI, the complexity and required digital processing speed become exorbitant, and this TDE approach becomes unattractive [92]. Therefore, for channels with severe delay spread, equalizazion in the frequency domain might be more convenient since the receiver complexity can be kept low. In fact, as for the Orthogonal Frequency Division Multiplexing (OFDM), equalization is performed on a block of data at a time, and the operations on this block involve an efficient Fast Fourier Transform (FFT) operation and a simple channel inversion operation.

An SC system transmits a single carrier, modulated, for example, with QAM, at a high symbol rate. Linear FDE in an SC system is simply the frequency analog of what is done by a conventional time domain equalizer. For channels with severe delay spread, Single Carrier-Frequency Domain Equalization (SCFDE) is computationally simpler than corresponding time domain equalization for the same reason OFDM is simpler: because equalization is performed on a block of data at a time, and the operations on this block involve an efficient FFT operation and a simple channel inversion operation. Sari *et al.* [12, 13] pointed out that when combined with FFT processing and the use of a Cyclic Prefix (CP), an SC system with FDE (SCFDE) has essentially the same performance and low complexity as an OFDM system. It is worth noting that a frequency domain receiver processing SC modulated data shares a number of common signal processing functions with an OFDM receiver. In fact, as pointed out in Section 5.1.3, SC and OFDM modems can easily be configured to coexist, and significant advantages may be obtained through such coexistence.

Figure 5.1 shows conventional linear equalization, using a transversal filter with N tap coefficients, but with filtering done in the frequency domain. The block length N is usually chosen in the range of 64–2048 for both OFDM and SCFDE systems.

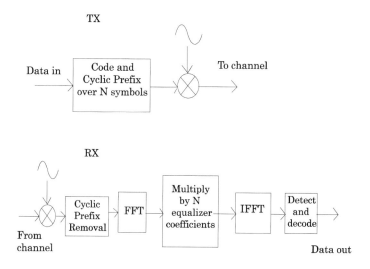

Fig. 5.1 SCFDE with linear FDE.

A Cyclic Prefix (CP) is appended to each block of N symbols, exactly as in OFDM. As an additional function, the CP can be combined with a training sequence for equalizer adaptation. An Inverse Fast Fourier Transform (IFFT) returns the equalized signal to the time domain prior to the detection of data symbols. Adaptation of the FDE's transfer function can be done with LMS, RLS, or Least Squares (LS) minimization techniques, analogously to adaptation of time domain equalizers [59, 93].

5.1.1 Single-Carrier vs Multi-Carrier, FDE vs TDE

Some recent studies have clearly shown that the basic issue is not OFDM vs SC but rather FDE vs TDE. FDE has several advantages over TDE in outdoor high mobility propagation environments (usually with long tail channel impulse response).

The conventional approach to digital communications over dispersive channels is single-carrier transmission with TDE. TDE covers the simple linear equalizers, decision-feedback equalizers, as well as maximum-likelihood sequence estimation. These techniques have been in use for decades in digital microwave radio, and more recently in mobile radio systems. Although FDE was originally introduced in the late 1970s, it was not pursued and quickly disappeared from the literature [94].

Let us consider a FDE with N_{taps} taps. The FFT operator which forms the first stage of the equalizer gives N_{taps} signal samples denoted $(Y_1, \ldots, Y_{N_{taps}})$. These samples are sent to a complex multiplier bank whose coefficients are denoted $(F_1, \ldots, F_{N_{taps}})$. The coefficient values which minimize signal distortion are

$$F_n = \frac{H_n^*}{|H_n|^2} = \frac{1}{H_n}. \tag{5.1}$$

Clearly, each coefficient is only a function of the channel frequency response at the corresponding frequency, and the equalizer is easily adapted to channel variations even if the number of taps is very large [94].

From the above discussions, SC-TDE is sufficient for channels with a small delay spread, because these channels can be equalized using a small number of taps. In contrast, SCFDE or OFDM are required on channels with a large delay spread, as these channels require a large number of taps and this leads to convergence and tracking problems with SC-TDE. Indeed, the normalized complexity of both OFDM and SCFDE is proportional to $\log(N_{taps})$, whereas the complexity of SC-TDE grows linearly with N_{taps}. For N_{taps} large, the complexity considerations clearly favor the use of frequency-domain techniques. In fact, these considerations indicate that the real problem is not OFDM vs SC, but instead FDE vs TDE [94].

5.1.2   Analogies and Differences Between OFDM and SCFDE

There is a strong analogy between OFDM and SCFDE. Analyzing the operation principle of OFDM, Sari *et al.* [13] noticed a striking resemblance to frequency-domain channel equalization for traditional single-carrier systems, a concept proposed more than three decades ago [95]. The motivation for frequency-domain equalization was due to the ability of this technique to accelerate the initial convergence of the equalizer coefficients. With a frequency-domain equalizer at the receiver, single-carrier systems can handle the same type of channel impulse responses as OFDM systems. In both cases, time/frequency and frequency/time transformations are made. The difference is that in OFDM systems, both channel equalization and receiver decisions are performed in the frequency domain, whereas in SCFDE systems the receiver decisions are made in the time domain, although channel equalization is performed in the frequency domain.

From a purely channel equalization capability standpoint, both systems are equivalent, assuming they use the same FFT block length. They have, however, an essential difference: *since the receiver decisions in uncoded OFDM are independently made on different carriers, those corresponding to carriers located in a region with a deep amplitude depression will be unreliable.* This problem does not exist for SCFDE, in fact *once the channel is equalized in the frequency domain, the signal is transformed back to the time domain, and the receiver decisions are based on the signal energy transmitted over the entire channel bandwidth.* In other words, the SNR value that dictates performance (assuming that residual ISI is negligible) corresponds to the average SNR of the channel. In fact, as noted in [13], the effect of the deep nulls in the channel frequency response is spread out over all symbols by the IFFT operation. Consequently, the performance degradation due to a deep notch in the signal spectrum remains small with respect to that suffered by OFDM.

The foregoing analysis indicates that with FDE, SC transmission is substantially superior to OFDM signaling. Without channel coding, OFDM is in fact not usable on fading channels, as deep notches in the transmitted signal spectrum lead to an irreducible BER. In order to work satisfactorily, *OFDM requires Error Correction Coding (ECC) with frequency-domain interleaving so as to scatter the signal samples falling in a spectral notch.* In this case, the interleaver uniformly distributes the low-SNR samples over the channel bandwidth. *In contrast, SCFDE can work without ECC.*

The main hardware difference between OFDM and SCFDE is that for SCFDE the transmitter's IFFT block is moved to the receiver. The complexities are the same. *Both OFDM and SCFDE can be enhanced by* adaptive modulation and *space diversity* [96].

The use of SC modulation and FDE by processing the FFT of the received signal has several attractive features:

- SC modulation has reduced Peak to Average Power Ratio (PAPR) requirements with respect to OFDM, thereby allowing the use of less costly power amplifiers.
- Its performance with FDE is similar to that of OFDM, even for very long channel delay spread.
- Frequency domain receiver processing has a similar complexity reduction advantage with respect to that of OFDM: complexity is proportional to *log* of multipath spread.
- Coding, while desirable, is not necessary for combating frequency selectivity, while it is needed in nonadaptive OFDM.
- SC modulation is a well-proven technology in many existing wireless and wireline applications, and its RF system linearity requirements are well known.

Comparable SCFDE and OFDM systems would have the same block length and CP lengths. The CP at the beginning of each block (Figure 5.2), used in both SCFDE and OFDM systems, has two main functions:

- It prevents contamination of a block by ISI from the previous block.
- It makes the received block appear to be periodic of period N, which is essential to the proper functioning of the FFT operation.

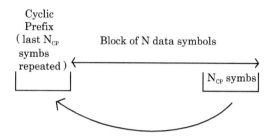

Fig. 5.2 Block processing in FDE.

If the first and the last N_{CP} symbols are identical unique word sequences of training symbols, the overhead fraction is $\frac{2N_{CP}}{N+2N_{CP}}$.

The following considerations provide a further support to the thesis of the similarity between SCFDE and OFDM. Precoding in OFDM disperses the energy of symbols over the channel bandwidth, i.e. it restores the frequency diversity broken by the IFFT operator. A common precoding matrix is the Walsh–Hadamard matrix which uniformly spreads the symbol energy across the channel bandwidth using orthogonal spreading sequences. Another precoding matrix that uniformly spreads the symbol energy over the channel bandwidth is the FFT matrix. This matrix cancels the IFFT matrix that generates the OFDM signal and the system reduces to SCFDE. From this discussion, it is clear that precoded OFDM mimics SCFDE. That is, *by precoding OFDM in order to restore frequency diversity, we get a SCFDE-type system* [94].

5.1.3 Interoperability of SCFDE and OFDM

Figure 5.3 shows block diagrams for OFDM and SC systems with linear FDE. It is evident that the two types of systems differ mainly in the placement of the IFFT operation: in OFDM it is placed at the transmitter to multiplex the data into parallel sub-carriers; in SC it is placed at the receiver to convert FDE signals back into time domain symbols. The signal processing complexities of these two systems are essentially the same for equal FFT block lengths [92]. A dual-mode system, in which a software radio modem can be reconfigured to handle either SC or OFDM signals, could be implemented by switching the IFFT block between the transmitter and the receiver at each end of the link, as suggested in Figure 5.4. *There may actually be an advantage in operating*

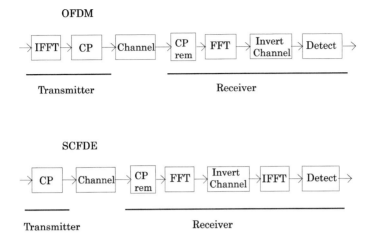

Fig. 5.3 OFDM and SCFDE signal processing similarities and differences.

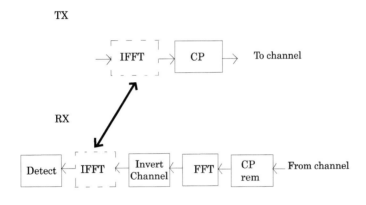

Fig. 5.4 Potential interoperability of SCFDE and OFDM: a "convertible" modem.

a dual mode system, wherein the base station uses an OFDM transmitter and an SC receiver, and the subscriber modem uses an SC transmitter and an OFDM receiver, as illustrated in Figure 5.5. This arrangement — OFDM in the downlink and SC in the uplink — has two potential advantages [92]:

- Concentrating most of the signal processing complexity at the hub or base station. The hub has two IFFTs and one FFT, while the subscriber has just one FFT.

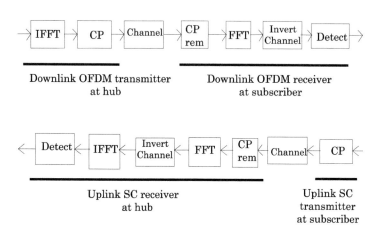

Fig. 5.5 Coexistence of SCFDE and OFDM: uplink/downlink asymmetry.

• The subscriber transmitter is SC, and thus inherently more efficient in terms of power consumption due to the reduced power backoff requirements of the SC mode. This may reduce the cost of a subscriber's power amplifier.

5.2 Single Carrier FDMA

5.2.1 Transceiver Architecture

Single Carrier FDMA is an extension of SCFDE which allows multi-user access. SC-FDMA can also be seen as a modified OFDMA with an extra Discrete Fourier Transform (DFT) at the transmitter and an extra Inverse Discrete Fourier Transform (IDFT) block at the receiver (see Figures 5.6 and 5.7) [97]. Since it has an inherent single carrier structure, as SCFDE also SC-FDMA has the advantage of lower PAPR with respect to OFDMA [98]: this is important especially in Uplink (UL) where lower PAPR means an higher power efficiency for the mobile terminal. Another problem with OFDMA in cellular uplink transmissions derives from the offset in frequency references among the different terminals that transmit simultaneously. This frequency

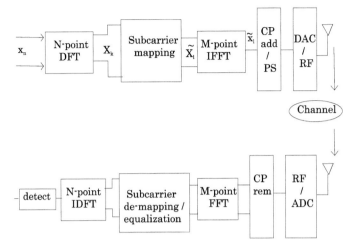

Fig. 5.6 Transceiver structure of SC-FDMA system.

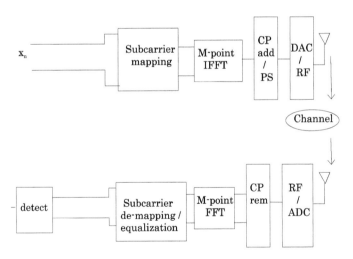

Fig. 5.7 Transceiver structure of OFDMA system.

offset destroys the orthogonality of the transmissions, introducing MAI [98]. As a consequence of these inherent advantages of SC-FDMA over OFDMA, SC-FDMA is a working assumption for UL multiple access scheme in The Third Generation Partnership Project (3GPP) LTE and Evolved-UMTS Terrestrial Radio Access (E-UTRA) [99, 100, 101].

As in OFDMA, in SC-FDMA one transmits the information symbols over different orthogonal frequencies, so-called sub-carriers, but differently from OFDMA, the sub-carriers are transmitted sequentially, and not in parallel: this implies a considerable reduction in the envelope fluctuations of the transmitted waveform with respect to OFDMA, and thus a lower PAPR [98].

As partially mentioned above, SC-FDMA can be regarded as DFT-spread OFDMA, where time domain data symbols are transformed to frequency domain by DFT before going through OFDMA modulation.

In Figures 5.6 and 5.7 [97, p. 23] one can see the SC-FDMA and OFDMA transceiver architectures. At the transmitter, one groups the modulation symbols $\{x_n\}$ into block each containing N symbols. Then an N-point DFT is used to produce a frequency domain representation X_k of the input time-domain symbols. Then the sub-carrier mapping takes place, associating each of the N DFT outputs to one of the M $(> N)$ orthogonal sub-carriers. If $N = M/Q$ and all terminals transmit N symbols per block, the system can handle Q simultaneous transmissions without co-channel interference, where Q is the bandwidth expansion factor of the symbol sequence. The result of the sub-carrier mapping is the set $\widetilde{X}_l$, $l = 0, \ldots, M - 1$ of complex sub-carrier amplitudes, where N of the amplitudes are nonzero. As in OFDMA, an M-point IFFT transforms the sub-carrier amplitudes to a complex time domain signal $\widetilde{x}_m$, which are then transmitted sequentially [97].

At this point, at the trasnmitter one inserts the CP, whose aim is to prevent Inter-Block Interference (IBI) due to multi-path propagation. The transmitter also performs a linear filtering referred to as pulse shaping in order to reduce out-of-band signal energy. One of the commonly used pulse shaping filter is the raised-cosine filter. The frequency domain and the time domain representations of the filter are as follows [97]:

$$P(f) = \begin{cases} T, & 0 \leq \mid f \mid \leq \frac{1-\alpha}{2T} \\ \frac{T}{2}\left\{1 + \cos\left[\frac{\pi T}{\alpha}\left(\mid f \mid -\frac{1-\alpha}{2T}\right)\right]\right\}, & \frac{1-\alpha}{2T} \leq \mid f \mid \leq \frac{1+\alpha}{2T} \\ 0, & \mid f \mid \geq \frac{1+\alpha}{2T} \end{cases} \tag{5.2}$$

$$p(t) = \frac{\sin(\pi t/T)}{\pi t/T} \frac{\cos(\pi \alpha t/T)}{1 - 4\alpha^2 t^2/T^2}, \tag{5.3}$$

where T is the symbol period and α is the roll-off factor. Roll-off factor α changes from 0 to 1 and It controls the amount of out-of-band radiation; $\alpha = 0$

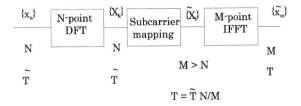

Fig. 5.8 SC-FDMA transmit symbols generation.

generates no out-of-band radiation and as α increases, the out-of-band radiation increases. In time domain, the pulse has higher side lobes when α is close to 0 and this increases the peak power for the transmitted signal after pulse shaping [97].

Figure 5.8 details the generation of SC-FDMA transmit symbols [97]. Among the M sub-carriers, N are occupied by input data. Seeing the situation from time-domain, the input data symbol has symbols duration of T and this duration is compressed to $\widetilde{T} = \frac{N}{M}T$ after going through the SC-FDMA modulation [97].

As already discussed for SCFDE, the receiver transforms the received signal into the frequency domain via FFT, performs the sub-carriers de-mapping, and afterwards the FDE. Being SC-FDMA a single-carrier modulation-based technique, it suffers from ISI and thus to combat this phenomenon equalization is needed. After equalization, the symbols are transformed back to the time domain via IDFT, and then detection and decoding are performed [97].

5.2.2 Sub-carrier Mapping

The approaches to mapping transmission symbols X_k to SC-FDMA sub-carriers are broadly divided into two categories: distributed and localized (see Figure 5.9 [98]). In the distributed sub-carrier mapping mode, DFT outputs of the input data are allocated over the entire bandwidth with zeros occupying the unused sub-carriers resulting in a noncontinuous comb-shaped spectrum. Interleaved SC-FDMA (IFDMA) is an important special case of distributed SC-FDMA. In contrast with IFDMA, consecutive sub-carriers are occupied by the DFT outputs of the input data in the localized sub-carrier mapping mode (so called Localized SC-FDMA (LFDMA)), resulting in a continuous spectrum that occupies a fraction of the total available bandwidth [98].

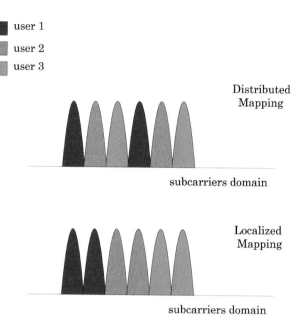

Fig. 5.9 Sub-carrier allocation methods for the case of 6 sub-carriers, 3 users and 2 sub-carriers per user.

Sub-carrier mapping methods are further divided into static and Channel Dependent Scheduling (CDS) methods. CDS assigns sub-carriers to users according to the channel frequency response of each user. For both scheduling methods, distributed sub-carrier mapping provides frequency diversity because the transmitted signal is spread over the entire bandwidth. With distributed mapping, CDS incrementally improves performance. By contrast CDS is of great benefit with localized sub-carrier mapping because it provides significant multi-user diversity [98].

When referring to mappings of N symbols in each block onto the $M > N$ transmission sub-carriers, already with $M = 256$ in a practical system the number of possible mappings is far too large for practical scheduling algorithms to consider. Therefore, to reduce the complexity of the mapping, sub-carriers are grouped into chunks and all of the sub-carriers in a chunk are assigned together. Figure 5.10 shows an example of SC-FDMA transmit symbols in the frequency domain for $N = 2$, $Q = 3$, and $M = 6$. After sub-carrier mapping, the frequency data is transformed back to the time domain by applying M-point IDFT. Different users occupy different orthogonal sub-carriers.

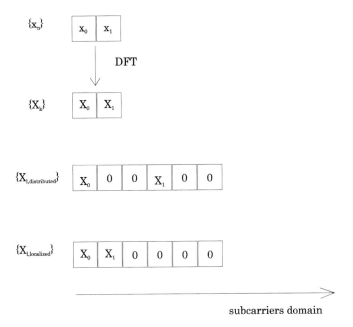

$\{x_n\}$

X_0 | X_1

DFT

$\{X_k\}$

X_0 | X_1

$\{X_{l,distributed}\}$

X_0 | 0 | 0 | X_1 | 0 | 0

$\{X_{l,localized}\}$

X_0 | X_1 | 0 | 0 | 0 | 0

subcarriers domain

Fig. 5.10 Example of SC-FDMA transmit symbols in frequency domain for $N = 2$ sub-carriers/users, $Q = 3$ users, and $M = 6$ sub-carriers in the system. Both distributed and localized sub-carriers allocations are shown.

For IFDMA, time symbols are simply a repetition of the original input symbols with a systematic phase rotation applied to each symbol in the time domain [102]. Therefore, the PAPR of IFDMA signal is the same as in the case of a conventional single carrier signal. In the case of LFDMA, the time signal has exact copies of input time symbols in N sample positions. The other $M-N$ time samples are weighted sums of all the symbols in the input block [103]. Figure 5.11 [98] shows an example of IFDMA and LFDMA signals, where $*_m = \sum_{k=0}^{1} c_{k,m} x_k$, with $c_{k,m}$ complex weights.

5.2.3 Relation Between SC-FDMA, OFDMA, and DS-CDMA/FDE

Relation between SC-FDMA and OFDMA: OFDMA transmitter has much in common with SC-FDMA. The only difference is the presence of the DFT in SC-FDMA. For this reason SC-FDMA is sometimes referred to as DFT-spread of DFT-precoded OFDMA. Other similarities between the two include: block-based data modulation and processing, division of the transmission bandwidth

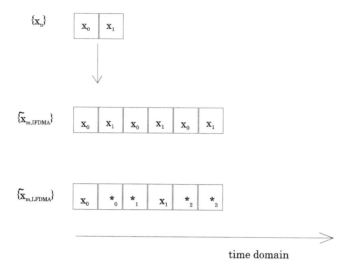

Fig. 5.11 Example of SC-FDMA transmit symbols in time domain for $N = 2$ sub-carriers/users, $Q = 3$ users, and $M = 6$ sub-carriers in the system. Both cases correspondent to distributed and localized sub-carriers allocations are shown.

into narrower sub-bands, frequency domain channel equalization process, and the use of CP [97].

However, there are distinct differences that make the two systems perform differently. In terms of data detection at the receiver, OFDMA performs it on a per-sub-carrier basis, while SC-FDMA does it after additional IDFT operation (see Figure 5.12 [97]). Because of this difference, OFDMA is more sensitive to a null in the channel spectrum and it requires channel coding or power/rate control to overcome this deficiency [97].

Another difference is that in OFDMA the modulated time symbols are expanded with parallel transmission of the data block during the elongated time period, while SC-FDMA modulated symbols are compressed into smaller chips with serial transmission of the data block, similarly to a DS-CDMA system [97] (see Figure 5.13 [97], where for SC-FDMA a bandwidth spreading factor of 4 is assumed).

Relation between SC-FDMA and DS-CDMA/FDE: DS-CDMA with FDE is a technique that replaces the rake combiner, commonly used in the conventional DS-CDMA, with the frequency domain equalizer [104]. A rake receiver consists of a bank of correlators, each of which correlate to a particular

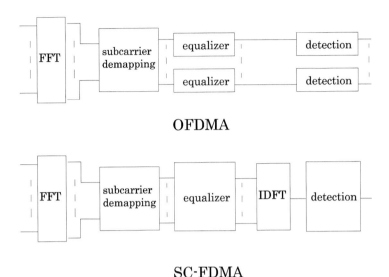

OFDMA

SC-FDMA

Fig. 5.12 Detection process dissimilarities between OFDMA and SC-FDMA.

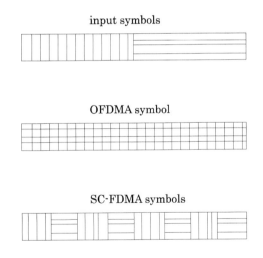

Fig. 5.13 Modulated symbol durations dissimilarities between OFDMA and SC-FDMA.

multi-path component of the desired signal. As the number of multi-paths increase, the frequency selectivity in the channel increases too, and so does the rake combiner's complexity, since more correlators are needed. This complexity problem in DS-CDMA can be alleviated by the use of FDE

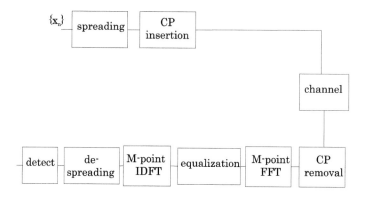

Fig. 5.14 Transceiver chain of DS-CDMA with FDE.

instead of rake combining. The block diagram of DS-CDMA/FDE is shown in Figure 5.14 [97].

The transmitter of DS-CDMA/FDE is the same as for the conventional DS-CDMA, but for the addition of CP. The FDE at the receiver removes the channel distorsion from the received chip symbols to recover ISI-free chip symbols. More insights about the DS-CDMA/FDE system can be found in [104].

SC-FDMA is similar to DS-CDMA/FDE in terms of the following aspects [97]:

- Both spread narrow-band data into broader band.
- They achieve processing gain or spreading gain from spreading.
- They both maintain low PAPR because of single carrier transmission.

An interesting relation between DS-CDMA and IFDMA is that by exchanging the roles of spreading sequence and data sequence, DS-CDMA modulation becomes IFDMA modulation [105, 106] (see Figures 5.15 and 5.16 [97]).

One advantage of SC-FDMA over DS-CDMA/FDE is that channel dependent resource scheduling is possible to exploit the frequency selectivity of the channel.

5.3 Chapter Summary

Single-carrier techniques using CP and FDE have been described and the benefits of FDE over TDE for future wireless systems are outlined.

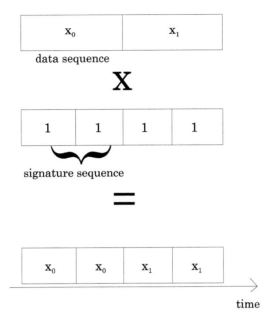

Fig. 5.15 Conventional spreading; spreading signature of (1,1) and block size of 2.

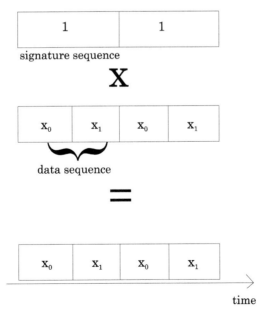

Fig. 5.16 Exchanged spreading; spreading signature of (1,1) and block size of 2.

The analogies and differences between single-carrier and multi-carrier systems operating with FDE, i.e. SCFDE and OFDM and their potential interoperability are discussed. Also, another version of single-carrier transmission scheme with FDE, SC-FDMA is presented and its relation with other FDE based schemes, such as OFDMA and DS-CDMA/FDE is described.

6

Synchronization in Time and Frequency Domain

Till now the OFDM transmitter–receiver system is shown without considering channel impairments and nonideal receiver conditions. The primary channel impairment is multi-path fading. This is taken care by the channel estimation and equalization module of the receiver which is described later. There are several nonideal conditions of the receiver. The nonideal conditions considered here are the synchronization issues. The prime issue under synchronization is the timing synchronization. Timing synchronization means alignment of the frame window. In OFDM systems several modulated symbols are combined through the FFT at the transmitter to generate the data part of an OFDM symbol. It is prefixed with a Cyclic Prefix (CP). The data part of the OFDM symbol together with the CP forms an OFDM symbol. In order to decode the transmitted information the receiver must first find the alignment of the OFDM symbol window. This is because the FFT operation performed in the received must contain signal from the data part of one complete OFDM symbol. If such is not the case then last half of one OFDM symbol and the first half of the next OFDM symbol is fed into the FFT module. This results in ISI. Under such situations the decoded signal is completely in error. Therefore, the task of the timing synchronization module is to find the OFDM symbol window.

The issue of frequency synchronization is a very important one especially in OFDM systems. The local oscillator which is used to generate the carrier frequency at the transmitter and the receiver do not generate exactly the same frequency. Then the local oscillator generated carrier frequency is used to convert down the received RF signal to the base band range, then the mismatch between the center frequencies of the transmitter and the receiver leads to receiver Carrier Frequency Offset (CFO). Due to such a condition the orthogonality of the sub-carriers in OFDM is lost giving rise to Inter-Carrier Interference (ICI). Such loss in orthogonality between the sub-carriers leads to severe degradation in performance.

127

A related problem is phase noise; a practical oscillator does not produce a carrier at exactly one frequency, but rather a carrier that is phase modulated by random phase jitter. As a result, the frequency, which is the time derivative of the phase, is never perfectly constant, thereby causing ICI in an OFDM receiver. For single-carrier systems, phase noise and frequency offsets only give degradation in the received signal-to-noise ratio (SNR), rather than introducing interference. This is the reason that the sensitivity to phase noise and frequency offset are often mentioned as disadvantages of OFDM relative to single-carrier systems. Although it is true that OFDM is more susceptible to phase noise and frequency offset than single-carrier systems, the following sections show that degradation can be kept to a minimum.

The sampling clock between the transmitter and the receiver is also not perfectly matched. Such difference in sampling clock frequencies lead to drift in the sampling timing position. This also leads to orthogonality loss. The other important error introduced due to the sampling clock offset is either missed sample or over sampled. The earlier case occurs when the receiver clock runs slower than the transmitter clock, while the latter occurs when the receiver clock is faster than that of the transmitter.

This chapters presents the detailed analysis of the synchronization errors which are very important for realizing OFDM systems.

6.1   Introduction

In an OFDM link, the sub-carriers are perfectly orthogonal only if the transmitter and the receiver use exactly the same frequencies. Any frequency offset immediately results in ICI. A related problem is phase noise; a practical oscillator does not produce a carrier at exactly one frequency, but rather a carrier that is phase modulated by random phase jitter. As a result, the frequency, which is the time derivative of the phase, is never perfectly constant, thereby causing ICI in an OFDM receiver. For single-carrier systems, phase noise and frequency offsets only give a degradation in the received SNR, rather than introducing interference. This is the reason that the sensitivity to phase noise and frequency offset are often mentioned as disadvantages of OFDM relative to single-carrier systems. Although it is true that OFDM is more susceptible to phase noise and frequency offset than single-carrier systems, the following sections show that degradation can be kept to a minimum. They describe techniques to achieve symbol timing and frequency synchronization by using the CP or special OFDM training symbols. They also demonstrate that OFDM is rather insensitive to timing offsets, although such offsets do reduce the delay spread robustness. An optimal symbol timing technique is derived that maximizes the delay spread robustness.

6.2 Sensitivity to Phase Noise

The issue of phase noise in OFDM systems has been the subject of many studies [66, 107, 108, 109, 110]. In [66], the power density spectrum of an oscillator signal with phase noise is modeled by a *Lorentzian* spectrum, which is equal to the squared magnitude of a first order lowpass filter transfer function. The single-sided spectrum $S_s(f)$ is given by

$$S_s(f) = \frac{\frac{2}{\pi f_l}}{1 + \frac{f^2}{f_l^2}}. \tag{6.1}$$

Here, f_l is the $-3\,\mathrm{dB}$ linewidth of the oscillator signal. In practice, only double-sided spectra are measured, which are equal to mirrored versions of the one-sided spectrum around the carrier frequency f_c. Further, because the bandwidth is doubled, the spectrum is divided by 2 in order to keep the total power normalized to 1. Hence, the double-sided phase noise spectrum is given by

$$S_s = \frac{\frac{1}{\pi f_l}}{1 + \frac{|f - f_c^2|}{f_l^2}}. \tag{6.2}$$

Figure 6.1 shows an example of a Lorentzian phase noise spectrum with a single-sided $-3\,\mathrm{dB}$ linewidth of 1 Hz. The slope of $-20\,\mathrm{dB}$ per decade of this model agrees with measurements in [108], which shows measured phase noise spectra for two oscillators are at 5 and 54 GHz. Phase noise basically has two effects. First, it introduces a random phase variation that is common to all sub-carriers. If the oscillator linewidth is much smaller than the OFDM symbol rate, which is usually the case, then the common phase error is strongly correlated from symbol to symbol, so tracking techniques or differential detection can be used to minimize the effects of this common phase error. The second and more disturbing effect of phase noise is that it introduces ICI, because the sub-carriers are no longer spaced at exactly $\frac{1}{T}$ in the frequency domain. In [66],

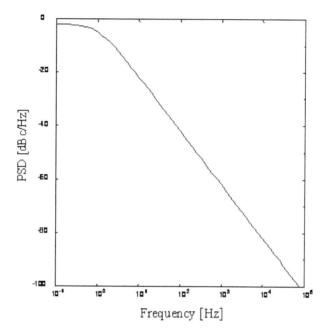

Fig. 6.1 Phase noise power spectral density (PSD) with a single-sided $-3\,\mathrm{dB}$ linewidth of 1 Hz and a $-100\,\mathrm{dBc/Hz}$ density at 100 kHz offset.

the amount of ICI is calculated and translated into a degradation in SNR that is given as

$$D_{\text{phase}} \cong \frac{11}{6\ln 10} 4\pi\beta T \frac{E_s}{N_o}. \tag{6.3}$$

Here, β is the $-3\,\mathrm{dB}$ one-sided bandwidth of the power density spectrum of the carrier. The phase noise degradation is proportional to βT, which is the ratio of the linewidth and sub-carrier spacing $\frac{1}{T}$. Figure 6.2 shows the SNR degradation in dB as a function of the normalized linewidth T. Curves are shown for three different $\frac{E_s}{N_o}$ values, corresponding to the required values to obtain a BER of 10^{-6} for uncoded QPSK, 16-QAM, and 64-QAM, respectively. The main conclusion that we can draw from this figure is that for a negligible SNR degradation of less than 0.1 dB, the $-3\,\mathrm{dB}$ phase noise bandwidth has to be about 0.1% to 0.01% of the sub-carrier spacing, depending on the modulation. For instance, to support 64-QAM in an OFDM link with a sub-carrier spacing of 300 kHz, the $-3\,\mathrm{dB}$ linewidth should be 30 Hz at most. According to (4.2),

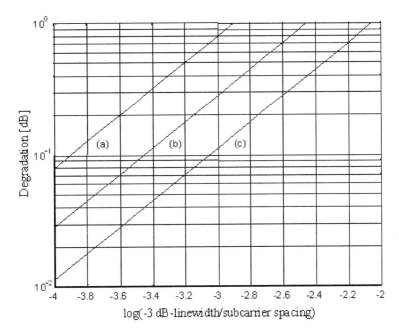

Fig. 6.2 SNR degradation in dB vs the $-3\,\text{dB}$ bandwidth of the phase noise spectrum for (a) 64-QAM $\left(\frac{E_s}{N_0} = 19\,\text{dB}\right)$, (b) 16-QAM $\left(\frac{E_s}{N_0} = 14.5\,\text{dB}\right)$, (c) QPSK $\left(\frac{E_s}{N_0} = 10.5\,\text{dB}\right)$.

this means that at a distance of 1 MHz from the carrier frequency, the phase noise spectral density has to have a value of approximately $-110\,\text{dBc/Hz}$.

The phase noise analysis in [66] assumed a free-running Voltage Controlled Oscillator (VCO). In practice, however, normally a Phase Locked Loop (PLL) is used to generate a carrier with a stable frequency. In a PLL, the frequency of a VCO is locked to a stable reference frequency, which is usually produced by a crystal oscillator. The PLL is able to track the phase jitter of the free-running VCO for jitter frequency components that fall within the tracking loop bandwidth of the loop. As a result, for frequencies below the tracking loop bandwidth the phase noise of the PLL output is determined mainly by the phase noise of the reference oscillator, which is usually smaller than the VCO phase noise, while for frequencies larger than the tracking loop bandwidth, the phase noise is dominated by the VCO phase noise. In this case, a typical phase noise spectrum will have a shape as depicted in Figure 6.3. The loop bandwidth of this example is around 100 Hz. For such phase noise spectra, the above analysis does not directly apply. We can, however, use the above

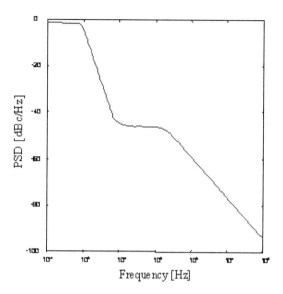

Fig. 6.3 Example of a PLL phase noise spectrum.

results to get some requirements for a practical phase noise spectrum. For example, a heuristic approach is to require that the total power in the range of a minimum frequency offset of 10% of the sub-carrier spacing to a maximum offset equal to the total bandwidth of the OFDM signal is equal to that of the Lorentzian model. For instance, suppose we have an OFDM system with a sub-carrier spacing of 300 kHz and a bandwidth of 20 MHz. For the above-mentioned example of a 30 Hz linewidth, the total power in the range of 30 kHz to 20 MHz is $(\pi/2)(a\tan(2.10^6/30)-a\tan(3.10^4/30)) \cong -32$ dBc. In fact, for this case, the exact value of the total bandwidth does not matter much, as the amount of phase noise power for frequency offsets larger than 20 MHz is negligible. The value of -32 dBc means that the total amount of phase noise for frequency offsets larger than 10% of the sub-carrier spacing is less than 0.1% of the total carrier power. For a practical PLL, the phase noise spectrum can be measured and integrated over the same frequency interval to check whether the total phase noise power meets the requirement.

6.3 Sensitivity to Frequency Offset

It is explained before that all OFDM sub-carriers are orthogonal if they all have a different integer number of cycles within the FFT interval. If there is a frequency offset, then the number of cycles in the FFT interval is not an integer anymore, with the result that ICI occurs after the FFT. The FFT output for each sub-carrier will contain interfering terms from all other sub-carriers, with an interference power that is inversely proportional to the frequency spacing. The amount of ICI for sub-carriers in the middle of the OFDM spectrum is approximately twice as large as that for sub-carriers at the band edges, because the sub-carriers in the middle have interfering sub-carriers on both sides, so there are more interferers within a certain frequency distance. In [66], the degradation in SNR caused by a frequency offset that is small relative to the sub-carrier spacing is approximated as

$$D_{\text{freq}} \cong \frac{10}{3 \ln 10} (\pi \Delta f T)^2 \frac{E_s}{N_o}. \tag{6.4}$$

This degradation is depicted in Figure 6.4 as a function of the frequency offset, normalized to the sub-carrier spacing, and for three different E_s/N_o values. Note that for a negligible degradation of about 0.1 dB, the maximum tolerable frequency offset is less than 1% of the sub-carrier spacing. For instance, for

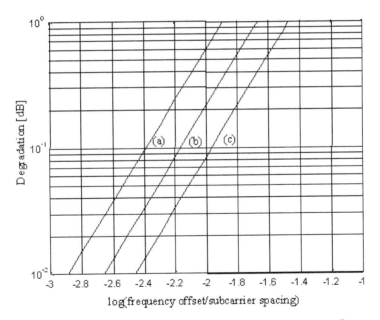

Fig. 6.4 SNR degradation in dB vs the normalized frequency offset for (a) 64-QAM ($\frac{E_s}{N_o} = 19$ dB), (b) 16-QAM ($\frac{E_s}{N_o} = 14.5$ dB), (c) QPSK ($\frac{E_s}{N_o} = 10.5$ dB).

an OFDM system at a carrier frequency of 5 GHz and a sub-carrier spacing of 300 kHz, the oscillator accuracy needs to be 3 kHz or 0.6 ppm. The initial frequency error of a low-cost oscillator will normally not meet this requirement, which means that a frequency synchronization technique has to be applied before the FFT. Examples of such synchronization techniques are described further in this chapter.

6.4 Sensitivity to Timing Errors

The previous section explained that frequency offset and phase jitter introduce a certain amount of ICI. With respect to timing offsets, OFDM is relatively more robust; in fact, the symbol timing offset may vary over an interval equal to the guard time without causing ICI or ISI, as depicted in Figure 6.5. ICI and ISI occur only when the FFT interval extends over a symbol boundary or extends over the rolloff region of a symbol. Hence, OFDM demodulation is quite insensitive to timing offsets. To achieve the best possible multi-path robustness, however, there exists an optimal timing instant as explained in later sections. Any deviation from this timing instant means that the sensitivity to delay spread increases, so the system can handle less delay spread than the value it was designed for. To minimize this loss of robustness, the system should be designed such that the timing error is small compared with the guard interval.

An interesting relationship exists between symbol timing and the demodulated sub-carrier phases [111]. Looking at Figure 6.5, we can see that as the timing changes, the phases of the sub-carriers change. The relation between the phase ϕ_i of sub-carrier i and the timing offset τ is given by

$$\phi_i = 2\pi f_i \tau. \tag{6.5}$$

Here, f_i is the frequency of the ith sub-carrier before sampling. For an OFDM system with N sub-carriers and a sub-carrier spacing of $1/T$, a timing delay of one sampling interval of T/N causes a significant phase shift of $2\pi(1 - 1/N)$ between the first and last sub-carrier. These phase shifts add to any phase shifts that are already present because of multi-path propagation. In a coherent OFDM receiver, channel estimation is performed to estimate these phase shifts for all sub-carriers, which is described in the next chapter. Figure 6.6 shows an example of the QPSK constellation of a received OFDM signal with

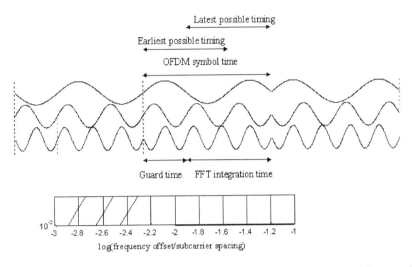

Fig. 6.5 Example of an OFDM signal with three sub-carriers, showing the earliest and latest possible symbol timing instants that do not cause ISI or ICI.

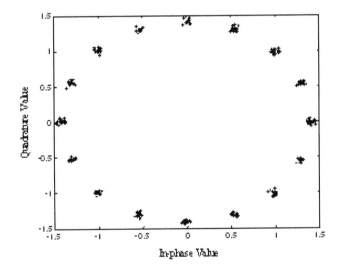

Fig. 6.6 Constellation diagram with a timing error of $T/16$ before phase correction.

48 sub-carriers, an SNR of 30 dB, and a timing offset equal to 1/16 of the FFT interval. The timing offset translates into a phase offset of a multiple of $2\pi/16$ between the sub-carriers. Because of this phase offset, the QPSK constellation points are rotated to 16 possible points on a circle. After estimation and

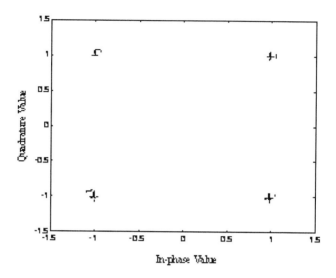

Fig. 6.7 Constellation diagram with a timing error of $T/16$ and after phase correction.

correction of the phase rotations, the constellation diagram in Figure 6.7 is obtained. In the above analysis, we implicitly assumed that there is an error only in the timing offset and not in the sampling frequency. An error in the sampling frequency has two effects [112]. First, it gives a time-varying timing offset, resulting in time-varying phase changes that have to be tracked by the receiver. Second, it causes ICI because an error in the sampling frequency means an error in the FFT interval duration, such that the sampled sub-carriers are not orthogonal anymore. Fortunately, for practically achievable sampling offsets of 10 ppm, the amount of ICI is rather small, about 0.01 dB at an E_s/N_o of 20 dB, as shown in [112].

6.5 Synchronization Using Cyclic Extension

Because of the CP, the first T_G seconds part of each OFDM symbol is identical to the last part. This property can be exploited for both timing and frequency synchronization by using a synchronization system like depicted in Figure 6.8. Basically, this device correlates a T_G long part of the signal with a part that is T seconds delayed [113, 114]. The correlator output can be written as

$$x(t) = \int_0^{T_G} r(t - \iota)_r (\iota - \iota - T)d\,\iota. \tag{6.6}$$

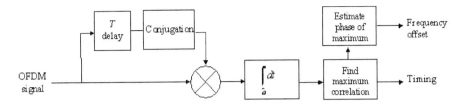

Fig. 6.8 Synchronization using the cyclic prefix.

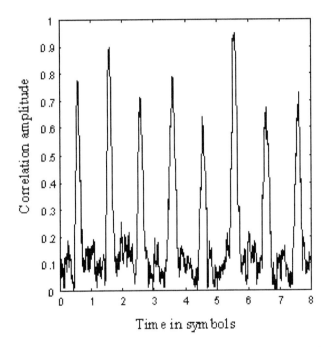

Fig. 6.9 Example of correlation output amplitude for eight OFDM symbols with 192 sub-carriers and a 20% guard time.

Two examples of the correlation output are shown in Figures 6.9 and 6.10 for eight OFDM symbols with 192 and 48 sub-carriers, respectively. These figures illustrate a few interesting characteristics of the cyclic extension correlation method. First, both figures clearly show eight peaks for the eight different symbols, but the peak amplitudes show a significant variation. The reason for this is that although the average power for a T-seconds interval of each OFDM symbol is constant, the power in the guard time can substantially

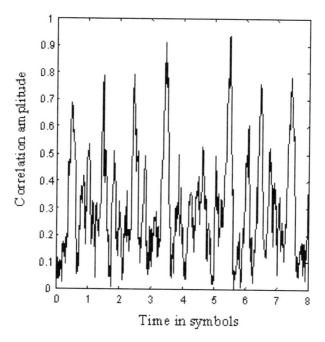

Fig. 6.10 Example of correlation output amplitude for eight OFDM symbols with 48 sub-carriers and a 20% guard time.

vary from this average power level. Another effect is the level of the undesired correlation sidelobes between the main correlation peaks. These sidelobes reflect the correlation between two pieces of the OFDM signal that belong partly or totally to two different OFDM symbols. Because different OFDM symbols contain independent data values, the correlation output is a random variable, which may reach a value that is larger than the desired correlation peak. The standard deviation of the random correlation magnitude is related to the number of independent samples over which the correlation is performed. The larger the number of independent samples, the smaller the standard deviation is. In the extreme case where the correlation is performed over only one sample, the output magnitude is proportional to the signal power, and there is no distinct correlation peak in this case. In the other extreme case where the correlation is performed over a very large number of samples, the ratio of sidelobes-to-peak amplitude will go to zero. Because the number of independent samples is proportional to the number of sub-carriers, the cyclic extension correlation technique is only effective when a large number of sub-carriers are used, preferably more than 100. An exception to this is the case where instead of random data symbols, specially designed training symbols are used [115]. In this case, the integration can be done over the entire symbol duration instead of the guard time only. The level of undesired correlation sidelobes can be minimized by a proper selection of the training symbols.

Notice that the undesired correlation sidelobes only create a problem for symbol timing. For frequency offset estimation, they do not play a role. Once symbol timing is known, the cyclic extension correlation output can be used to estimate the frequency offset. The phase of the correlation output is equal to the phase drift between samples that are T seconds apart. Hence, the frequency offset can simply be found as the correlation phase divided by $2\pi T$. This method works up to a maximum absolute frequency offset of half the sub-carrier spacing. To increase this maximum range, shorter symbols can be used, or special training symbols with different PN sequences on odd and even sub-carrier frequencies to identify a frequency offset of an integer number of sub-carrier spacings [116]. The noise performance of the frequency offset estimator is now determined for an input signal $r(t)$ that consists of an OFDM signal $s(t)$ with power P and additive Gaussian noise $n(t)$ with a one-sided noise power spectral density of N_o within the bandwidth of the OFDM signal:

$$r(t) = s(t) + n(t). \tag{6.7}$$

The frequency offset estimator multiplies the signal by a delayed and conjugated version of the input to produce an intermediate signal $y(t)$ given by

$$y(t) = r(t)\dot{r}(t - T) = ||s(t)||^2 e^{(j\varphi)} + n(t)\dot{s}(t - T) + \dot{n}(t - T)s(t)$$
$$+ n(t)\dot{n}(t - T). \tag{6.8}$$

The first term in the right-hand side of (6.8) is the desired output component with a phase equal to the phase drift over a T-second interval and a power equal to the squared signal power. The next two terms are products of the signal and the Gaussian noise. Because the signal and noise are uncorrelated, and because noise samples separated by T seconds are uncorrelated, the power of the two terms is equal to twice the product of signal power and noise power. Finally, the power of the last term of (6.8) is equal to the squared noise power. If the input SNR is much larger than one, the power of the squared noise component becomes negligible compared with the power of the other two noise terms. For practical OFDM systems, the minimum input SNR is about 6 dB, so the signal power is four times the noise power. In this case, the power of the squared noise component is eight times smaller than the power of the two signal-noise product terms. The frequency offset is estimated by averaging $y(t)$ over an interval equal to the guard time T_G and then estimating the phase of $y(t)$. Because the desired output component of 6.8 is a constant vector, averaging reduces the noise that is added to this vector. Assuming that the squared noise component may be neglected, the output SNR is approximated as

$$\text{SNR}_o = \frac{P^2}{2PN_o/T_G} = \frac{PT_G}{2N_o}. \tag{6.9}$$

Figure 6.11 shows a vector representation of the phase estimation, where the noise is divided into in-phase and quadrature components, both having a noise power of N_o/T_G. The phase error θ is given by (6.10), where the approximation has been made that n_i and n_q are small compared with the signal amplitude $\sqrt{P}$.

$$\theta = \tan^{-1}\left(\frac{n_q}{\sqrt{P} + n_i}\right) \simeq \frac{n_q}{\sqrt{P}}. \tag{6.10}$$

Because the frequency offset estimation error is equal to the phase error θ divided by $2\pi T$, the standard deviation of the frequency error is given by

$$\sigma_f = \frac{1}{2\pi T}\sqrt{\frac{N_o}{PT_G}} = \sqrt{\frac{1}{2\pi T}\frac{1}{E_s/N_o}\frac{T_G}{T_s}}. \tag{6.11}$$

Here, T_s is the symbol interval and E_s/N_o is the symbol-to-noise energy ratio, defined as

$$\frac{E_s}{N_o} = \frac{PT_s}{N_o}. \tag{6.12}$$

E_s/N_o is equal to the bit energy-to-noise density, E_b/N_o multiplied by the number of bits per symbol. Because OFDM typically has a large number of

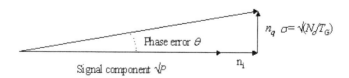

Fig. 6.11 Vector representation of phase drift estimation.

bits per symbol and E_b/N_o is larger than 1 for successful communications, typical E_s/N_o values are much larger than 1. For instance, with 48 sub-carriers using 16-QAM and rate 1/2 coding, there are 96 bits per OFDM symbol. In this case, E_s/N_o is about 20 dB larger than E_b/N_o. So for typical E_b/N_o values around 10 dB, typical E_s/N_o values are around 30 dB. Figure 6.12 shows the frequency estimation error versus E_s/N_o for three different T_G/T_s ratios. The frequency error is normalized to the sub-carrier spacing $1/T$, so a value of 0.01 means 1% of the sub-carrier spacing. The solid lines represent calculated values according to Equation (6.11), while the dotted lines are derived from simulations. The difference between the two set of curves show the effect of the simplifications made in the derivation of (6.11). For E_s/N_o values of 30 dB or more, the difference is negligible, but around 20 dB, the simulated errors are about 50% larger than the calculated values. Earlier it is explained that the frequency error preferably had to be less than 1% of the sub-carrier spacing

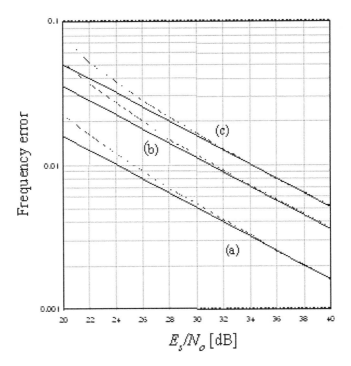

Fig. 6.12 Frequency estimation error normalized to the sub-carrier spacing. Solid lines are calculated; dotted lines are simulated. (a) $T_G/T_s = 1$, (b) $T_G/T_s = 0.2$, (c) $T_G/T_s = 0.1$.

to have a negligible performance degradation. From Figure 6.12, we can learn that such an error level can be achieved at an E_s/N_o value of 26, 31, and 34 dB for a T_G/T_s ratio of 1, 0.2, and 0.1, respectively. A lower T_G/T_s ratio means that a smaller fraction of an OFDM symbol is used for synchronization, hence more SNR is required to attain the same performance as for a larger T_G/T_s value.

If the required E_s/N_o value for an acceptable frequency error level is too large, then averaging the vector $y(t)$ in (6.8) over multiple OFDM symbols can be used to increase the effective signal-to-noise ratio. For averaging over K symbols, the frequency error standard deviation becomes

$$\sigma_f = \frac{1}{2\pi T} \sqrt{\frac{1}{K E_s/N_o} \frac{T_G}{T_s}}. \tag{6.13}$$

Averaging over K symbols has the effect that the curves in Figure 6.12 shift to the left by $10\log K$ dB. For instance, when averaging over four OFDM symbols, a 1% frequency error is achieved at an E_s/N_o value of 28 dB for a T_G/T_s ratio of 0.1 instead of 34 dB without averaging.

Notice that a T_G/T_s ratio of one — curve (a) in Figure 6.12 — is a special case where the guard time is equal to the symbol period. For normal OFDM data symbols, this is not possible, as it would mean that the FFT interval is zero. It does, however, correspond to the interesting case where two identical OFDM symbols are used to estimate the frequency offset. In this case, all samples of a symbol can be used to estimate the phase difference with the corresponding samples of the other symbol. Hence, (6.11) applies with T_G/T_s set to 1, although T_G is not really a guard time in this case.

6.6 Synchronization Using Special Training Symbols

The synchronization technique based on the cyclic extension is particularly suited to tracking or to blind synchronization in a circuit-switched connection, where no special training signals are available. For packet transmission, however, there is a drawback because an accurate synchronization needs an averaging over a large (> 10) number of OFDM symbols to attain a distinct correlation peak and a reasonable SNR. For high-rate packet transmission, the synchronization time needs to be as short as possible, preferably a few OFDM symbols only. To achieve this, special OFDM training symbols can be used for

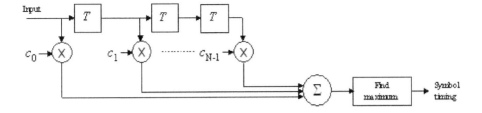

Fig. 6.13 Matched filter that is matched to a special OFDM training symbol.

which the data content is known to the receiver [116, 117, 118]. In this way, the entire received training signal can be used to achieve synchronization, whereas the cyclic extension method only uses a fraction of each symbol. Figure 6.13 shows a block diagram of a matched filter that can be used to correlate the input signal with the known OFDM training signal. Here, T is the sampling interval and c_i are the matched filter coefficients, which are the complex conjugates of the known training signal. From the correlation peaks in the matched filter output signal, both symbol timing and frequency offset can be estimated, as will be explained in this section. Notice that the matched filter correlates with the OFDM time signal, before performing an FFT in the receiver. Hence, this technique is very similar to synchronization in a direct-sequence spread-spectrum receiver, where the input signal is correlated with a known spreading signal. In fact, the latter approach of using a single-carrier training signal can also be combined with OFDM, as proposed in [119], but here we will assume that the training signal consists of normal OFDM data symbols. Figure 6.14 shows an example of the matched filter output for an OFDM training symbol with 48 sub-carriers. The training signal for this case consisted of five identical OFDM symbols without a guard time. Alternatively, it could be stated that there is only one OFDM symbol with a guard time equal to four IFFT intervals, because the IFFT output is repeated four times. The reason for having the training symbol interval equal to the IFFT interval is that this gives the best possible cyclic autocorrelation properties in terms of low undesired sidelobes. This can be seen in Figure 6.14, which shows the undesired sidelobes to be at least 20 dB lower than the main correlation peaks. An exception to this occurs at the beginning of the correlation. The reason for this is that at this point, an aperiodic correlation is performed instead of a cyclic correlation, because the matched filter is partly filled with zero

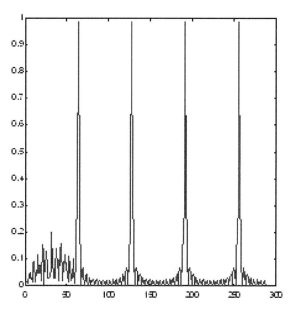

(a) Zero fractional timing offset between input signals and matched filter coefficients.

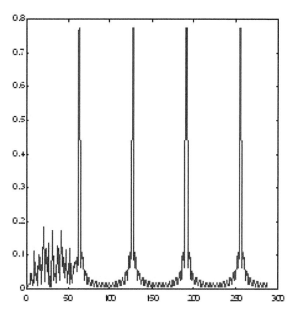

(b) Worst case fractional offset of half a sample between input signal and reference pulse.

Fig. 6.14 Matched filter output vs sample number for 4 training symbols, using 48 sub-carriers and 64 samples per symbol.

values until a full OFDM symbol has been received. A similar effect happens at the end of the training not shown in the picture when the matched filter will partly correlate with samples from the following OFDM data symbol that is different from the training symbol. Hence, to avoid undesired partial correlations, the matched filter outputs during the first and last symbol intervals should be skipped. The values in between can be used to detect the main correlation peak, which gives the desired symbol timing information. The correlation function of Figure 6.14 was made for the case of a zero fractional timing offset between the input signal and the known training signal. This means that the matched filter tap values, which are equal to the conjugated training signal samples, are exactly equal to the conjugated sample values of the incoming OFDM signal. This ideal situation does not occur when there is a timing offset of some fraction of a sample interval between the input signal and the known training signal. To see the effect of a fractional timing offset, Figure 6.14 shows the correlation output for the worst case timing offset of half a sampling interval. In this case, instead of one main peak per symbol interval, there are two equally strong peaks with a slightly smaller amplitude than the single peak in the case of no timing offset. However, the relative level of undesired correlation sidelobes is still 20 dB below the main peaks. The plots of Figure 6.14 assumed unquantized input signals and tap values. In practice, it is desirable to have a low number of quantization bits to keep the implementation simple. Figure 6.15 shows the correlation output where the matched filter tap values are quantized to $\{-1, 0, 1\}$ values for both the real and imaginary parts. This reduces the complexity of the multiplications in the matched filter to additions, having a relatively low hardware complexity. As we can see from Figure 6.15, the correlation output looks different from the unquantized case in Figure 6.14, but the undesired sidelobe level is still about 20 dB below the main peak. Such good correlation properties cannot be achieved with any arbitrary quantized OFDM signal. To minimize the effects of the quantization, the best results are obtained with OFDM signals that have minimum amplitude fluctuations. In the case of Figure 6.15, the OFDM symbol consists of the IFFT of a length 48 complementary code, which results in a signal with peak amplitude fluctuations that are no more than 3 dB larger than the root mean square value. More details about these complementary codes can be found in the chapter which deals with the OFDM peak-to-average power issue.

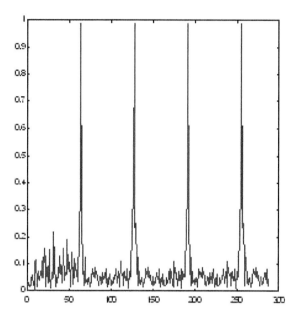

(a) Zero fractional timing offset between input signals and matched filter coefficients.

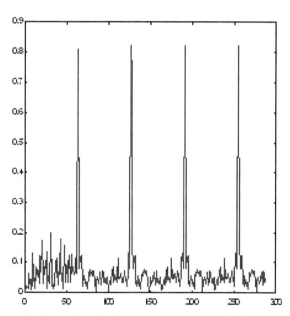

(b) Worst case fractional offset of half a sample between input signal and reference pulse.

Fig. 6.15 Matched filter output vs sample number with $\{1, -1, 0\}$ values for in-phase and quadrature coefficients.

6.7 Optimal Training in Presence of Multi-path

The task of OFDM symbol timing is to minimize the amount of ISI and ICI. This type of interference is absent when the FFT is taken over the flat part of the signaling window, which is shown in Figure 6.16. This window is the envelope of the transmitted OFDM symbols. Within the flat part of the window, all subchannels maintain perfect orthogonality. In the presence of multi-path, however, orthogonality is lost if the multi-path delays exceed the effective guard time, which is equal to the duration of the flat window part minus the FFT period. The effect of multi-path propagation on ISI and ICI is illustrated in Figure 6.17. It shows the windowing envelopes of three OFDM symbols. The radio channel consists of two paths with a relative delay of almost half a symbol and a relative amplitude of 0.5. The receiver selects the FFT timing such that the FFT is taken over the flat envelope part of the strongest path. Because the multi-path delay is larger than the guard time, however, the FFT period cannot at the same time cover a totally flat envelope part of the weaker signal. As a result, the nonflat part of the symbol envelope causes ICI. At the same time, the partial overlap of the previous OFDM symbol in the FFT period causes ISI. The solution to the timing problem is to find the delay

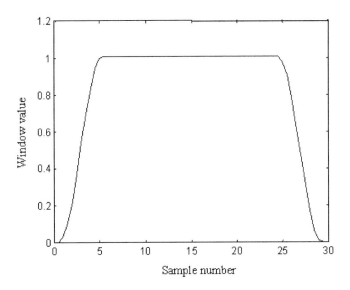

Fig. 6.16 Raised cosine window.

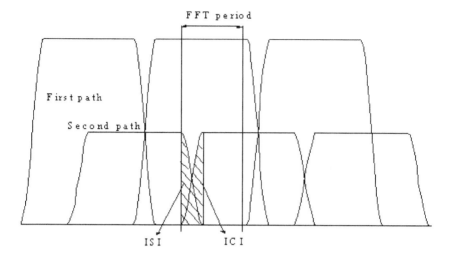

Fig. 6.17 ISI/ICI caused by multi-path signals.

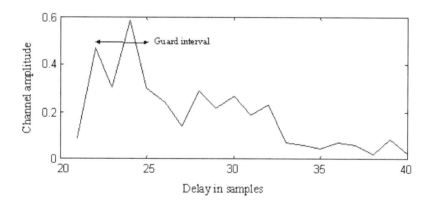

Fig. 6.18 Example of channel impulse response.

window with a width equal to the guard time that contains maximum signal power. The optimal FFT starting time, then, is equal to the starting delay of the found delay window, plus the delay that occurs between a matched filter peak output from a single OFDM pulse and the delay of the last sample on the flat part of the OFDM signal envelope, minus the length of the FFT interval. Figure 6.18 clearly shows the advantage of looking for maximum power in the whole guard interval, rather than looking for the maximum correlation output only. If the latter is applied to the example of Figure 6.18, then sample 23 would be chosen instead of sample 21. As a result, the multi-path power at samples 21 and 22 would cause extra ISI and ICI, while the useful signal power of samples 24 to 27 would be less than the power of the samples 21 to 24. Hence, the signal-to-interference ratio can easily be degraded by several dB if the suboptimal maximum peak detection is used. We now prove that maximizing power in a certain delay window actually maximizes the signal-to-interference ratio. Figure 6.19 shows the OFDM symbol structure, where T is the time needed by the FFT. If a multi-path signal is introduced with a relative delay exceeding T_{g1}, it will cause ISI and ICI. Similarly, multi-path signals with relative delays less than $-T_{g2}$ cause ISI and ICI. The timing problem is now to choose T_{g1} and T_{g2} such that the amount of ICI and ISI after the FFT is minimized.

From the above, it is clear that ISI and ICI are caused by all multi-path signals, which delays fall outside a window of $T_g = T_{g1} + T_{g2}$. All multi-path signals within this delay window contribute to the effectively used signal power. Hence, the optimal timing circuit maximizes the signal-to-(ISI + ICI) ratio (SIR), given by

$$\text{SIR} = \frac{S_u}{S_t - S_u}, \ S_u = \int_{T_o}^{T_o+T_g} ||h(\tau)||^2 d\tau, \ S_t = \int_{-\infty}^{\infty} ||h(\tau)||^2 d\tau. \quad (6.14)$$

Here, $T_o = -T_{g2}$ is the timing offset of the guard time window T_g. S_t denotes the total received signal power and S_u is the useful signal power. Because only S_u depends on the timing offset T_o, the SIR is maximized by maximizing S_u; that is, choosing the T_o value that contains the largest power of $h(\tau)$ in the interval $\{T_o, T_o + T_G\}$.

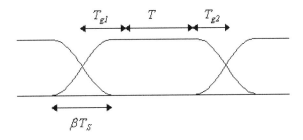

Fig. 6.19 OFDM symbol structure.

7

Channel Estimation and Equalization

7.1 Introduction

In an Orthogonal Frequency Division Multiplexing (OFDM) link, the data bits are modulated on the sub-carriers by some form of Phase Shift Keying (PSK) or Quadrature Amplitude Modulation (QAM). To estimate the bits at the receiver, knowledge is required about the reference phase and amplitude of the constellation on each sub-carrier. In general, the constellation of each sub-carrier shows a random phase shift and amplitude change, caused by carrier frequency offset, timing offset, and frequency selective fading. To cope with these unknown phase and amplitude variations, two different approaches exist. The first one is coherent detection, which uses estimates of the reference amplitudes and phases to determine the best possible decision boundaries for the constellation of each sub-carrier. The main issue with coherent detection is how to find the reference values without introducing too much training overhead. To achieve this, several channel estimation techniques exist that will be described in the next section. The second approach is differential detection, which does not use absolute reference values, but only looks at the phase and/or amplitude differences between two QAM values. Differential detection can be done both in the time domain or in the frequency domain; in the first case, each sub-carrier is compared with the sub-carrier of the previous OFDM symbol. In the case of differential detection in the frequency domain, each sub-carrier is compared with the adjacent sub-carrier within the same OFDM symbol.

7.2 Coherent Detection

Figure 7.1 shows a block diagram of a coherent OFDM receiver. After down-conversion and analog-to-digital conversion, the Fast Fourier Transform (FFT)

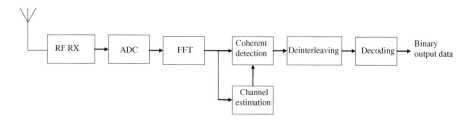

Fig. 7.1 Block diagram of an OFDM receiver with coherent detection.

is used to demodulate the N sub-carriers of the OFDM signal. For each symbol, the FFT output contains N QAM values. However, these values contain random phase shifts and amplitude variations caused by the channel response, local oscillator drift, and timing offset. It is the task of the channel estimation block to learn the reference phases and amplitudes for all sub-carriers, such that the QAM symbols can be converted to binary soft decisions. The next subsections present several techniques to obtain the channel estimates that are required for coherent detection.

7.2.1 Two-Dimensional Channel Estimators

In general, radio channels are fading both in time and in frequency. Hence, a channel estimator has to estimate time-varying amplitudes and phases of all sub-carriers. One way to do this is to use a two-dimensional channel estimator that estimates the reference values based on a few known pilot values. This concept is demonstrated in Figure 7.2, which shows a block of 9 OFDM symbols with 16 sub-carriers. The gray sub-carrier values are known pilots. Based on these pilots, all other reference values can be estimated by performing a two-dimensional interpolation [120, 121, 122, 123].

To be able to interpolate the channel estimates both in time and frequency from the available pilots, the pilot spacing has to fulfill the Nyquist sampling theorem, which states that the sampling interval must be smaller than the inverse of the double-sided bandwidth of the sampled signal. For the case of OFDM, this means that there exist both a minimum sub-carrier spacing and a minimum symbol spacing between pilots. By choosing the pilot spacing much smaller than these minimum requirements, a good channel estimation can be made with a relatively easy algorithm. The more pilots are used, however, the

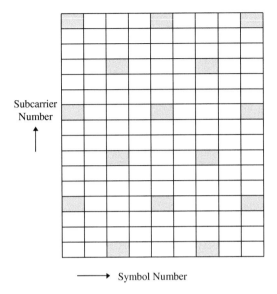

Fig. 7.2 Example of pilots (marked gray) in a block of 9 OFDM symbols with 16 sub-carriers.

smaller the effective Signal-to-Noise Ratio (SNR), becomes that is available for data symbols. Hence, the pilot density is a trade-off between channel estimation performance and SNR loss.

To determine the minimum pilot spacing in time and frequency, we need to find the bandwidth of the channel variation in time and frequency. These bandwidths are equal to the Doppler spread B_d in the time domain and the maximum delay spread $\tau_{\max}$ in the frequency domain [44]. Hence, the requirements for the pilot spacings in time and frequency s_t and s_f are

$$s_t < \frac{1}{B_d} \tag{7.1}$$

$$s_f < \frac{1}{\tau_{\max}}. \tag{7.2}$$

Assume now the available pilot values are arranged in a vector $\hat{\mathbf{p}}$ and the channel values that have to be estimated from $\hat{\mathbf{p}}$ are in a vector $\mathbf{h}$. Notice that we use bold letters to distinguish vectors from scalar variables. It is assumed that any known modulation of the pilots is removed before the estimation; for instance, if the transmitter applies some phase shifts to the pilots, then the receiver has to back-rotate those known phases. The channel estimation problem is now to find the channel estimates $\hat{\mathbf{h}}$ as a linear combination of the pilot estimates $\hat{\mathbf{p}}$. According to [124], the minimum mean square error estimate for this problem is given by

$$\hat{\mathbf{h}} = \mathbf{R}_{h\hat{p}} \mathbf{R}_{\hat{p}\hat{p}}^{-1} \hat{\mathbf{p}}. \tag{7.3}$$

$\mathbf{R}_{h\hat{p}}$ is the cross-covariance matrix between $\mathbf{h}$ and the noisy pilot estimates $\hat{\mathbf{p}}$, given by

$$\mathbf{R}_{\hat{p}\hat{p}} = E\left\{\mathbf{h}\hat{\mathbf{p}}^{H}\right\}. \tag{7.4}$$

$\mathbf{R}_{\hat{p}\hat{p}}$ is the auto-covariance matrix of the pilot estimates:

$$\mathbf{R}_{\hat{p}\hat{p}} = E\left\{\hat{\mathbf{p}}\hat{\mathbf{p}}^{H}\right\} = \mathbf{R}_{pp} + \sigma_{n}^{2}\left(\mathbf{p}\mathbf{p}^{H}\right)^{-1}. \tag{7.5}$$

Assuming the pilots all have the same power, which is the case if all pilots are for instance known Quadrature Phase Shift Keying (QPSK) symbols, then the pilots' auto-covariance matrix can be rewritten as

$$\mathbf{R}_{\hat{p}\hat{p}} = E\left\{\hat{\mathbf{p}}\hat{\mathbf{p}}^{H}\right\} = \mathbf{R}_{pp} + \frac{1}{\mathrm{SNR}}\mathbf{I}, \tag{7.6}$$

where SNR is the signal-to-noise ratio per pilot and $\mathbf{R}_{pp}$ is the auto-covariance matrix of the noiseless pilots. With this, the channel estimates can be written as

$$\hat{\mathbf{h}} = \mathbf{R}_{h\hat{p}}\left(\mathbf{R}_{pp} + \frac{1}{\mathrm{SNR}}\mathbf{I}\right)^{-1}\hat{\mathbf{p}}. \tag{7.7}$$

The above Equation (7.7) basically gives the desired channel estimates as the multiplication of an interpolation matrix with the pilot estimates. Notice that the interpolation matrix does not depend on the received symbols; it only depends on the position of the pilots and the number of pilots and channel estimates. Hence, the interpolation matrix can be designed as a constant matrix, avoiding the need to do matrix inversions in the OFDM receiver.

The elements covariance matrices $\mathbf{R}_{h\hat{p}}$ and $\mathbf{R}_{\hat{p}\hat{p}}$ can be calculated as follows. Both matrices contain correlation values between sub-carrier values for different time and frequency spacings. If k and l are the sub-carrier number and OFDM symbol number, respectively, the correlation values are given by

$$E\left\{h_{k,l}\hat{p}_{k',l'}^{H}\right\} = E\left\{p_{k,l}p_{k',l'}^{*}\right\} = r_{f}(k - k')r_{t}(l - l'). \tag{7.8}$$

Here, $r_t(l)$ and $r_f(k)$ are the correlation functions in time and frequency, respectively. For an exponentially decaying multi-path power delay profile, $r_f(k)$ is given by

$$r_f(k) = \frac{1}{1 + j2\pi \tau_{\text{rms}}k/T}. \tag{7.9}$$

Here, $1/T$ is the sub-carrier spacing, which is the inverse of the FFT interval T. For a time-fading signal with a maximum Doppler frequency f_{max} and a Jakes spectrum, the time correlation function $r_t(l)$ is given as

$$r_t(l) = J_0(2\pi f_{\text{max}} l T_s), \tag{7.10}$$

where $J_0(x)$ is the zeroth order Bessel function of the first kind and T_s is the OFDM symbol duration, which is the FFT interval T plus the guard time. To illustrate the channel estimation technique described above, an example will be given for the case of five pilots in a block of five OFDM symbols with five sub-carriers. Four pilots are located at the corners of the block and one in the middle, being the third sub-carrier of the third OFDM symbol. Using the Jakes fading channel model, an OFDM signal was generated that experienced fading both in time and frequency. Table 7.1 lists the 25 reference channel values for the five sub-carriers in each of the five symbols for this example. Of course, an OFDM receiver does not know these reference values; it only has knowledge about the five pilot values, which are located at row and column numbers $\{1, 1\}, \{1, 5\}, \{3, 3\}, \{5, 1\}$, and $\{5, 5\}$. In this particular example, no noise is added to the pilot values. Each column represents one symbol with five sub-carrier values. We can see from the table that there is fading both in frequency — across the rows — and in time.

Table 7.2 gives the elements of the pilot auto-covariance matrix. This matrix is independent of the received signal, so it can be precalculated using (7.8), (7.9), and (7.10). The first row of $\mathbf{R_{pp}}$ consists of the correlations between the first pilot and all five pilots. The first value is 1, as this is the correlation

Table 7.1 Example channel values for a block of five OFDM symbols and five sub-carriers.

$1.0386 - 0.2468i$	$1.1333 - 0.2441i$	$1.1777 - 0.2491i$	$1.1693 - 0.2617i$	$1.1048 - 0.2761i$
$0.8938 - 0.4782i$	$0.9798 - 0.4821i$	$1.0172 - 0.4842i$	$1.0040 - 0.4847i$	$0.9386 - 0.4778i$
$0.6726 - 0.6302i$	$0.7479 - 0.6398i$	$0.7802 - 0.6402i$	$0.7675 - 0.6316i$	$0.7099 - 0.6082i$
$0.4173 - 0.6794i$	$0.4809 - 0.6913i$	$0.5093 - 0.6897i$	$0.5007 - 0.6750i$	$0.4567 - 0.6421i$
$0.1794 - 0.6258i$	$0.2321 - 0.6349i$	$0.2578 - 0.6303i$	$0.2548 - 0.6125i$	$0.2258 - 0.5774i$

Table 7.2 Example of covariance matrix $\mathbf{R_{pp}}$.

1.0000	0.4720	$0.7967 - 0.2589i$	$0.7568 - 0.5499i$	$0.3572 - 0.2595i$
0.4720	1.0000	$0.7967 - 0.2589i$	$0.3572 - 0.2595i$	$0.7568 - 0.5499i$
$0.7967 - 0.2589i$	$0.7967 - 0.2589i$	1.0000	$0.7967 - 0.2589i$	$0.7967 - 0.2589i$
$0.7568 - 0.5499i$	$0.3572 - 0.2595i$	$0.7967 - 0.2589i$	1.0000	0.4720
$0.3572 - 0.2595i$	$0.7568 - 0.5499i$	$0.7967 - 0.2589i$	0.4720	1.0000

of the first pilot with itself. The second value of the first row is the correlation between the first pilot at the first sub-carrier with the pilot of the first sub-carrier of the fifth symbol. Because these pilots are both on the same sub-carrier, the frequency correlation component $r_f(k)$ is equal to 1, and the correlation value is purely determined by the time fading component $r_t(k)$. This explains why the imaginary component of the second correlation value is zero, because $r_t(k)$ (7.10) is a strictly real function. Notice that from the matrix $\mathbf{R_{pp}}$ the matrix is formed by adding an identity matrix multiplied by the inverse of the SNR (7.6). In practice, the SNR is not known *a priori*, so an expected value is used. In our example, we will use an SNR of 10 dB. Using a large SNR value gives a relatively large weight on $\mathbf{R_{pp}}$ in (7.7).

Table 7.3 gives the example matrix $\mathbf{R_{h\hat{p}}}$. This matrix contains the correlation values between all 25 channel values — in the block of five symbols by five sub-carriers — with each of the five pilots. Hence, the matrix has 25 rows and 5 columns. Because the pilots are also part of the 25 channel values, the rows of $\mathbf{R_{pp}}$ are all part of $\mathbf{R_{h\hat{p}}}$; for instance, the first row of $\mathbf{R_{h\hat{p}}}$ is the same as the first row of $\mathbf{R_{pp}}$.

The above matrices can be combined to the final interpolation matrix listed in Table 7.4. The matrix has 25 rows and 5 columns; when multiplied by the row vector containing five measured pilots, 25 channel estimates are obtained. Figure 7.3 shows the estimation errors for the channel example of Table 7.1. We can see that the relative estimation error is between 3% and 14%. The largest errors are located at the edges of the block, which is a typical phenomenon of this kind of interpolation. It suggests that to minimize the interpolation error, pilots should be used that surround the channel positions that have to be estimated.

7.2.2 One-Dimensional Channel Estimators

The channel estimation technique described in the previous section basically performed a two-dimensional interpolation to estimate points on a

Table 7.3 Example of covariance matrix $\mathbf{R}_{h\hat{p}}$.

1.0000	0.4720	0.7967 − 0.2589i	0.7568 − 0.5499i	0.3572 − 0.2595i
0.9836 − 0.1558i	0.4643 − 0.0735i	0.8377 − 0.1327i	0.8584 − 0.4374i	0.4052 − 0.2064i
0.9355 − 0.3040i	0.4416 − 0.1435i	0.8516	0.9355 − 0.3040i	0.4416 − 0.1435i
0.8584 − 0.4374i	0.4052 − 0.2064i	0.8377 − 0.1327i	0.9836 − 0.1558i	0.4643 − 0.0735i
0.7568 − 0.5499i	0.3572 − 0.2595i	0.7967 − 0.2589i	1.0000	0.4720
0.9618	0.6820	0.8998 − 0.2924i	0.7279 − 0.5289i	0.5161 − 0.3750i
0.9461 − 0.1498i	0.6708 − 0.1062i	0.9461 − 0.1498i	0.8256 − 0.4207i	0.5854 − 0.2983i
0.8998 − 0.2924i	0.6380 − 0.2073i	0.9618	0.8998 − 0.2924i	0.6380 − 0.2073i
0.8256 − 0.4207i	0.5854 − 0.2983i	0.9461 − 0.1498i	0.9461 − 0.1498i	0.6708 − 0.1062i
0.7279 − 0.5289i	0.5161 − 0.3750i	0.8998 − 0.2924i	0.9618	0.6820
0.8516	0.8516	0.9355 − 0.3040i	0.6445 − 0.4683i	0.6445 − 0.4683i
0.8377 − 0.1327i	0.8377 − 0.1327i	0.9836 − 0.1558i	0.7310 − 0.3725i	0.7310 − 0.3725i
0.7967 − 0.2589i	0.7967 − 0.2589i	1.0000	0.7967 − 0.2589i	0.7967 − 0.2589i
0.7310 − 0.3725i	0.7310 − 0.3725i	0.9836 − 0.1558i	0.8377 − 0.1327i	0.8377 − 0.1327i
0.6445 − 0.4683i	0.6445 − 0.4683i	0.9355 − 0.3040i	0.8516	0.8516
0.6820	0.9618	0.8998 − 0.2924i	0.5161 − 0.3750i	0.7279 − 0.5289i
0.6708 − 0.1062i	0.9461 − 0.1498i	0.9461 − 0.1498i	0.5854 − 0.2983i	0.8256 − 0.4207i
0.6380 − 0.2073i	0.8998 − 0.2924i	0.9618	0.6380 − 0.2073i	0.8998 − 0.2924i
0.5854 − 0.2983i	0.8256 − 0.4207i	0.9461 − 0.1498i	0.6708 − 0.1062i	0.9461 − 0.1498i
0.5161 − 0.3750i	0.7279 − 0.5289i	0.8998 − 0.2924i	0.6820	0.9618
0.4720	1.0000	0.7967 − 0.2589i	0.3572 − 0.2595i	0.7568 − 0.5499i
0.4643 − 0.0735i	0.9836 − 0.1558i	0.8377 − 0.1327i	0.4052 − 0.2064i	0.8584 − 0.4374i
0.4416 − 0.1435i	0.9355 − 0.3040i	0.8516	0.4416 − 0.1435i	0.9355 − 0.3040i
0.4052 − 0.2064i	0.8584 − 0.4374i	0.8377 − 0.1327i	0.4643 − 0.0735i	0.9836 − 0.1558i
0.3572 − 0.2595i	0.7568 − 0.5499i	0.7967 − 0.2589i	0.4720	1.0000

time–frequency grid based on several pilots. Instead of directly calculating the two-dimensional solution, it is also possible to separate the interpolation into two one-dimensional interpolations, as illustrated by Figure 7.4 [125]. With this technique, first an interpolation in the frequency domain is performed for all symbols containing pilots. Then, for each sub-carrier an interpolation in the time domain is performed to estimate the remaining channel values.

7.2.3 Special Training Symbols

The channel estimation techniques from the previous sections were designed to estimate a channel that varied both in time and frequency. These techniques are especially suitable for continuous transmission systems such as Digital Audio Broadcasting (DAB) or Digital Video Broadcasting (DVB), which are both described in Chapter 10. They are not very suited, however, for packet-type communications for two reasons. First, in many packet transmission systems, such as Wireless Local Area Network (WLAN), the packet

Table 7.4 Example of interpolation matrix $\mathbf{R}_{h\hat{p}}\left(\mathbf{R}_{pp} + \frac{1}{\text{SNR}}\mathbf{I}\right)^{-1}$.

0.9044 + 0.0902i	0.0331 − 0.0034i	0.0333 − 0.0749i	0.0145 − 0.0630i	−0.0241 + 0.0562i
0.6009 + 0.0022i	−0.0908 − 0.0346i	0.4006 + 0.0861i	0.1446 − 0.0753i	−0.1200 − 0.0040i
0.2760 − 0.0836i	−0.2100 − 0.0647i	0.7484 + 0.2459i	0.2760 − 0.0836i	−0.2100 − 0.0647i
0.1446 − 0.0753i	−0.1200 − 0.0040i	0.4006 + 0.0861i	0.6009 + 0.0022i	−0.0908 − 0.0346i
0.0145 − 0.0630i	−0.0241 + 0.0562i	0.0333 − 0.0749i	0.9044 + 0.0902i	0.0331 − 0.0034i
0.7534 + 0.0819i	0.2917 + 0.0323i	0.0409 − 0.1125i	0.0040 − 0.0267i	−0.0165 + 0.0364i
0.4658 + 0.0002i	0.0993 − 0.0193i	0.4555 + 0.0678i	0.0816 − 0.0547i	−0.0587 − 0.0169i
0.1620 − 0.0797i	−0.0955 − 0.0697i	0.8483 + 0.2478i	0.1620 − 0.0797i	−0.0955 − 0.0697i
0.0816 − 0.0547i	−0.0587 − 0.0169i	0.4555 + 0.0678i	0.4658 + 0.0002i	0.0993 − 0.0193i
0.0040 − 0.0267i	−0.0165 + 0.0364i	0.0409 − 0.1125i	0.7534 + 0.0819i	0.2917 + 0.0323i
0.5412 + 0.0619i	0.5412 + 0.0619i	0.0435 − 0.1257i	−0.0067 + 0.0077i	−0.0067 + 0.0077i
0.2920 − 0.0072i	0.2920 − 0.0072i	0.4746 + 0.0614i	0.0111 − 0.0344i	0.0111 − 0.0344i
0.0333 − 0.0749i	0.0333 − 0.0749i	0.8830 + 0.2483i	0.0333 − 0.0749i	0.0333 − 0.0749i
0.0111 − 0.0344i	0.0111 − 0.0344i	0.4746 + 0.0614i	0.2920 − 0.0072i	0.2920 − 0.0072i
−0.0067 + 0.0077i	−0.0067 + 0.0077i	0.0435 − 0.1257i	0.5412 + 0.0619i	0.5412 + 0.0619i
0.2917 + 0.0323i	0.7534 + 0.0819i	0.0409 − 0.1125i	−0.0165 + 0.0364i	0.0040 − 0.0267i
0.0993 − 0.0193i	0.4658 + 0.0002i	0.4555 + 0.0678i	−0.0587 − 0.0169i	0.0816 − 0.0547i
−0.0955 − 0.0697i	0.1620 − 0.0797i	0.8483 + 0.2478i	−0.0955 − 0.0697i	0.1620 − 0.0797i
−0.0587 − 0.0169i	0.0816 − 0.0547i	0.4555 + 0.0678i	0.0993 − 0.0193i	0.4658 + 0.0002i
−0.0165 + 0.0364i	0.0040 − 0.0267i	0.0409 − 0.1125i	0.2917 + 0.0323i	0.7534 + 0.0819i
0.0331 − 0.0034i	0.9044 + 0.0902i	0.0333 − 0.0749i	−0.0241 + 0.0562i	0.0145 − 0.0630i
−0.0908 − 0.0346i	0.6009 + 0.0022i	0.4006 + 0.0861i	−0.1200 − 0.0040i	0.1446 − 0.0753i
−0.2100 − 0.0647i	0.2760 − 0.0836i	0.7484 + 0.2459i	−0.2100 − 0.0647i	0.2760 − 0.0836i
−0.1200 − 0.0040i	0.1446 − 0.0753i	0.4006 + 0.0861i	−0.0908 − 0.0346i	0.6009 + 0.0022i
−0.0241 + 0.0562i	0.0145 − 0.0630i	0.0333 − 0.0749i	0.0331 − 0.0034i	0.9044 + 0.0902i

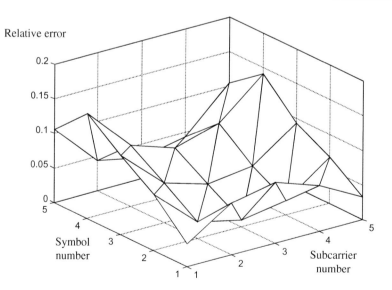

Fig. 7.3 Example of relative channel estimation errors vs sub-carrier number and symbol number.

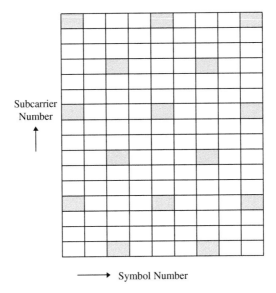

Fig. 7.4 Channel estimation with separable filters in frequency (1) and time (2) direction.

length is short enough to assume a constant channel during the length of the packet. This means there is no need to estimate time fading, which greatly simplifies the channel estimation problem. Second, using pilots scattered over several OFDM data symbols introduces a delay of several symbols before the first channel estimates can be calculated. Such a delay is undesirable in packet transmission like in an IEEE 802.11 WLAN, which requires an acknowledgment to be sent after each packet transmission. Any delay in the reception of a packet will also delay the acknowledgment and hence decrease the effective throughput of the system. An additional disadvantage is the fact that the receiver needs to buffer several OFDM symbols, thereby requiring extra hardware.

For the specific problem of channel estimation in packet transmission systems, the most appropriate approach seems to be the use of a preamble consisting of one or more known OFDM symbols. This approach is sketched in Figure 7.5. The figure shows the time–frequency grid with sub-carriers on the vertical axis and symbols on the horizontal axis. All gray sub-carriers are pilots. The packet starts with two OFDM symbols for which all data values are known. These training symbols can be used to obtain channel estimates,

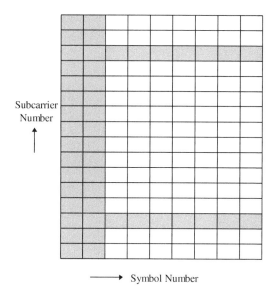

Fig. 7.5 Example of a packet with two training symbols for channel estimation and two pilot sub-carriers used for frequency synchronization.

as well as a frequency offset estimate, as was explained in Chapter 7. After the first two training symbols, Figure 7.5 shows two pilot sub-carriers within the data symbols. These pilots are not meant for channel estimation, but for tracking the remaining frequency offset after the initial training. Because this frequency offset affects all sub-carriers in a similar way, there is no need to have many pilots with a small frequency spacing as in the case of channel estimation. This type of pilot structure was first mentioned in a proposal for the IEEE 802.11 OFDM standard [126].

The choice of the number of training symbols is a trade-off between a short training time and a good channel estimation performance. Using two training symbols is a reasonable choice, because it gives a 3 dB-lower noise level in the channel estimates by a simple averaging of the two training symbols, and a minimum of two training symbols is convenient anyway to estimate the frequency offset by comparing the phase shift of the two identical symbols.

If multiple training symbols are used, there is actually no need to repeat an entire OFDM symbol including guard time. A more efficient way is to repeat the Inverse Fast Fourier Transform (IFFT) interval and to keep a single guard time, as depicted in Figure 7.6. This can also be viewed as extending

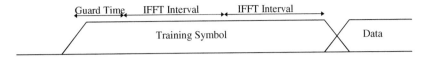

Fig. 7.6 Extended training symbol for channel estimation with a single guard time and multiple IFFT intervals.

the guard time to twice the original guard time plus an extra time equal to the IFFT interval. The advantage of this approach is that it makes the training extra robust to multi-path; a receiver can perform a channel estimation by taking the FFT of the averaged IFFT intervals of the long training symbol. This channel estimate will be free from Inter-Symbol Interference (ISI) and Inter Carrier Interference (ICI) as long as the relative multi-path delays are smaller than the guard time of the training symbol, which is now doubled relative to the guard time of the OFDM data symbols.

One of the main assumptions when using pilot symbols only at the start of a packet is that channel variations during the rest of the packet are negligible. Whether this is a valid assumption depends on the packet duration and the Doppler bandwidth. This issue is discussed further in Section 7.3.1. This section describes a related differential detection technique that relies on the same assumption of a channel that is nearly constant in time.

7.2.4 Decision-Directed Channel Estimation

The previously described coherent detection techniques are all based on pilots to estimate the channel. A disadvantage of those pilots is that they cost a certain percentage of the transmitted power. To avoid this loss, decision-directed channel estimation can be used. Here, instead of pilots, data estimates are used to remove the data modulation from the received sub-carriers, after which all sub-carriers can be used to estimate the channel. Of course, it is not possible to make reliable data decisions before a good channel estimate is available. Therefore, only decisions from previous symbols are used to predict the channel in the current symbol [127]. This is in contrast to the pilot methods, where the channel for a certain symbol is estimated from pilots within, before, and after that particular symbol. If the channel is relatively slowly varying in time, however, such that there is a large correlation between adjacent symbols, then there is a negligible impact on performance if only earlier symbols are used to estimate the channel for a particular OFDM symbol.

To start the decision-directed channel estimation, at least one known OFDM symbol must be transmitted. This enables the receiver to attain channel estimates for all sub-carriers, which are then used to detect the data in the following OFDM symbol. Once data estimates are available for a symbol, these estimates are used to remove the data modulation from the sub-carriers, after which those sub-carrier values can be used as pilots in exactly the same way as described in Sections 7.2.1 and 7.2.2.

7.3 Differential Detection

The key idea behind all coherent detection techniques discussed in the previous subsections is that they somehow estimate the channel to obtain an absolute reference phase and amplitude for each sub-carrier in each OFDM symbol. In contrast to this, differential detection does not perform any channel estimation, thereby saving both complexity and pilots at the cost of a somewhat reduced SNR performance. A general block diagram of an OFDM receiver using differential detection is shown in Figure 7.7. Instead of using an absolute reference, differential detection compares each subsymbol with another subsymbol, which can be a previous sub-carrier in the same OFDM symbol, or the same sub-carrier of a previous OFDM symbol. The next sections explain both variations of differential detection and show how differential detection can even be applied to multi-amplitude modulation.

7.3.1 Differential Detection in the Time Domain

If differential detection is applied in the time domain, then each sub-symbol is compared to the sub-symbol on the same sub-carrier of the previous OFDM symbol, as depicted in Figure 7.8.

To make differential detection possible, the transmitter has to apply differential encoding. For a PSK signal with input phases ϕ_{kj}, the differentially

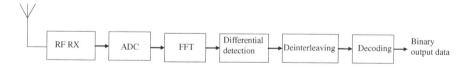

Fig. 7.7 Block diagram of an OFDM receiver with differential detection.

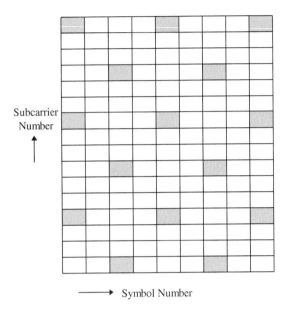

Subcarrier Number

Symbol Number

Fig. 7.8 Differential detection in the time domain. Gray sub-carriers are pilots that are needed as initial phase references.

encoded phases θ_{ij} are

$$\theta_{ij} = \sum_{k=0}^{i} \phi_{kj} \mathrm{mod}(2\pi), \qquad (7.11)$$

where i and k are the symbol number and j is the sub-carrier number. Differential detection is essentially applied to each sub-carrier separately. Because at the start of a transmission no previous symbol values are yet available, the sub-carrier values of the first symbol are chosen to be some arbitrary values.

At the receiver, the FFT output of symbol i and sub-carrier j can be written as

$$x_{ij} = a_{ij} e^{\theta_{ij} + \beta_{ij}} + n_{ij}, \qquad (7.12)$$

where a_{ij}, β_{ij}, and n_{ij} are the channel amplitude, channel phase, and additive noise component of symbol i and sub-carrier j. A differential phase detection in the time domain is performed by multiplying each FFT output with the conjugated FFT output of the same sub-carrier from the previous OFDM symbol.

$$
\begin{aligned}
y_{ij} &= x_{ij}x_{i-1,j}^* \\
&= a_{ij}a_{i-1,j}e^{\phi_{ij}+\beta_{ij}-\beta_{i-1,j}} + n_{ij}a_{i-1,j}e^{\theta_{i-1,j}+\beta_{i-1,j}} \\
&\quad + n_{i-1,j}a_{ij}e^{\theta_{ij}+\beta_{ij}} + n_{ij}n_{i-1,j}^*.
\end{aligned} \tag{7.13}
$$

The first term of (7.13) has the desired phase ϕ_{kj}, but it also has an undesired phase disturbance $\beta_{ij} - \beta_{i-1,j}$, which is the channel phase shift on sub-carrier j from symbol $i - 1$ to i. The latter disturbance depends only on the time fading, so to have a negligible impact on the phase detection, the OFDM symbol duration has to be small relative to the channel coherence time.

Figure 7.9 shows the correlation between signal samples as a function of the normalized time difference $f_{max}T_s$, where f_{max} is the maximum Doppler spread and T_s is the OFDM symbol duration. The correlation is calculated according to (7.10), with the number of OFDM symbols l set to one. Figure 7.9 can be used to determine the maximum tolerable Doppler spread, depending

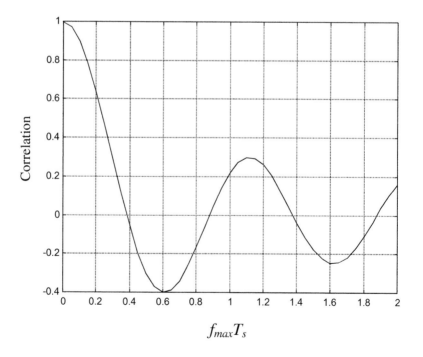

Fig. 7.9 Correlation between symbols vs the normalized time distance $f_{max}T_s$.

on the allowable phase error and the OFDM symbol duration. The maximum tolerable level of channel estimation errors can be related to the correlation between two OFDM symbols by writing the channel value y_{ij} as a function of the channel value $y_{i-1,j}$ of the previous symbol

$$y_{ij} = r_t y_{i-1,j} + \sqrt{\left(1 - r_t^2\right)} q_{ij}. \tag{7.14}$$

Here, r_t is the correlation between channel values of two OFDM symbols on the same sub-carrier, as given by (7.10). q_{ij} is a randomly distributed component with unity power. The difference between the channel values $y_{i-1,j}$ and y_{ij} is given by

$$y_{i-1,j} - y_{ij} = y_{i-1,j}(1 - r_t) + \sqrt{\left(1 - r_t^2\right)} q_{ij}. \tag{7.15}$$

This difference in channel values between two OFDM symbols determines the loss in SNR performance. For a negligible loss of performance, the Signal-to-Distortion Ratio (SDR) should be much larger than the SNR that is needed to achieve a certain maximum Bit Error Rate (BER) or Packet Error Rate (PER).

Assuming $y_{i-1,j}$ and q_{ij} are uncorrelated, the distortion power is the sum of the powers of both components in (7.15). Because both $y_{i-1,j}$ and q_{ij} have unity power, the SDR can be written as

$$\text{SDR} = \frac{1}{2(1 - r_t)}. \tag{7.16}$$

The required SNR values depend on coding rate and type of modulation. For instance, an SNR of 4 dB is required to get a BER of 10^{-5} using QPSK and rate 1/2 convolutional coding. For a loss in SNR performance of less than 1 dB, the distortion power should be at least 6 dB lower than the noise power, so the SDR needs to be 10 dB or more. Using (7.16), an SDR of 10 dB requires a correlation value r_t of 0.95. From Figure 7.10 — which is a zoom-in on the first part of Figure 7.9 — we can see that a correlation value of 0.95 corresponds to a normalized time distance $f_{\max} T_s$ of approximately 0.07. Hence, the maximum allowable Doppler frequency in this case is $0.07/T_s$. The Doppler frequency can be related to the maximum allowable velocity v as

$$f_{\max} = f_c \frac{v}{c}, \tag{7.17}$$

where f_c is the carrier frequency and c the speed of light. For example, for a carrier frequency of 5 GHz and a symbol duration of 4 s, a maximum

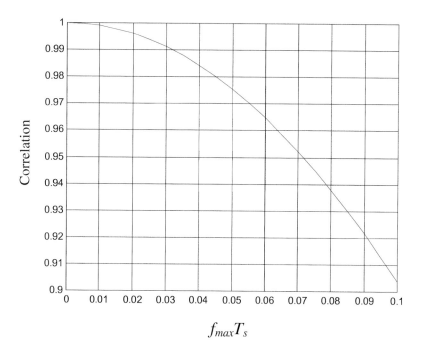

Fig. 7.10 Correlation between symbols vs the normalized time distance $f_{max}T_s$, zoom-in of Figure 7.9.

Doppler frequency of $0.07/T_s = 17.5\,$kHz leads to a maximum allowable user velocity of $1050\,$m/s. Because this speed is orders of magnitudes above practical values, we can conclude that for normal speeds, the channel change between two OFDM symbols is negligible for the parameters of the above example.

An interesting analogy exists between differential detection in the time domain and the coherent detection method using pilot symbols at the beginning of a packet, which was described in Section 7.2.3. Both methods rely on the fact that the channel is relatively constant in time. When the first symbols of a packet are used as a coherent reference for the rest of the packet, a nonzero Doppler bandwidth will cause the channel estimation errors to grow with the packet length. Hence, the maximum possible packet length is determined by the allowable level of channel estimation errors. To calculate the errors in the channel estimation, the same equations can be used as for differential detection in the time domain, with the only difference that T_s is replaced by the packet duration T_p. For instance, for the parameters of the previous example, it is required that $f_{max} T_p$ is approximately 0.07. Because f_{max} is equal to $f_c v/c$, the maximum possible packet duration T_p becomes $0.07c/f_c v \approx 2.8\,\text{ms}$ at a walking speed of $v = 1.5\,\text{m/s}$. At a vehicle speed of $30\,\text{m/s}$, however, the maximum packet duration is limited to about $0.1\,\text{ms}$.

The remaining three terms of (7.13) are noise components. If the difference between the amplitudes $a_{i-1,j}$ and a_j is neglected, the power of the second and third term is equal to $2P_s P_n$, where P_s and P_n are signal and noise power, respectively. The power of the last term in (7.13) is equal to the squared noise power P_n^2. For an SNR that is much larger than one, the cross products dominate the squared noise component, and the output SNR can be written as

$$\text{SNR}_y = \frac{P_s^2}{P_n^2 + 2P_s P_n} \approx \frac{P_s}{2P_n} = \frac{\text{SNR}_x}{2}. \tag{7.18}$$

Hence, the SNR after differential detection is approximately $3\,\text{dB}$ worse than the input SNR. This $3\,\text{dB}$ is the worst case SNR loss of differential detection relative to coherent detection. In practice, coherent detection also has an SNR loss because of imperfect channel estimates and because a part of the signal power is spent on pilots. This typically reduces the difference between differential and coherent detection from 3 to about 1 to $2\,\text{dB}$.

7.3.2 Differential Detection in the Frequency Domain

Differential detection can also be applied across sub-carriers instead of symbols. In this case, for a PSK signal with input phases ϕ_{kj}, the differentially encoded phases θ_{ij} are

$$\theta_{ij} = \sum_{k=0}^{j} \phi_{ik} \,\text{mod}(2\pi), \tag{7.19}$$

where i is the symbol number and j and k are sub-carrier numbers. Differential detection is now applied to each symbol separately, as depicted in Figure 7.11. The first sub-carrier of each symbol is a known pilot value that is needed to provide an initial value to start the differential detection process.

A differential phase detection in the frequency domain is performed by multiplying each FFT output with the conjugated FFT output of the same

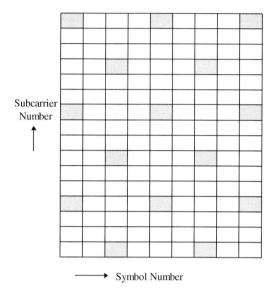

Fig. 7.11 Differential detection in the frequency domain. Gray sub-carriers are pilots.

symbol from the previous sub-carrier:

$$\begin{aligned}
y_{ij} &= x_{ij}x_{i,j-1}^* \\
&= a_{ij}a_{i,j-1}e^{\phi_{ij}+\beta_{ij}-\beta_{i,j-1}} + n_{ij}a_{i,j-1}e^{\theta_{i,j-1}+\beta_{i,j-1}} \\
&\quad + n_{i,j-1}a_{ij}e^{\theta_{ij}+\beta_{ij}} + n_{ij}n_{i,j-1}^*.
\end{aligned} \tag{7.20}$$

Here, x_{ij} is the FFT output at the receiver as defined by (7.12). Equation (7.20) has exactly the same structure as (7.13), which described differential detection in the time domain. Because of this, a similar signal-to-noise analysis can be made, showing that differential detection in the frequency domain also has an SNR loss of 3 dB compared with ideal coherent detection. The main difference between differential detection in frequency and time is the phase disturbance component in the first term of (5.20). This first term contains the desired phase ϕ_{kj}, but also an undesired phase $\beta_{ij} - \beta_{i,j-1}$, which is the channel phase shift on symbol i from sub-carrier $j-1$ to j.

The influence of the phase disturbance can be analyzed by looking at the correlation between adjacent sub-carriers as a function of the normalized frequency difference $\tau_{\max}/T$. This correlation can be calculated from (5.9) by

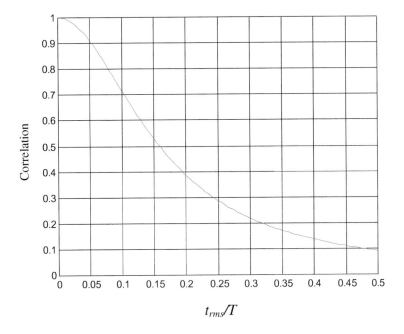

Fig. 7.12 Correlation between sub-carriers vs the normalized sub-carrier spacing τ_{rms}/T.

setting the number of sub-carriers k to one. A plot of the correlation versus the normalized frequency difference is shown in Figure 7.12.

Similar to the case of correlation in the time domain as described in the previous section, we can deduce that the SDR for a correlation value r_f is given by

$$\text{SDR} = \frac{1}{2(1 - r_f)} \tag{7.21}$$

If we use the same example as in the previous section, which required an SDR of 10 dB, then the correlation r_f has to be 0.95. From Figure 7.12 we can deduce that the normalized sub-carrier spacing τ_{rms}/T has to be approximately 0.03. This means that the sub-carrier spacing has to be $0.03/\tau_{rms}$ at most. Equivalently, it can be stated that the maximum tolerable delay spread is 3% of the FFT period T. The latter means that the delay spread robustness of differential detection in the frequency domain is generally significantly worse than other detection techniques, where the delay spread robustness is related only to the guard time and not to the FFT period.

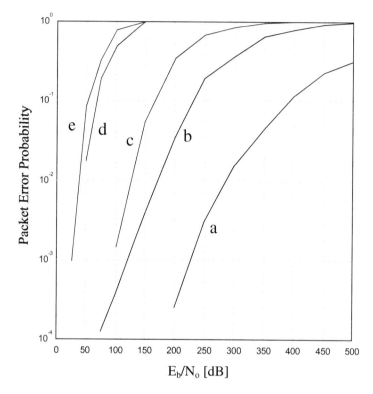

Fig. 7.13 Irreducible packet error ratio versus rms delay spread, simulated for an exponentially decaying power delay profile with Rayleigh fading paths. (a) 16 Mbps with coherent QPSK and rate 1/2 coding, (b) 32 Mbps with coherent 16-QAM and rate 1/2 coding, (c) 16 Mbps with differential QPSK (in frequency domain) and rate 1/2 coding, (d) 24 Mbps with differential 8-PSK (in frequency domain) and rate 1/2 coding, (e) 32 Mbps with differential 8-PSK (in frequency domain) and rate 2/3 coding.

Figure 7.13 presents some simulation results which compare coherent demodulation and differential detection for an OFDM system for WLAN applications. Table 7.5 lists the main system parameters. The data rates for this system are variable, dependent on the coding rate and the modulation type. Some possible data rates for this system are 32 Mbps with 16-QAM and rate 1/2 coding, or 8-PSK with rate 2/3 coding, 24 Mbps with QPSK and rate 3/4 coding, or 8-PSK with rate 1/2 coding, and 16 Mbps with QPSK and rate 1/2 coding.

Figure 7.13 shows the irreducible packet-error probabilities versus rms delay spread for an exponentially decaying multi-path delay profile. No noise was present in the simulation, so all errors are caused by ISI and ICI because

Table 7.5 Main parameters of the simulated OFDM system.

Number of sub-carriers	48
OFDM symbol duration	$3\,\mu s$
Guard interval	$600\,ns$
T_{prefix}: Pre-guard interval	$600\,ns$
$T_{postfix}$: Post-guard interval	$75\,ns$
Sub-carrier spacing	$416.666\,kHz$
Roll-off factor β	0.025
Channel spacing	$25\,MHz$
Occupied $-3\,dB$ bandwidth	$20\,MHz$

of multi-path components with relative delays extending the guard time of the OFDM symbols. To obtain channel estimates for coherent detection, a training symbol was present at the start of each packet. All sub-carrier values of this training symbol are known to the receiver, so this is the channel estimation method discussed in Section 7.2.3.

Clearly, coherent demodulation [curves (a) and (b)] performs much better than differential detection in the frequency domain. For the same data rate and packet-error probability, coherent demodulation can tolerate about three times as much delay spread as differential detection. The reason for the relatively poor performance of differential in frequency detection is the significant phase fluctuation between sub-carriers. Differential detection in the frequency domain assumes that there is a negligible phase difference between two adjacent sub-carriers. For delay spreads around $50\,ns$, however, a significant percentage of channels show several phase changes exceeding $pi/8$ within the 48 OFDM sub-carriers per symbol. Differential 8-PSK will generate two erroneous sub-carriers if the phase changes more than $\pi/8$, and that explains why the error curves for 8-PSK quickly converge to 1 for delay spreads exceeding $50\,ns$. Differential QPSK is more robust, but still worse than coherent 16-QAM, which operates at twice the data rate. Coherent demodulation is not affected by phase changes across the sub-carriers, because it uses training symbols to estimate reference phases and amplitudes of all sub-carriers. The same holds for differential detection in the time domain, which will have approximately the same delay spread robustness as coherent detection.

7.3.3 Differential Amplitude and Phase Shift Keying

Traditionally, differential detection is applied to phase-modulated systems only, as it is not obvious how differential detection can be applied to

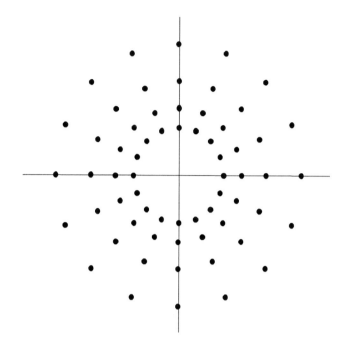

Fig. 7.14 64-DAPSK constellation.

amplitude-modulated systems. It is possible to use Differential Amplitude and Phase Shift Keying (DAPSK), however, by using a star constellation like depicted in Figure 7.14 [128, 129, 130].

The main advantage of DAPSK is that it does not require pilot symbols to estimate a time-varying channel or a remaining local oscillator offset. It only requires a single pilot symbol to initialize the differential detection, as depicted in Figures 7.8 and 7.11. The disadvantage, however, is a loss in SNR performance, because the minimum distance for the DAPSK constellation is clearly worse than for the corresponding square QAM constellation. Added to this is the loss of doing differential versus coherent detection, although this loss is partly compensated by the fact that DAPSK uses fewer pilot sub-carriers.

8

High Power Amplifier and PAPR in OFDM

8.1 Introduction

In contrast to its appealing properties, OFDM suffers from high Peak to Average Power Ratio (PAPR) problem. When the signal with high PAPR passes through the High Power Amplifier (HPA), which has nonlinear transfer function, spectral regrowth occurs that can be separated as in band intermodulation and out band intermodulation [131]. This in turn causes Block Error Rate (BLER) degradation and Adjacent Channel Interference (ACI) [11, 45].

To reduce the nonlinear distortion effect, signal power is reduced before applying to the HPA input so that the probability of the signal peaks to be amplified in the linear region of the amplifier transfer function increases. This reduction of signal power is referred to as Back Off (BO). The larger the BO, higher the reduction of the distortion effect. However, it reduces the total available transmit power which eventually reduces the coverage area of the system. Therefore it is aimed that minimum BO is used while optimizing the BLER performance by limiting the signal distortion due to Peak to Average Power Ratio (PAPR) problem. It is also found that the amount of BO needed for optimum performance varies for different modulation schemes [132]. Since LA employs varying modulation, power and coding rate, fixed value of power BO will not optimize the performance. So, it becomes important to analyze the impact of nonlinearity in Orthogonal Frequency Division Multiplexing (OFDM) system using rate, modulation, and power adaptation.

8.2 HPA Models

Several HPA models are available in the literature mainly for two types of amplifier. One is relatively older and is known as Traveling Wave Tube Amplifier (TWTA) and another is Solid State Power Amplifier (SSPA).

The models developed so far can mainly be divided into two categories. One exhibits nonlinear distortion in both amplitude (AM/AM) and phase (AM/PM). Other exhibits nonlinear distortion in amplitude (AM/AM) only. A brief description of two models that are widely employed for wireless communication related studies has been given below.

8.2.1　TWTA Model

The most widely used model for TWTA is known as Saleh model. It considers nonlinear distortion in both amplitude (AM/AM) and phase (AM/PM). The model is extensively used in nonlinear distortion analysis related to OFDM system [131, 133].

In Saleh model, input signal is defined as [134]

$$x(t) = r(t) \cdot \cos[w_0 t + \psi(t)], \tag{8.1}$$

where w_0 is the carrier frequency, $r(t)$ and $\psi(t)$ are modulated envelope and phase, respectively.

The corresponding output can be written as [134]

$$y(t) = A[r(t)] \cdot \cos\{w_0 t + \psi(t) + \Phi[r(t)]\}, \tag{8.2}$$

where $A(r)$ is an odd function of r, with a linear leading term representing AM/AM conversion and $\Phi(r)$ is an even function of r, with a quadratic leading term representing AM/PM conversion.

8.2.2　SSPA Model

For SSPA, nonlinear distortion has been analyzed using Rapp's Model [135]. This HPA model is simulated using the following relation considering distortion in amplitude only [11].

$$g(x) = \frac{|x|}{(1 + |x|^{2p})^{\frac{1}{2p}}}. \tag{8.3}$$

Here x is the signal amplitude and the variable p is used to tune the amount of nonlinearity. A good approximation of existing HPA can be obtained by choosing p in the range of 2 to 3. For large values of p, the model converges to a clipping amplifier and is perfectly linear until it reaches the maximum output power level. This is, however very hard to achieve in practical system.

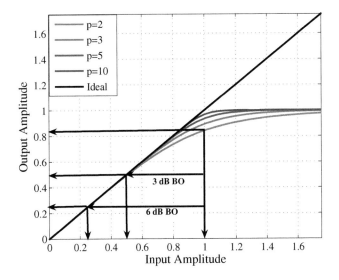

Fig. 8.1 Transfer function and of Rapp's model with BO scenario.

Available literatures suggests that the AM/PM distortion is small enough to be neglected [11].

Figure 8.1 represents the transfer function of the Rapp's model. It also explains the influence of BO on nonlinearity. When no BO is applied, the signal is more likely to be amplified in the nonlinear region of the transfer function. To increase the possibility of amplifying in the linear region, operating point needs to be shifted to the left of the horizontal axis of Figure 8.1. To shift the operating point 3 dB to the left, the signal needs to be reduced by a factor of half. In order to accommodate more signal peaks in the linear region, operating point needs to be shifted more left. To implement 6 dB BO, the signal should be reduced by a factor of $\frac{1}{4}$.

Thus BO always gives some performance improvement in terms of BLER but at the cost to coverage reduction due to reduced output power. The amount of distortion caused by the amplifier is estimated by measuring the difference in signal power at the input and output of the amplifier. The distortion versus BO relation is shown in Figure 8.2 which again proves that with increasing BO, distortion decreases, i.e. the signal operates more in the linear region.

Selection of HPA model: Both TWTA and SSPA has advantages as well as disadvantages over one another. However, there have been a long debate whether

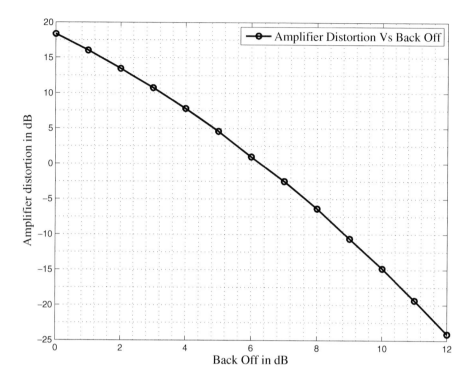

Fig. 8.2 Relation between amplifier distortion and BO power.

tube based or solid state amplifier is better for wireless applications. For this work SSPA have been chosen due to some of its distinct advantages [136]

- No warm up time required.
- Inherently good linear performance for multi-carrier, digital transmission.
- Built in soft-fail capabilities in case of single device or module failure.
- No expected RF section sparing requirements.
- High volume production capability since the industry is growing exponentially.

And Rapp's model is used to model the SSPA.

8.3 PAPR in OFDM

Orthogonal Frequency Division Multiplexing (OFDM) symbol is generated by superimposing several narrow band sub-carriers. These carriers may add up constructively resulting in high amplitude for some part of the signal. When power instead of amplitude of the signal is considered situation becomes even worse. This problem is widely known as PAPR problem. Large PAPR of a system makes the implementation of Digital-to-Analog Converter (DAC) and Analog-to-Digital Converter(ADC) extremely difficult. The design of RF amplifier also becomes more difficult with the increasing PAPR as it gives rise to the intermodulation in RF domain. Since the source of the problem is constructive addition of sub-carriers, PAPR is proportional to the number of sub-carrier in an OFDM symbol. PAPR can be defined mathematically as

$$\text{PAPR} = \frac{\max |x(n)|^2}{E[|x(n)|^2]}. \tag{8.4}$$

Several methods have been proposed to reduce nonlinear distortion due to high PAPR. Some techniques directly deal with nonlinearity of the amplifier such as predistortion, negative feedback, linear amplification with nonlinear component (LINC), feed forward etc. While some other techniques involve reduction of PAPR so that operating point does not fluctuate very much. Three category of techniques have been found in literature to reduce PAPR, namely *Signal Distortion Techniques*, *Coding Techniques*, and finally the *Scrambling Technique*. A good comparison of these technique can be found in [137].

8.3.1 CDF of PAPR

Orthogonal Frequency Division Multiplexing (OFDM) signal is characterized by

$$x(t) = \frac{1}{\sqrt{N}} \sum_{n=1}^{N} a_n e^{j w_n t}. \tag{8.5}$$

Here a_n is the modulating signal. For large number of a_n both the real and imaginary parts tend to be Gaussian distributed, So the amplitude of the OFDM symbol has a Rayleigh distribution, while the power distribution is central chi squared with two degrees of freedom and zero mean with a cumulative distribution function given by:

$$F(z) = \int_0^z \frac{1}{2 \cdot \sigma^2} \cdot e^{-\frac{u}{2 \cdot \sigma^2}} du = 1 - e^{-z}. \qquad (8.6)$$

Assuming that the samples z to be mutually uncorrelated and the cumulative distribution function for the peak power per OFDM symbol is given by [138]:

$$P(\text{PAPR} \leq z) = F(z)^N = (1 - e^{-z})^N, \qquad (8.7)$$

where N is the number of sub-carriers in one OFDM symbol.

PAPR and number of sub-carrier: As stated earlier, high value of PAPR originates from the constructive addition of sub-carriers, so PAPR increases with the increase in number of sub-carriers. Figure 8.3 shows the theoretical and simulated Cumulative Distribution Function (CDF) of PAPR for different number of sub-carriers. Theoretical curve is generated using Equation (8.7). It is obvious that there is no difference between theoretical and simulated values except for the very low CDF which probably occurs due to not enough number of simulation samples at that range. Here CDF for 128, 512, and 1024 sub-carriers are shown and it shows that PAPR increases with the increase in number of sub-carriers.

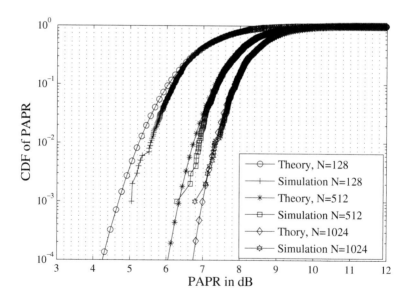

Fig. 8.3 Comparison of theoretical and simulated CDF of PAPR.

The CDF of PAPR reaches 90 percentile value, which is around 9 dB, very quickly and rest of the values span over wide range of SNR. There is very small difference between 90 percentile value and median value. Therefore, it can be concluded that high PAPR occurs very rarely. This, in fact, gives very useful information for system design. Since high PAPR occurs very rarely, for acceptable performance it is quite enough to accommodate high peaks up to median value in the linear region during amplification.

Effect modulation and coding schemes on PAPR: The Figure 8.4 shows the CDF of PAPR when different modulation order were used while keeping number of sub-carrier and code rate fixed to 512 and $\frac{1}{2}$, respectively. All the curve completely overlapped each other which proves that PAPR is completely independent of modulation scheme.

Figure 8.5 exhibits a very well known fact that PAPR does not depends on coding rate. Here modulation and number of sub-carrier were fixed to 16QAM

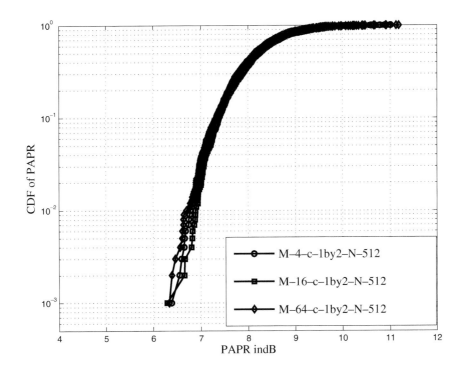

Fig. 8.4 Effect of different modulation scheme on CDF of PAPR.

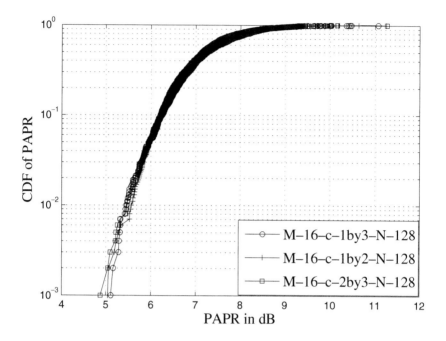

Fig. 8.5 Effect of FEC on CDF of PAPR.

and 128QAM respectively while coding rate was varied. All the CDF curves again overlapped each other. So, it can be concluded that PAPR only depends on how the sub-carriers are added together and their number. It is not influenced by the parameters like modulation, FEC coding etc.

8.4 Effect of HPA and BO Power

8.4.1 Effect on Constellation Points

Figure 8.6 shows the received constellation points for 16QAM modulation scheme when no power amplifier was used. The received constellation points for subsequent symbols are quite close to each other. However, when power amplifier is plugged into the system the received constellation points become affected by nonlinearity and therefore scattered.

Figure 8.7 shows the received constellation diagram for 3 dB and 6 dB BO. It can be seen from the figures that for 3 dB BO, received constellation points are more scattered than that of 6 dB BO. Since with higher modulation

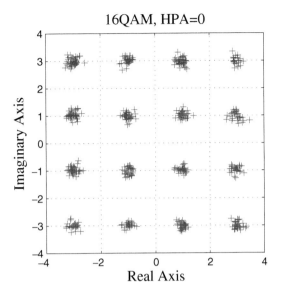

Fig. 8.6 16QAM basic constellation points.

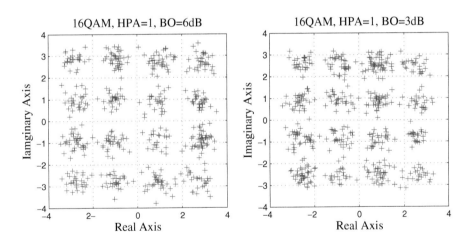

Fig. 8.7 Effect of BO on 16QAM constellation points.

order, decision boundaries in constellation diagram become more dense and so the impact of HPA and BO will surely become more severe. On the other hand, for low modulation rate like 4QAM, impact is much less since decision boundaries have comparatively wider space.

8.4.2 Effect on Power Spectrum

To investigate the effect of HPA on power spectrum, during generation of an OFDM symbol, sub-carrier with high frequency components are forcefully set to zero to avoid ACI. Figure 8.8 shows the sub-carrier arrangement at FFT output. The spectrum spans from $-\frac{N}{2}$ to $\frac{N}{2} - 1$. Frequencies higher than $-p$ and $p-1$ are set to zero, where p is chosen in a way that around 60% of the sub-carriers are loaded. The 0 frequency is also set to zero to avoid unnecessary power wastage by DC transmission. The power spectrum of OFDM signal measured at the output of the power amplifier is shown in Figure 8.9 for

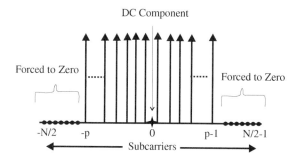

Fig. 8.8 Sub-carrier organization for spectrum plot.

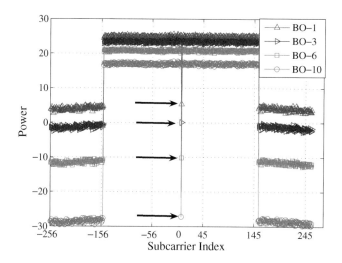

Fig. 8.9 Spectrum plot of OFDM signal.

different BO. The number of sub-carriers were fixed to 512. It shows that with increasing BO power, out of band emission decreases, i.e. power leakage reduces. If a small BO of 1 dB is applied, it can be seen from the figure that high frequency leakage is few dBi. However, with 10 dB of BO, out of band emission reduces to nearly −30 dB. Moreover, though DC component was set to zero, after power amplifier, DC value rises to −30, −12, and 3 dB for BO power of 10, 6, and 1 dB respectively.

8.4.3 SDNR Plot

The parameter Signal to Distortion plus Noise Ratio (SDNR) helps to identify the actual reduction in signal quality. SDNR can be measured both in the transmitter side and in the receiver side. If it is measured in the transmitter side then it will only capture nonlinear distortion due to HPA and white noise. On the other hand, if it is measured on the receiver side it will also include channel noise, multi-path fading effect etc.

SDNR at the receiver end can be expressed as follows:

$$\text{SDNR} = \frac{1}{N} \sum_{i=0}^{N-1} \frac{|X_i|^2}{|X_i - \hat{X}_i|^2}. \tag{8.8}$$

Here, X_i is the transmitted signal,
$\hat{X}_i$ is the received signal,
N is the number of sub-carrier.

Amount of SDNR does not depend on the modulation scheme used. However, the decision boundary for higher modulation order in the constellation diagram is smaller as compared to the lower modulation order. Hence the probability of correct decision for the symbol will be much lower for higher modulation order than that of lower modulation order for the same amount of distortion. Therefore for the same amount of SDNR, the effect on BLER performance of different modulation will be different.

Since LA system usually selects modulation adaptively, it is more likely that at low SNR 4QAM modulation will be selected and then with increasing input SNR higher modulation will be selected gradually. So, even though the SDNR plots is independent of modulation, they are represented individually for each modulation in both Additive White Guassian Noise (AWGN) and

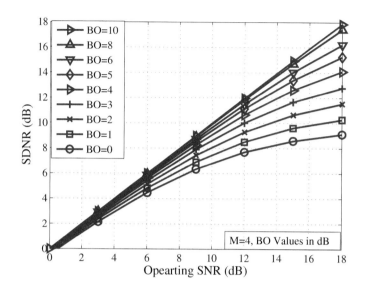

Fig. 8.10 SDNR plot for 4QAM modulation in AWGN channel.

fading channel with emphasis on the different SNR range where they are more likely to operate. Thus for 4QAM modulation, SDNR is plotted against SNR range of 0 to 18 dB, while for 16QAM it ranges from 5 to 30 dB and for 64QAM the range is between 12 and 40 dB.

Figures 8.10–8.12 shows plots for the modulation scheme of 4, 16, and 64QAM, respectively, in AWGN channel while Figures 8.13–8.15 show the SDNR plots for 4, 16, and 64QAM, respectively, in Fading channel. For higher modulation level, it can be seen that, when the BO is reduced, then the SDNR is reaching the highest values quite speedily, i.e. with increasing transmit SNR, the SDNR increases until around 20 dB, after that the SDNR starts to saturate. This means increasing SNR after certain value will not give much gain.

Also it is evident from the figure that though the nature of SDNR curve is same for both AWGN and fading channel, the amount of SDNR for fading channel is much lower than that of AWGN channel since the SDNR for fading channel includes the distortion due to multi-path fading. For example, in 4QAM system with BO power of 8 dB, if the input power is 18 dB then the output power is 17 dB and 9 dB for AWGN and fading channel, respectively. Similarly, for modulation scheme of 16QAM with the same BO power as above,

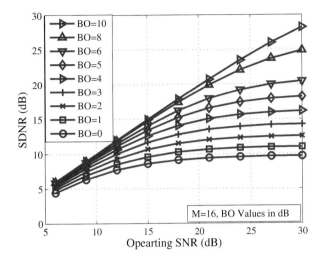

Fig. 8.11 SDNR plot for 16QAM modulation in AWGN channel.

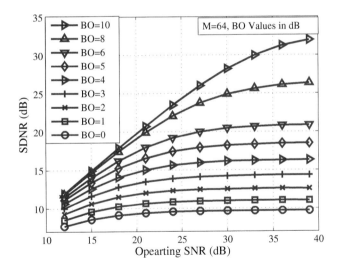

Fig. 8.12 SDNR plot for 64QAM modulation in AWGN channel.

if the input power is 24 dB then the out power is 22 dB and 15 dB for AWGN and fading channel respectively. This clearly shows the effect of multipath.

Signal to Distortion plus Noise Ratio (SDNR) plots shows that with the increasing BO, distortion decreases and after 8 dB of BO power, performance

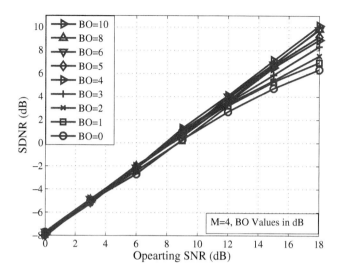

Fig. 8.13 SDNR plot for 4QAM modulation in fading channel.

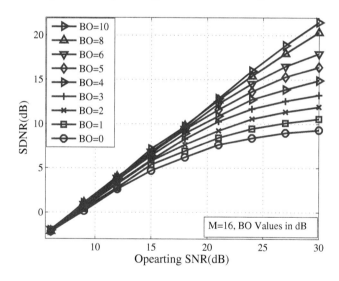

Fig. 8.14 SDNR plot for 16QAM modulation in fading channel.

is more of less same since 8 dB BO power is enough to accommodate most of the peaks in the linear region. So SDNR curve shows the optimum BO power as well. From the SDNR curve it is also possible to obtain the value at which it starts to saturate. Thus it helps to decide the operating SNR as well as the

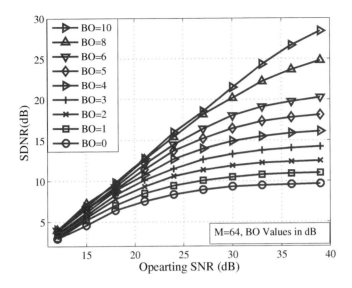

Fig. 8.15 SDNR plot for 64QAM modulation in fading channel.

maximum SNR after which power will be wasted for different modulation scheme and thereby ensures optimum use of power.

8.5 Performance of Different Modulation and Coding

In this section system performance is presented for basic system with different modulation and coding rate. Number of sub-carriers per OFDM symbol used in the simulation is 512. All the system parameters are selected as per 3GPP-LTE specification. Performance is studied initially for AWGN channel and then is extended for fading channel to realize how it will be affected in practical scenarios. Modulation schemes under investigation are 4QAM, 16QAM and 64QAM. BO power employed in the system simulation is from 0 to 10 dB. And as seen from the CDF curves presented in Section 8.3, 10 dB back off will accommodate almost all the high peaks and nonlinearity effect is mostly mitigated. Therefore, performance with 10 dB back off can also be considered as basic system performance. The doppler frequency for all the investigation was fixed at 50 Hz.

Block interleaver has been used in both AWGN and fading channel. However, it has no effect in AWGN channel but gives quite a good performance improvement in fading channel. Block interleaver spreads the errors in time and thereby converts frequency selective channel to the flat fading channel. So, even if the channel goes through deep fade at one instant, error due to this deep fade is distributed over time by the de-interleaver which makes the task of FEC decoder to recover the erroneous bit much easier.

Effect of HPA and BO power on all the modulation schemes are studied with FEC coding rate of $\frac{1}{3}$, $\frac{1}{2}$, and $\frac{2}{3}$. However, in order to avoid repetitive information, only results for FEC $= \frac{1}{2}$ is presented here.

The main parameters, that are usually used to characterize the impact of nonlinearities on digital communication systems are: EVM (Error Vector Magnitude), ACPR (Adjacent Channel Protection Ratio), power spectrum and BER or BLER. In this study the performance is shown in terms of BER for uncoded system. Since in all practical system FEC coding is employed and performance is measured in terms of Block Error Rate (BLER), it is used instead of BER to represent the performance of the coded system in order to keep our investigation close to the reality.

It is worth mentioning that for convenience of using results obtained here in link adapted system, all the performance is plotted in terms of *POST-SNR* which is defined as the actual SNR measured at the receiver. For AWGN channel, there is no difference between *PRE-SNR* or the transmit SNR and *POST-SNR* or received SNR. However, in fading channel they are quite different due to the addition of multi-path effect which reduces the SNR at the receiver.

8.5.1 Performance in AWGN Channel

Uncoded system: Figures 8.16–8.18 show the BER vs SNR performance for uncoded modulation of 4QAM, 16QAM, and 64QAM, respectively. For 4QAM modulation, backing off does not give that much performance improvement. However, impact of BO increases gradually with increasing modulation rate. For 4QAM, to achieve BER of 10^{-3}, the difference in required SNR with 10 dB and 5 dB BO is around 1 dB only. However, if the same scenario is considered for 16QAM, the required SNR will be 5 dB. For 64QAM it is even impossible to reach that BER level if 5 dB BO is used. This shows the importance of considering BO values carefully for different scenarios.

Coded system: The Figures 8.19–8.21 show the BLER vs SNR performance for 4QAM, 16QAM, and 64QAM modulation, respectively with FEC $= \frac{1}{2}$.

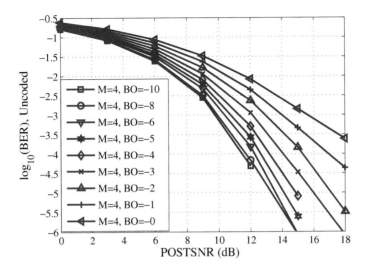

Fig. 8.16 BER vs SNR curve for $M = 4$, uncoded system in AWGN.

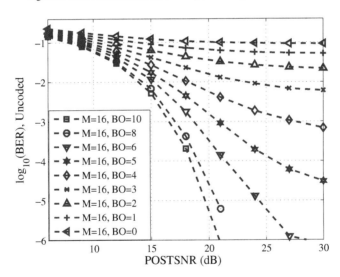

Fig. 8.17 BER vs SNR curve for $M = 16$, uncoded system in AWGN.

There is some performance improvement due to FEC coding gain. For 4QAM modulation, no significant difference in performance is observed between 5 dB and 10 dB back off to reach the BLER threshold of 10^{-2}. Even though there is some difference in required SNR to reach the BLER level stated above,

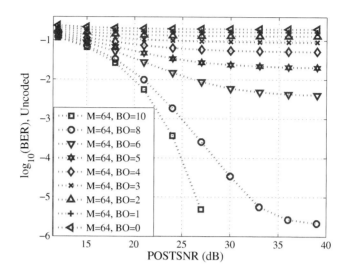

Fig. 8.18 BER vs SNR curve for $M = 64$, uncoded system in AWGN.

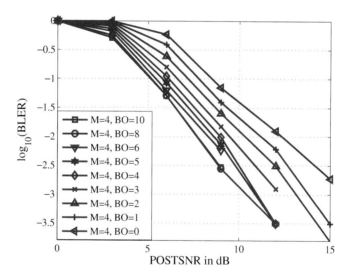

Fig. 8.19 BLER vs SNR curve for $C = \frac{1}{2}$ and $M = 4$ in AWGN.

Spectral Efficiency (SE) curves in Figure 8.22 shows that if the SNR is above 8 dB, there is no difference at all in spectral efficiency.

For the same scenario with 16QAM, the required SNR difference is around 2.5 dB while the SE performance saturates after 15 dB of SNR

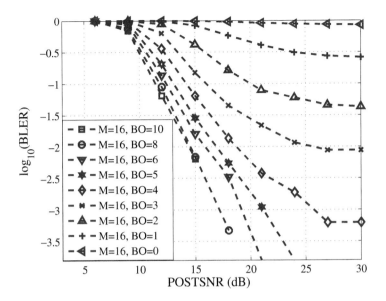

Fig. 8.20 BLER vs SNR curve for $C = \frac{1}{2}$ and $M = 16$ in AWGN.

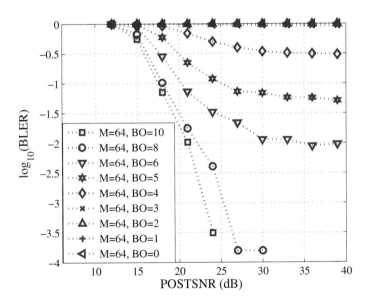

Fig. 8.21 BLER vs SNR curve for $C = \frac{1}{2}$ and $M = 64$ in AWGN.

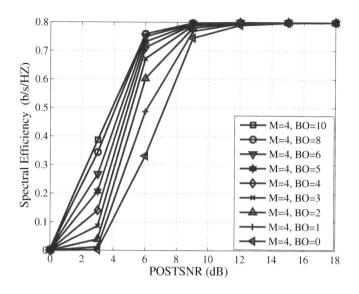

Fig. 8.22 Spectral efficiency vs SNR curve for $C = \frac{1}{2}$ and $M = 4$ in AWGN.

(see Figure 8.23). In case of 64QAM, the performance is worse and with 5 dB BO power, it is almost impossible to reach the selected level and significant difference is observed for SE performance (see Figure 8.24).

Even if the differences in BO between 10 and 6 dB and between 6 and 2 dB are same (4 dB), the required SNR to reach the threshold level of 10^{-1} is not the same. For 16QAM required SNR is 1 dB only for the first case while it is around 10 dB for the later case, i.e. the same amount of BO will not always give same improvement. So BO should be carefully selected to get the optimum performance. As it is observed in the CDF curve, median and 90 percentile value of the PAPR is very close and concentrated around some value. Beyond that level, the rest of the high peaks span over a wide area of SNR values. However, from the BLER curves it is obvious that the peaks with level higher than the median value do not have significant impact on BLER performance since implementing BO higher than 8 dB (which is the median value of CDF of PAPR for OFDM symbol used) do not give any performance improvement. Thus no performance improvement is achieved by using 10 dB of BO instead of 8 dB. Since the amount of BO is inversely proportional to the cell coverage, it is very important to keep BO as minimum as possible.

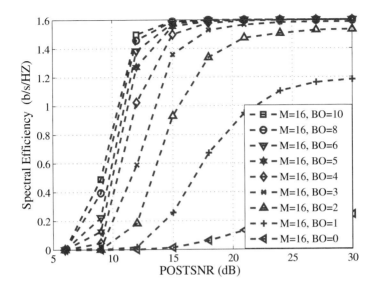

Fig. 8.23 Spectral efficiency vs SNR curve for $C = \frac{1}{2}$ and $M = 16$ in AWGN.

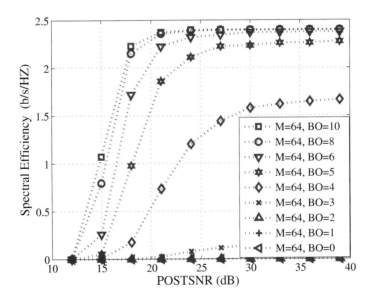

Fig. 8.24 Spectral efficiency vs SNR curve for $C = \frac{1}{2}$ and $M = 64$ in AWGN.

Total degradation plot: Modulation and coding rate shows different behavior at various BO. So the choice of suitable BO or different modulation and coding are different. Also it is highly complex, if not impossible, to change the BO power frequently with the change in modulation and coding rate in link adapted system. So finding an optimum BO point considering modulation with coding rate becomes necessary for optimum performance.

Total Degradation (TDe) for a certain BLER threshold is defined as the amount of BO plus the SNR degradation due to nonlinearity in BLER performance as compared to the performance in basic system [131].

Total Degradation (TDe) curve is very useful to find this optimum operating BO. Table 8.1 shows how the TDe values are calculated. For a certain value of BER or BLER, the corresponding required SNR to reach that level for ideal system (i.e system without HPA) is noted and used as the basic reference point as shown in the second column of the table. Then required SNR to reach the same BLER threshold for the system with HPA at various BO is noted. The difference between these values and basic reference values represent the degradation. Finally these degradation values are added to the corresponding BO values to get the TDe. Third row of Table 8.1 shows the TDe values.

Figure 8.25 and Figure 8.26 represent the TDe curve for BLER threshold of 0.1 and 0.05, respectively. From this curve the minimum TDe point can easily be obtained. For 4QAM, 16QAM, and 64QAM, the best BO points are 0, 4, and 7 dB, respectively.

The amount of degradation can vary for different modulation, coding as well as for BLER threshold. From Figure 8.25, we can see that for 64QAM, 16QAM, and 4QAM, optimum BO is 6, 4, and 0 dB, respectively. So TDe Curve at different modulations and coding for different threshold becomes crucial to select optimum BO.

8.5.2   Performance in Fading Channel

With the knowledge of the system performance in AWGN channel, the system is now investigated in the fading channel to see its behavior under practical

Table 8.1 Table for calculation of total degradation in dB.

BO Values	Ideal	10	8	6	5	4
Measured SNR	5.1	5.67	5.7	5.79	5.91	6.33
Degradation	0.0	0.57	0.6	0.69	0.91	1.33
TDe	—	10.57	8.60	6.69	5.91	5.33

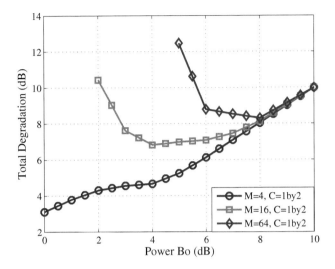

Fig. 8.25 TDe plot for $FEC = \frac{1}{2}$ with BLER threshold $= 0.1$ in AWGN.

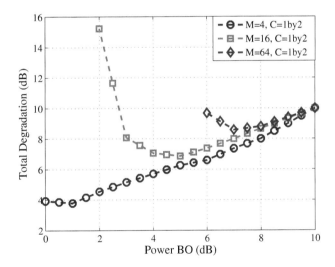

Fig. 8.26 TDe plot for $FEC = \frac{1}{2}$ with BLER threshold $= 0.05$ in AWGN.

channel impairments. The nature of the performance of fading channel is very much similar except additional degradation due to multi-path fading.

In the following sections the results for fading channel with uncoded system is presented which followed by the results for the coded system with $FEC = \frac{1}{2}$.

The performance is shown in terms of BLER versus POST-SNR. For each transmitted SNR, due to channel variation in the fading channel, different channels go through different fade, and thus received SNR for different subchannel varies even though the transmit SNR remains same. So if BLER performance is plotted against *PRE-SNR*, it will represent avarage performance of subchannels which experienced different received SNR during that transmission. Clearly, it will not exhibit the appropriate behavior of the system. To get the performance based on the received SNR, in the simulator, some beans were created for different SNR values. For example, if the received SNR ranges from 5 to 25 dB, then it can be subdivided into 5 different equally spaced beans. Each bean is represented by a central value of that bean, i.e. if a bean is spaced between 5 to 11 dB then it is represented by 8 dB. For each transmission received SNR is measured for each subchannel and necessary information is saved in the bean to which it belongs to. Finally BLER is calculated for each bean and plotted. Obviously this technique will give much more realistic view. However, to ensure enough number of samples in each bean, number of simulations required was quite high. The details of the performance is described below.

Uncoded system: In practical systems uncoded systems are used vary rarely. However, the BER performance is presented so that performance improvement due to coding system can be compared if needed.

Figures 8.27–8.29 show the BER performance for 4QAM, 16QAM, and 64QAM modulation scheme, respectively. It can be seen from the figures that for 4QAM, a little gain is possible by using power BO. By introducing 10 dB BO power, only around 2.5 dB of BER performance gain is achieved as compared to 1 dB of BO. However, for 16QAM, there is quite a good performance improvement for adding BO power until 5 dB. After 5 dB no significant improvement is found. Therefore, 64QAM modulation is most severely affected by the HPA and performance improvement is possible until 8 dB of BO power implementation.

So, the amount of BO required for optimum performance varies with modulation scheme. However, even in worse case, employing BO equal to the value of median of cdf of PAPR is enough to restore the performance.

Coded system: As mentioned earlier, coded system performance is shown in terms of BLER instead of BER in order to make it closer to the practical systems. In this section performance in terms of BLER and spectral efficiency

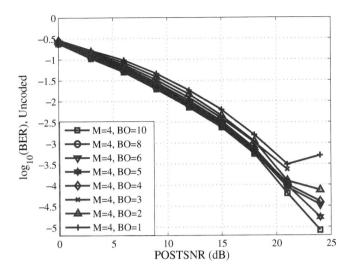

Fig. 8.27 BER vs SNR curve for $M = 4$, uncoded system in fading channel.

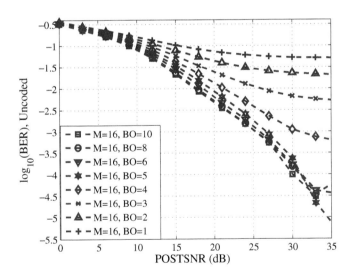

Fig. 8.28 BER vs SNR curve for $M = 16$, uncoded system in fading channel.

has been shown and analyzed for all three modulation rate at different BO power. $FEC = \frac{1}{2}$ is used in all cases.

Figures 8.30 and 8.31 show the BLER performance and corresponding SE performance for 4QAM modulation. Here almost no effect of BO is visible.

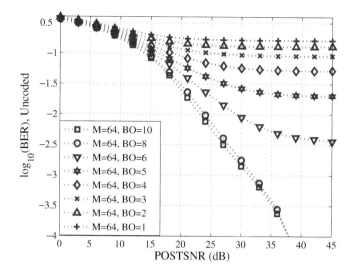

Fig. 8.29 BER vs SNR curve for $M = 64$, uncoded system in fading channel.

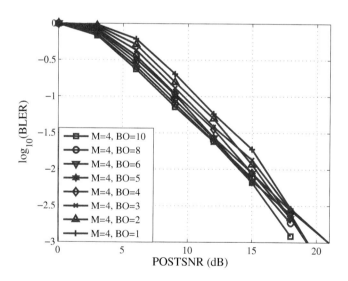

Fig. 8.30 BLER vs SNR curve for $C = \frac{1}{2}$ and $M = 4$ in fading channel.

Since constellation points are quite far from one another, with 1 dB of BO it requires only 11 dB SNR to reach the BLER level of 10^{-1}. Whereas with 10 dB BO, 8 dB of SNR is required to reach the same level. Thus a BO difference of 9 dB is required to gain 3 dB of BLER performance gain at threshold 10^{-1}

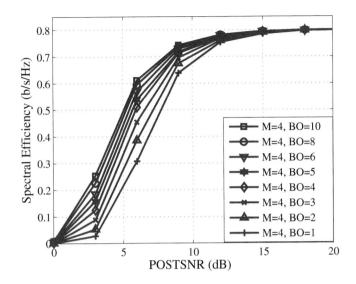

Fig. 8.31 Spectral efficiency vs SNR curve for $C = \frac{1}{2}$ and $M = 4$ in fading channel.

which is not practical at all. If the result is compared with the performance in AWGN channel, it can be seen that the nature is same, i.e. not much gain is achievable by employing large BO, but the performance degrades due to multi-path fading. To reach BLER of 10^{-1}, 5 and 8 dB of SNR is required in AWGN and fading channel, respectively. However, to reach BLER of 10^{-2}, around 8 and 14 dB of SNR is required in AWGN and fading channel, respectively. With decreasing BLER level, impact of multi-path fading increases. As it is obvious from the uncoded scenarios, the BO is not giving any improvement, it is better to use minimum BO and since coding reduces the spectral efficiency significantly, minimum coding rate should be chosen to achieve the high spectral efficiency for 4QAM modulation.

Figures 8.32 and 8.33 represent the BLER and SE performance for 16QAM modulation. It can be seen from the figures that BLER performance is more dependant on BO compared to the 4QAM modulation. Impact of BO is not that much significant if it is 4 dB or more. However, if BO is less than 4 dB then impact is quite significant. To reach the BLER level of 10^{-1}, difference in required SNR for the system with 2 and 3 dB BO is 5 dB. Situation is even worse if 1 dB BO is considered. And SE also severely degrades in case of 1 dB BO power.

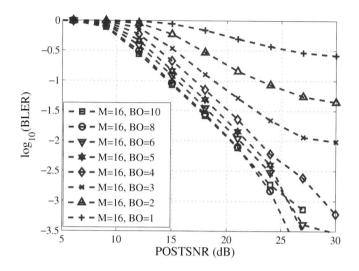

Fig. 8.32 BLER vs SNR curve for $C = \frac{1}{2}$ and $M = 16$ in fading channel.

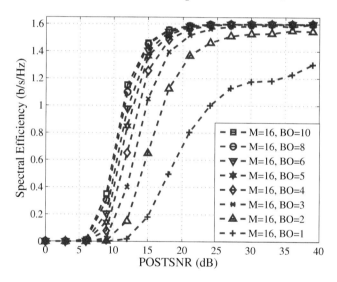

Fig. 8.33 Spectral efficiency vs SNR curve for $C = \frac{1}{2}$ and $M = 16$ in fading channel.

Like AWGN channel, most severely affected modulation among the schemes considered in this study is 64QAM. Figures 8.34 and 8.35 represent the performance of 64QAM modulation. Effect of HPA can be suppressed only if BO equal to or greater than the median value of CDF of PAPR is applied.

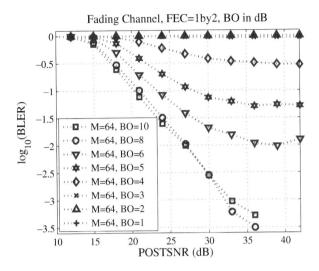

Fig. 8.34 BLER vs SNR curve for $C = \frac{1}{2}$ and $M = 64$ in fading channel.

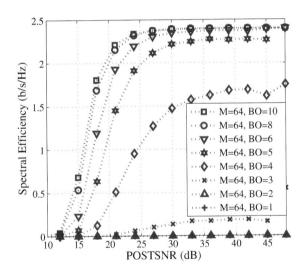

Fig. 8.35 Spectral efficiency vs SNR curve for $C = \frac{1}{2}$ and $M = 64$ in fading channel.

In order to reach BLER level of 10^{-1}, at least 5 dB BO is necessary. Based on the level of BLER needs to be reached, some BO scheme even become unusable. Since the curves show that to reach the level of 10^{-2}, we need BO power of at least more than 6 dB, the BO of 1 to 5 dB is not usable in this case. The BO has severe impact on SE performance as well. For BO of 5 and 10 dB, difference in SE is nearly $1\,b/s/Hz$ at 30 dB of SNR. If the amount of BO employed is equal to or less than 3 dB, almost no SE is achievable. So, to achieve lower BLER and higher SE, around 8 dB of BO needs to be employed which is again the median value. However, employing this amount of BO will cause a huge reduction of cell coverage.

Total degradation plot: Figures 8.36 and 8.37 show the TDe curve for BLER threshold of 0.1 and 0.05, respectively. As mentioned earlier, this curve is very useful in determining which BO is to be selected for optimum performance. Similar to AWGN channel, optimum BO is different for different modulation level. Even it can change depending on the level BLER to maintain. For 64QAM modulation, in both cases optimum BO is 8 dB which is the mean of the CDF of PAPR for the OFDM symbol with 512 sub-carrier. Interestingly, for 16QAM modulation to get minimum total degradation different BO is needed for different BLER threshold. From the figures it can be seen that required BO is 4 and 5 dB to get the minimum degradation of BLER threshold 0.1 and 0.05, respectively. For 4QAM, no BO is required at all.

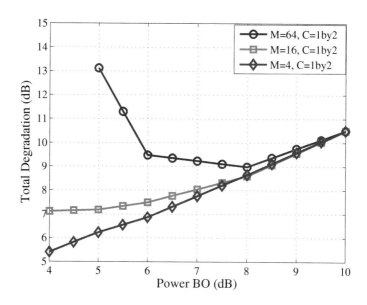

Fig. 8.36 TDe plot for $FEC = \frac{1}{2}$ with BLER threshold $= 0.1$ in fading channel.

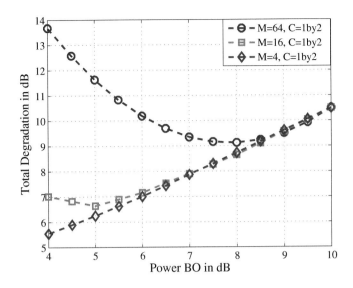

Fig. 8.37 TDe plot for $FEC = \frac{1}{2}$ with BLER threshold $= 0.05$ in fading channel.

8.6 Conclusion

In this chapter, the impact of HPA has been analyzed from different angles. Beginning with HPA model, CDF of PAPR is generated to see the nature of distribution of peaks in the symbol which also gives us the preliminary idea about the influence of BO required to reduce the nonlinearity due to HPA. The impact on signal level was investigated by generating spectrum plot. Then the impact on constellation diagram was also shown. SDNR plots gives a further insight to the problem. It helps us to decide about range of Signal-to-Noise Ratio (SNR) that are suitable for use.

In order to investigate the effect of HPA on system performance initially AWGN channel is studied with above mentioned impairments. The study is finally extended to the fading channel to include the multi-path effect and also to see the performance under real environment scenarios. It is found that there is significant impact of HPA on system performance which varies with modulation and coding rate. Though the impact is not very significant for low modulation rate, it becomes severe for higher modulation rates. BO is employed to reduce the effect of nonlinearity. However, the amount of BO power required to reduce the effect of nonlinearity to an acceptable level varies with modulation and coding and since BO reduces coverage careful selection of the combination of BO with modulation and coding rate becomes very important. And since LA system changes modulation and coding rate adaptively, investigation of LA system is also necessary. The result for LA system is presented later.

9

Multi-antenna Gains

Multi-antenna is one of the key technologies for mitigating the negative effects of the wireless channel, providing better link quality and/or higher data rate without consuming extra bandwidth or transmitting power. The usage of multiple antennas at either receiver, transmitter or at both locations provides different benefits, namely *array gain*, *interference reduction*, *diversity gain*, and/or *multiplexing gain* [139]. The combination of multi-antenna techniques with Orthogonal Frequency Division Multiplexing (OFDM) can be very beneficial, due to the fact that OFDM greatly simplifies the multi-channel equalization problem in multi-antenna systems, thus making implementation of multi-antenna techniques practical in frequency-selective fading channels. Numerous research works are published on utilizing multiple antennas for OFDM-based 4th Generation (4G) wireless systems, which can be broadly classified into three categories as illustrated in Figure 9.1:

- Smart array processing, such as Beamforming (BF), which aims at increasing receiving power and rejecting unwanted interference.
- Diversity combining, which is employed to mitigate fading and increase link reliability.
- Spatial multiplexing, which is used to increase data rate.

9.1 Gains Obtained by Exploiting the Spatial Domain

Let us start by examining the possible transceiver architectures configurations. Single Input Single Output (SISO) is referring to a system with one transmit and one receive antenna; Single Input Multiple Output (SIMO), refers to one antenna at the transmitter and multiple antennas at the receiver, while Multiple Input Single Output (MISO) is the converse; finally, Multiple Input Multiple

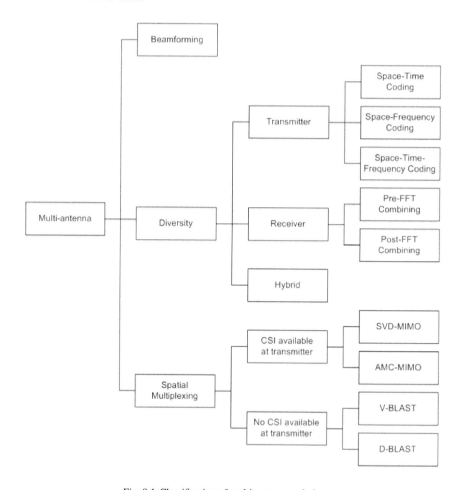

Fig. 9.1 Classification of multi-antenna techniques.

Output (MIMO) indicates an architecture with multiple antennas at the transmitter and at the receiver.

Let us now discuss the possible gains that are achievable through the use of multiple antennas in a wireless system. In the following, we will refer to the number of transmit antennas as N_T and to the number of receive antennas as N_R.

9.1.1 Array Gain

This term is indicating the average increase in SNR deriving from the coherent combining effect of multiple antennas in a MISO, SIMO, or MIMO system.

In case of SIMO, signals arriving at the receive array generally have different amplitudes and phases: in this case, the receiver can combine the incoming signals coherently, such that the resultant signal is enhanced. This results in an average increase in signal power at the receiver which is proportional to N_R. In case of channels with multiple antennas at the transmitter, to exploit array gain one requires channel knowledge at the transmitter [139].

9.1.2 Diversity Gain

In the wireless channel, the signal power is subject to fluctuations (phenomenon also known as fading); when the signal power decreases of a consistent amount, the channel is said to be in fade. The techniques designed to try to combat the fading problem are so called diversity techniques.

When dealing with a SIMO channel, receive diversity techniques can be employed [141]. The fact that these algorithms exploit is that different receive antennas see independently faded versions of the same signal, and the receiver then combines these signals in a way that the resultant signal has a reduced amplitude variation with respect to the signal at each receive antenna. The diversity gain, also called diversity order, is equal to the number of receive antennas in SIMO case.

In case of MISO, we can apply transmit diversity techniques, and obtaining diversity gain in these channels is possible in both cases of knowledge or non-knowledge of the channel at the transmitter. Several designs have been proposed in the literature to extract transmit diversity. Probably the most famous example are the Space–Time Coding (STC) schemes (e.g., [30, 144]), which code across the spatial domain (transmit antennas) and do not make use of any channel knowledge at the transmitter. If the links between all N_T transmit antennas to the receive antenna experience independent fading conditions, then the diversity order of the MISO channel is equal to N_T.

If dealing with MIMO, one can combine the above-mentioned transmit and receive diversity schemes. If the links between each transmit–receive antenna pair are fading independently, then the diversity order becomes $N_T N_R$ (product of the number of transmit and receive antennas) [139].

9.1.3 Multiplexing Gain

Spatial Multiplexing (SM) offers a linear increase in $\min(N_T, N_R)$ of the capacity, at no extra bandwidth or power expense: this gain is called multiplexing gain. Only with multiple antennas at the transmitter and at the receiver, i.e. in MIMO channels, SM is possible [145, 146, 147].

In the case of $N_T = 2$ and $N_R = 2$, i.e. 2×2 MIMO channel, the bit stream to be transmitted is demultiplexed into two half-rate substreams, simultaneously modulated and transmitted from each transmit antenna. Under certain favorable channel conditions, the spatial signatures of these signals induced at the two receive antennas are well separated, thus the receiver, since it has knowledge of the channel, can differentiate between the two co-channel signals and extract both signals. At this point, demodulation yields the original sub-streams, that can be combined giving back the original bit stream [139].

Transmit and receive diversity are means to combat fading. A different line of thought suggests that in a MIMO channel, fading can in fact be beneficial, through increasing the degrees of freedom available for communication [146, 147]. Essentially, if the path gains between individual transmit–receive antenna pairs fade independently, the channel matrix is well conditioned with high probability, in which case multiple parallel spatial channels are created. Therefore, in rich scattering environment, the independent spatial channels can be exploited to send multiple signals at the same time and frequency, resulting in higher spectral efficiency. By transmitting independent information streams in parallel through the spatial channels, the data rate can be increased. This effect is also called spatial multiplexing [148], and is particularly important for the high-Signal-to-Noise Ratio (SNR) regime where the system is degrees-of-freedom limited (as opposed to power limited). Foschini [146] has shown that in the high-SNR regime, the capacity of a channel with N_T transmit antennas, N_R receive antennas, and i.i.d. Rayleigh-faded gains between each antenna pair, is given by

$$C(\rho) = \min(N_T, N_R) \log \rho + O(1), \qquad (9.1)$$

where ρ is the SNR. The number of degrees of freedom is thus the minimum of N_T and N_R.

9.1.4   Interference Reduction

Frequency reuse in wireless channels gives rise to co-channel interference: one way to limit this interference is to use multiple antennas, differentiating between the spatial signature of the desired signal and the ones of the interfering co-channel signals. Some knowledge of the channel of the desired signal is required to perform interference reduction.

Also at the transmitter one can implement interference reduction or avoidance techniques, by maximizing the energy sent to the desired user, and minimizing the one sent to the co-channel users. Positive effects of interference reduction are the possibility to use aggressive reuse factors and the improvement of network capacity [139].

In next sections, we will go more into details concerning diversity, array and multiplexing gains.

9.2 Multi-antenna and Diversity

In a multi-path environment, the received signal is the sum of multiple scattering paths with random amplitudes and phases. The interaction between these paths causes channel gain variations in time and deep fades. These fades lead to degradation in the received signal, which might make it impossible for the receiver to correctly detect the transmitted signal. Diversity is one of the most effective techniques to combat fading in wireless communications.

The main idea behind diversity is to send multiple copies of the transmitted signal via multiple (presumably independent) channels to the receiver. When the channels have low, or ideally zero, cross-correlation, the probability that all of them fall into deep fading simultaneously is very low [140]. That means if one radio path undergoes a deep fade at a particular point in time and/or frequency and/or space, another uncorrelated path may have a strong signal at that point. By having more than one path to select from, both the instantaneous and average Signal to Noise Ratio (SNR) at the receiver can be greatly improved.

There are different types of diversity techniques, which can be categorized by the methods from which multiple versions of the received signal are introduced at the receiver.

9.2.1 Time Diversity

In time (or temporal) diversity, the transmitted signal is repeated several times at different time frames. Thus, the receiver will receive more than one version of the transmitted message at different time. The only requirement for time diversity technique is that the time separation between two repetitions be greater than the channel coherence time, so as to ensure that the received signals are uncorrelated. If N_b diversity branches are needed, the reception delay is at least N_b times the repetition interval. As a result, time diversity system would observe large delay, especially when the number of branches N_b increases or when the channel coherence time is large. When the receiver is stationary and the fading is not time-selective, time diversity will not bring the desired gain [141].

9.2.2 Frequency Diversity

Another method to obtain multiple versions of the incoming signal at the receiver is frequency (or spectral) diversity. This method utilizes two or more frequency bands to transmit the same message. The receiver will listen to these frequency bands to get N_b diversity branches. The requirement for frequency diversity is that those frequency bands must be sufficiently separated in order to ensure that the fading in different frequency bands are uncorrelated. The coherence bandwidth of the channel is a convenient quantity to use in describing the degree of correlation existing between different frequencies. For frequency separation of more than several times the coherence bandwidth, the signal fading would be essentially uncorrelated. For example, if the channel coherence bandwidth is 500 kHz for a certain mobile environment, the frequency separation should be more than 1–2 MHz [141]. In mobile radio, where frequency is a precious resource, this diversity technique is not always a preferable choice. Frequency diversity also requires separate transmitter chain for each of the branches.

9.2.3 Space Diversity

Space (or spatial) diversity is a method in which two or more antennas physically separated from each other are used to obtain independent versions of the received signal. These antennas can be located at the receiver (receive diversity), at the transmitter (transmit diversity) or at both. Figure 9.2 illustrates a scenario where receiver diversity is employed at the Base Station (BS) to compensate for low transmitting power at the Mobile Station (MS). This technique is now commonly used in the BS of the 2nd Generation (2G) wireless systems.

The level of cross-correlation between signals received by those antennas depends on the distance between antennas. In general, the minimum distance

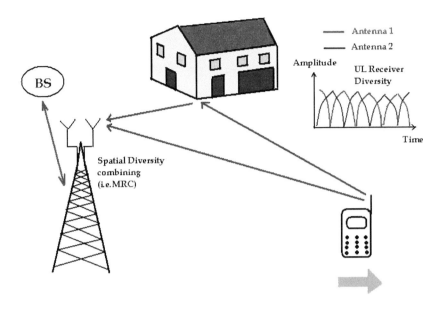

Fig. 9.2 Example of space diversity.

between antennas must be greater than that coherence distance of the channel so that the spatial fadings become uncorrelated [139]. Ideally, antenna separation of half of wavelength ($\lambda/2$) should be sufficient [141]. Space diversity is a convenient and attractive method of diversity for mobile radio. It is relatively simple to implement and it does not require additional frequency spectrum. If antenna separation is sufficient, spatial diversity can be employed to reduce fading, even if the channel is neither time- nor frequency-selective. The downside of space diversity is the cost, the size, the power consumption and the complexity of having additional antennas.

When multiple copies of the transmitted signal arrive at the receiver, they have to be pre-combined in a determined manner, in order to obtain the diversity gain.

9.3 Multi-antenna and Spatial Multiplexing

Parallel to diversity techniques, a recently new approach of multi-antenna transmission systems is spatial multiplexing. With respect to diversity technique, spatial multiplexing aims at increasing data rate of the system.

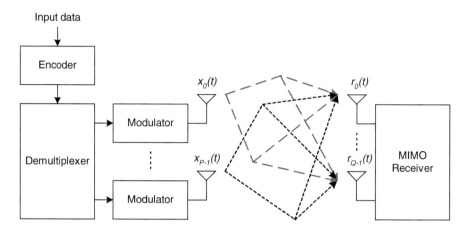

Fig. 9.3 An example of MIMO spatial multiplexing system.

Spatial multiplexing techniques require that multiple antennas be present at both the transmitter and the receiver. Such antenna arrangement is often referred to as Multiple Input Multiple Output (MIMO), as illustrated in Figure 9.3. The number of transmitting and receiving antennas are denoted by P and Q, respectively. At the transmitter, P independent data streams, $x_0(t)$, $x_1(t)$, ..., $x_{P-1}(t)$, are transmitted on different transmitting antennas simultaneously and in the same frequency band. For the sake of simplicity, flat fading channel is assumed between pairs of transmitting and receiving antennas. At the qth receiving antenna, the independent signals $\{x_0(t), x_1(t), \ldots, x_{P-1}(t)\}$ are linearly combined to produce the received signal $r_q(t)$, which is given by

$$r_q(t) = h_{q,0}x_0(t) + h_{q,1}x_1(t) + \cdots + h_{q,P-1}x_{P-1}(t) + \vartheta_q(t),$$
$$q = 0, 1, \ldots, Q - 1, \quad (9.2)$$

where $h_{q,p}$ is the envelop of flat fading channel between the pth transmitting antenna and the qth receiving antenna, and $\vartheta_q(t)$ is the thermal noise of the qth receiving antenna. Under a rich scattering environment, the channels between pairs of transmitting and receiving antennas undergo different multipaths and thus become independent from each other. The independence of the channels means that the receiver can estimate $\{x_0(t), x_1(t), \ldots, x_{P-1}(t)\}$ from $\{r_0(t), r_1(t), \ldots, r_{Q-1}(t)\}$ independently, provided that the channel responses are known [139]. The problem is equivalent to solving a set of Q linear equations with P unknown variables, and it can be solved if $Q \geq P$.

As a result, spatial multiplexing promises higher throughput without increasing system bandwidth or transmit power. It is shown in [142] that, under certain circumstances, system capacity increases linearly with the number of antennas used.

The main reason of using OFDM with MIMO is the fact that OFDM has the capability of turning a frequency-selective MIMO fading channel into multiple flat fading channels [143]. This renders the multi-channel equalization particularly simple, since for each OFDM sub-carrier only a constant matrix needs to be inverted.

9.4 Diversity Gain vs Coding Gain

Traditionally, multiple antennas have been used to better exploit the diversity due to the multi-path phenomenon. Each pair of transmit and receive antennas provides a signal path from the transmitter to the receiver. By sending signals that carry the same information through different paths, multiple independently faded replicas of the data symbol can be obtained at the receiver end; hence, more reliable reception is achieved. For example, in a slow Rayleigh-fading environment with one transmit and N_R receive antennas, the transmitted signal is passed through N_R different paths. It is well known that if the fading is independent across antenna pairs, a maximal diversity gain (advantage) of N_R can be achieved: the average error probability can be made to decay like $1/\rho^{N_R}$ at high SNR ρ, in contrast to the ρ^{-1} for the single-antenna fading channel [149]. The underlying idea is averaging over multiple path gains (fading coefficients) to increase the reliability. In a system with N_T transmit and N_R receive antennas, assuming the path gains between individual antenna pairs are independent and ideally identically distributed (i.i.d.) Rayleigh faded, the maximal diversity gain is $N_T N_R$, which is the total number of fading gains that one can average over.

While both diversity and coding gain improve system performance, i.e. they decrease error rate, these gains are very different in nature. Diversity gain acts on the Bit Error Rate (BER), increasing the magnitude of the slope of the curve, instead coding gain shifts the error rate curve to the left. The BER for a system employing both coding and diversity techniques at high SNR can be approximated by

$$\overline{P}_e \approx \frac{c}{(\gamma_c \, \rho)^M},\tag{9.3}$$

where c is a scaling constant dependent on the modulation employed and the channel, γ_c, $\gamma_c \geq 1$ indicates the coding gain, and M is the diversity order of the system. When increasing the diversity order, it increases also the SNR advantage, resulting in a lower error rate. Instead coding gain is in general constant at high SNR (see Figure 9.4 [139]).

It should be clear enough now that diversity is one of the main and most effective techniques to mitigate fading phenomenon in wireless systems. By using diversity techniques in time and frequency domains, we have to pay a price, i.e. delay in case of time diversity, and bandwidth in case of frequency diversity; furthermore it can be shown that the average received SNR by using these techniques is the same as for an Additive White Gaussian Noise (AWGN) channel [139].

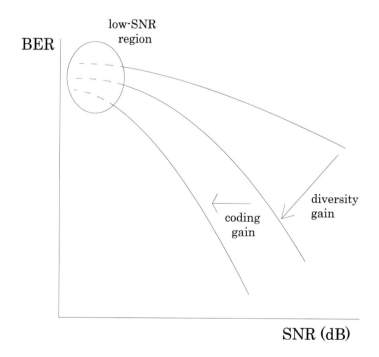

Fig. 9.4 Difference between diversity gain and coding gain.

An attractive alternative is spatial diversity, which does not imply a price to pay in terms of time and bandwidth, and provides diversity and array gains. We can have different multiple antennas to extract spatial diversity, depending on the transceiver configuration, i.e. SIMO, MISO, or MIMO and the study of some of these techniques will be the object of the next two chapters.

In Table 9.1 one shows the expected array gain and the diversity order for different multi-antenna configurations [139], where the channel knowledge is intended with respect to the transmitter. Without spatial diversity, i.e. $N_T = 1$, $N_R = 1$, the signal experiences and suffers from deep fades. Increasing the spatial diversity, i.e. increasing $N_T N_R$, the fades' depth reduces and the effective channel tightens: this effect has been demonstrated in real channels [150]. Moreover by increasing the diversity order, we also notice a progressive increase in the mean signal level: this comes from the array gain at the receiver and the transmitter side, assuming channel knowledge, which is usually the case at the receiver, and optionally can be there also at the transmitter.

We can quantify the degree of tightening, or "hardening" of the channel by its coefficient of variation μ_{var} [151, 152]. The coefficient of variation of a random variable is defined as the ratio of the standard deviation of the random variable to its mean. Assuming an i.i.d. (spatially white) MIMO channel, $\mathbf{H}_w$, its coefficient of variation is

$$\mu_{var} = \frac{1}{\sqrt{N_T N_R}}, \tag{9.4}$$

i.e. the degree of tightening of the effective channel is inversely proportional to the square root of the system diversity order. Asymptotically, i.e., when $N_T N_R \to \infty$ we can see from (9.4) that the coefficient of variation tends to zero: in other words, with infinite diversity order the channel is perfectly stabilized [139].

Table 9.1 Array gain and diversity order for different multi-antenna configurations.

Configuration	Expected array gain	Diversity order
SIMO (Channel unknown)	N_R	N_R
SIMO (Channel known)	N_R	N_R
MISO (Channel unknown)	1	N_T
MISO (Channel known)	N_T	N_T
MIMO (Channel unknown)	N_R	$N_T N_R$
MIMO (Channel known)	$E(\lambda_{max})$	$N_T N_R$

9.5   Capacity of MIMO Channels

The maximum error-free data rate that a channel can support is called the channel capacity. The derivation of the channel capacity for an AWGN channel is due to Claude Shannon [153]. In contrast to AWGN channels, ST channels experience fading and implicitly incorporate the spatial dimension.

9.5.1   Fundamentals on Channel Capacity

For a memoryless SISO system the capacity is given by

$$C_{\text{SISO}} = \log_2\left(1 + \rho \mid h \mid^2\right) \quad \left[\frac{bps}{Hz}\right], \tag{9.5}$$

where h is the normalized complex gain of a fixed wireless channel or that of a particular realization of a random channel, and ρ denotes the SNR at each receive antenna. By using multiple receive antennas, the statistics of the capacity improve and for a SIMO system with N_R receive antennas, the capacity is [154]

$$C_{\text{SIMO}} = \log_2\left(1 + \rho \sum_{i=1}^{N_R} \mid h_i \mid^2\right) \quad \left[\frac{bps}{Hz}\right], \tag{9.6}$$

where h_i is the channel gain for ith receive antenna. From Equation (9.6) one can see that the average capacity increase is only logarithmic with N_R.

In case have a MISO system with N_T transmit antennas employing transmit diversity, and the transmitter does not have channel knowledge, the capacity is [142]

$$C_{\text{MISO}} = \log_2\left(1 + \frac{\rho}{N_T} \sum_{i=1}^{N_T} \mid h_i \mid^2\right) \quad \left[\frac{bps}{Hz}\right], \tag{9.7}$$

where to ensure a fixed total transmitter power, ρ has been normalized over N_T. From (9.7) we can notice the absence of array gain, comparing with (9.6) where the channel energy could be combined coherently. Also we can see from (9.7) that for a MISO the capacity increases logarithmically with the number of transmit antennas, analogously to what we saw in (9.6) for a SIMO system.

If a MIMO $N_T \times N_R$ system is considered, employing transmit and receive diversity, we obtain the well-known capacity equation [142, 147, 155]

$$C_{\text{MIMO}} = \log_2 \left[\det \left(\mathbf{I}_{N_R} + \frac{\rho}{N_T} \mathbf{H}\mathbf{H}^H \right) \right] \quad \left[\frac{bps}{Hz} \right], \qquad (9.8)$$

where $\mathbf{H}$ is the $N_R \times N_T$ channel matrix. As in (9.7), also in (9.8) one uses the assumption of total power equally distributed among transmit antennas. In [142] and [155] it has been demonstrated that C_{MIMO} grows linearly with $\min(N_R, N_T)$, rather than logarithmically.

9.5.2 MIMO Channel Capacity: Information Theoretic Approach

Let us define the MIMO signal model

$$\mathbf{y} = \mathbf{Hs} + \mathbf{n}, \qquad (9.9)$$

where $\mathbf{y}$ is the $N_R \times 1$ received signal vector, $\mathbf{s}$ is the $N_T \times 1$ transmitted signal vector and $\mathbf{n}$ is an $N_R \times 1$ vector of additive noise terms, assumed i.i.d. complex Gaussian where each element has a variance equal to σ^2. The system equation (9.9) represents a Single User (SU)-MIMO system assuming AWGN channel, thus the only interference present is self-interference between the input streams to the MIMO system [154].

If we denote with $\mathbf{Q}$ the covariance matrix of $\mathbf{s}$, then the capacity of the system described by (9.9) is [147, 155]

$$C = \log_2 \left[\det \left(\mathbf{I}_{N_R} + \mathbf{HQH}^H \right) \right] \quad \left[\frac{bps}{Hz} \right], \qquad (9.10)$$

where $tr(\mathbf{Q}) \leq \rho$ holds providing a global power constraint. For equal power uncorrelated sources $\mathbf{Q} = \frac{\rho}{N_T} \mathbf{I}_{N_T}$ and (9.10) collapses to (9.8). This expression for the covariance matrix is optimal when $\mathbf{H}$ is unknown at the transmitter and the input distribution maximizing the mutual information is the Gaussian distribution [147, 155]. When channel feedback is available at the transmitter, the optimal $\mathbf{Q}$ is not proportional to the identity matrix but is constructed by a waterfilling approach [154].

Equation (9.8) can be rewritten as [155]

$$C = \sum_{i=1}^{m} \log_2 \left(1 + \frac{\rho}{N_T} \lambda_i \right) \quad \left[\frac{bps}{Hz} \right], \qquad (9.11)$$

where λ_i, $i = 1, \ldots, m$ are the nonzero eigenvalues of $\mathbf{W}$, $m = \min(N_R, N_T)$, and

$$\mathbf{W} = \begin{cases} \mathbf{HH}^H, & N_R \leq N_T \\ \mathbf{H}^H\mathbf{H}, & N_T < N_R. \end{cases} \qquad (9.12)$$

Now some exact capacity results will be shown and discussed, considering two main cases: when the channel known at the transmitter, and when it is unknown. Two main questions need to be answered [154], i.e. what is the effect of feedback and what is the impact of the channel?

Channel known at the transmitter: When channel is known at the transmitter (and at the receiver), $\mathbf{H}$ is known in (9.10) and capacity is optimized over $\mathbf{Q}$ subject to the power constraint $tr(\mathbf{Q}) \leq \rho$: in this case the optimal $\mathbf{Q}$ is well known [147, 155, 156, 157, 158, 159] and the resulting capacity is given by

$$C_{\mathrm{WF}} = \sum_{i=1}^{m} \log_2 (\mu\lambda_i)^+ \quad \left[\frac{bps}{Hz}\right], \qquad (9.13)$$

where μ is chosen to satisfy

$$\rho = \sum_{i=1}^{m} (\mu - \lambda_i^{-1})^+ \qquad (9.14)$$

and "+" takes into account only the positive terms. Being μ a complicated nonlinear function of λ_i, $i = 1, \ldots, m$ the distribution of C_{WF} appears intractable; anyway C_{WF} can be simulated using (9.13) and (9.14) for any given $\mathbf{W}$, and in this way the optimal capacity can be computed numerically for any channel.

Concerning impact of feedback, i.e. channel information being supplied to the transmitter, it has been shown in [159, 160, 161] that the gains of waterfilling over equal power uncorrelated sources are significant at low SNR but converge to zero as SNR increases. The fact that the gain from feedback reduces at higher SNR can be intuitively explained as follows. The main advantage of the knowledge of the transmit channel is the provision of array gain, while gains such as diversity gain and multiplexing gain can be achieved by blind schemes like Space–Time Block Code (STBC) and Vertical-Bell Labs Layered Space–Time Architecture (VBLAST) that do not require such kind of knowledge. But at high SNR the transmit array gain (that is practically a boost in average SNR) is marginal, i.e. the benefit of feedback reduces [154].

Channel unknown at the transmitter: In this case, the capacity expression is given by (9.8) and it was derived by Foschini and Telatar starting from different viewpoint [142, 147, 155]. For Ricean channels, capacity decreases as Line of Sight (LOS) strength increases [158, 162]. The impact of correlation is important too and various models and measurements have been used to quantify its impact [157, 163, 164].

9.5.3 Limiting Capacity Results

In Section 9.5.2 exact results for capacity have been shown, but all of these results are obtained assuming i.i.d. Rayleigh fading case. Hence, it may be useful to look for limiting results to cover a broader range of cases, to give simpler and more intuitive results, and to study the potential of very large scale systems [154]. In the literature many authors have considered the case where $N_T, N_R \to \infty$ and $\frac{N_R}{N_T} \to c$ for some constant c, i.e. number of antennas growing proportionally at both transmitter and receiver: under these assumptions, the limiting results are said to hold for "large systems."

Channel known at the transmitter: Analytical results in [159, 161] show that $\frac{C_{WF}}{N_R} \to \mu_{WF}$, with μ_{WF} constant, for large systems under both i.i.d. and correlated fading conditions. The remaining knowledge of large systems is mainly based on simulations, e.g. [160] shows that C_{WF} grows linearly for Ricean fading.

Channel unknown at the transmitter: In this case, the capacity expression is given by (9.8), and we can rename C as C_{EP} where "EP" stands for equal power uncorrelated sources. For large systems, the limiting mean capacity has been shown to be of the form $N_R \mu_{EP}$ [155], where μ_{EP} depends on N_T, N_R through the ratio $c = N_R/N_T$. A closed form expression for C_{EP} can be found in [165]. The limiting variance is a constant [166], once again dependent on the ratio c rather than on N_T and N_R individually. For a more general class of fading channels similar results hold and a central limit theorem can be stated as follows [167, 168]

$$\lim_{N_T, N_R \to \infty} \left(\frac{C_{EP} - E(C_{EP})}{\sqrt{Var(C_{EP})}} \right) = Z, \tag{9.15}$$

where $N_R/N_T \to c$ as $N_T, N_R \to \infty$ and $Z \sim N(0,1)$ is a standard Gaussian random variable. In [159] and [161], one shows that $\frac{C_{EP}}{N_R} \to \mu_{EP}$ for large systems, in both i.i.d. and correlated fading, and the correlation has a detrimental impact on the value of μ_{EP}.

Concerning asymptotic results in SNR, [159] gives both low and high SNR capacity results for large systems. It is shown that at high SNR, C_{EP} and C_{WF} are equivalent [154].

9.6 Trade-Off Between Spatial Multiplexing and Spatial Diversity

A MIMO system can provide several types of gains. In this section, the focus is on diversity gain and spatial multiplexing gain. There are schemes which switch between these two modes, depending on the instantaneous channel condition [32]. However, maximizing one type of gain may not necessarily maximize the other. In fact, each of the two design goals addresses only one aspect of the problem. This makes it difficult to compare the performance between diversity-based and multiplexing-based schemes.

In [149] a different viewpoint is put forth, i.e. given a MIMO channel, both gains can be simultaneously obtained, but there is fundamental trade-off between how much of each type of gain any coding scheme can extract: higher spatial multiplexing gain comes at the price of sacrificing diversity. The focus of [149] is on high-SNR regime. A scheme is said to have a spatial multiplexing gain r and a diversity advantage d if the rate of the scheme scales like $r \log \rho$ and the average error probability decays like $1/\rho^d$, where ρ is the SNR. Clearly, r cannot exceed the total number of degrees of freedom $\min(N_T, N_R)$ provided by the channel, and $d(r)$ (the diversity gain which the trade-off curve associates to the multiplexing gain r) cannot exceed the maximal diversity gain $N_T N_R$ of the channel. The trade-off curve bridges between these two extremes, as can be seen in Figure 9.5.

The specific study-cases that will address the multiplexing-diversity trade-off issue in this book are Joint Diversity and Multiplexing (JDM) schemes, in Chapters 12 and 13, and Linear Dispersion Codes (LDC) in Chapter 13. The trade-off studies are triggered based on the fact that none of the schemes mentioned above can become an absolute choice for best system performance in all scenarios. To be more precise, a user located very close to access point can better exploit the multiplexing benefits, while a user at farther location

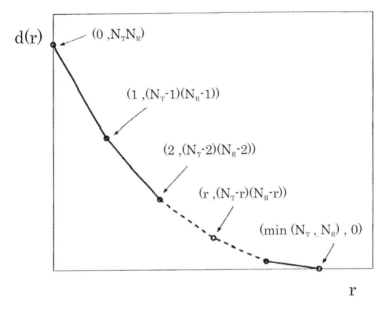

Fig. 9.5 Diversity-multiplexing trade-off.

may benefit more by using diversity schemes: to explore what is in between is one of the main research goals of this book.

9.7 Multi-antenna in OFDM

Most of the available MIMO techniques are effective in frequency flat scenarios [169, 170]. All these narrowband algorithms can be implemented on OFDM sub-carrier level, because OFDM converts a wideband frequency selective channel into a number of narrowband sub-carriers. In other words, OFDM turns a frequency-selective MIMO fading channel into a set of parallel frequency-flat MIMO fading channels. So, in wideband scenarios, OFDM can be combined with MIMO systems, for both diversity and multiplexing purposes [154]. In frequency selective environments, amalgamation of SM and OFDM techniques can be a potential source of high spectral efficiency at reasonable complexity. MIMO-OFDM drastically simplifies equalization in frequency-selective environments.

Let us assume an OFDM-MIMO system with N_T transmit antennas, N_R receive antennas and N sub-carriers. The input–output relation for the MIMO system for the kth tone, $k = 1, \ldots, N$ may be expressed as

$$\mathbf{Y}_k = \sqrt{\frac{P_T}{N_T}} \mathbf{H}_k \mathbf{s}_k + \mathbf{N}_k, \tag{9.16}$$

where $\mathbf{Y}_k$ and $\mathbf{N}_k$ are $N_R \times 1$ vectors, $\mathbf{H}_k$ is a $N_R \times N_T$ matrix, $\mathbf{s}_k$ is a $N_T \times 1$ vector, and P_T is the total transmit power. The matrix $\mathbf{H}_k$ is the frequency response of the matrix channel corresponding to the kth tone.

From Equation (9.16) we can see that, just as in SISO channels, OFDM-MIMO decomposes the otherwise frequency selective channel of bandwidth B into N orthogonal flat fading MIMO channels, each with bandwidth B/N (see Figure 9.6) [139]. MIMO signaling treats each OFDM tone as an independent narrowband frequency flat channel. We must take care to ensure that the modulation and demodulation parameters (carrier, phasing, FFT/IFFT, prefixes, etc.) are completely synchronized across all the transmit and receive antennas. With this precaution, every OFDM tone can be treated as a MIMO channel, and the tone index can be treated as a time index in the already existing ST techniques.

Multiple antenna systems for OFDM based 4G systems can be broadly classified into three categories:

- Diversity combining, which is employed to mitigate fading and increase link reliability;

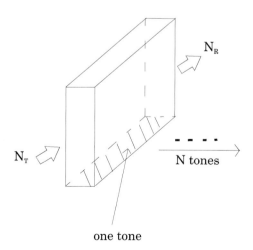

Fig. 9.6 Schematic of MIMO-OFDM. Each OFDM tone admits N_T inputs and N_R outputs.

- Spatial multiplexing, which is used to increase data rate;
- Smart array processing, such as BF, which aims at increasing receiving power and rejecting unwanted interference.

9.7.1 Space Diversity

Transmit Diversity (TD)-OFDM system model is shown in Figure 9.7. Space–Time (ST) diversity schemes map directly to OFDM by applying these coding techniques on a per tone basis across OFDM symbols in time exactly as in SC modulation. However, this requires that channel remains constant over T OFDM symbol periods. Since the duration of an OFDM symbol $\frac{N+N_{\mathrm{CP}}}{f_s}$ is usually large, this may be impractical (f_s is the sampling frequency).

For example, assuming a set of WiMAX parameters [171], i.e. $N = 128$, $N_{\mathrm{CP}} = 16$, $f_s = 1.429$ Msps, the OFDM symbol duration is $T_s = \frac{N+N_{\mathrm{CP}}}{f_s} = \frac{128+16}{1.429e6} = 100.77 \,\mu s$. At the speed of $v = 150$ Kmph$= 41.66$ m/s, the maximum Doppler shift $f_m = \frac{v}{c} f_c = \frac{41.66}{3e8} 3.5e9 = 486$ Hz, where c is the speed of light and f_c is the carrier frequency. Therefore, the coherence time [40] is $T_c = \frac{9}{16\pi f_m} = 368 \,\mu s$. To verify the Alamouti's assumption of channel constance over two symbol periods, we have to compare $2T_s$ with the coherence time T_c: since it is only $2T_s < T_c$ and not $2T_s << T_c$, it is risky to take for granted this assumption.

The possibility to access the frequency-domain in OFDM systems suggests to code also in space and frequency, not only in space and time: diversity techniques designed for Single Carrier (SC) modulation over flat fading channels are easily extended to OFDM modulation with the time index for

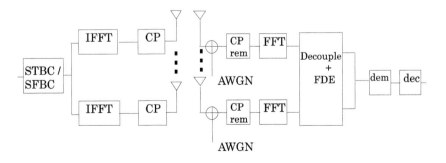

Fig. 9.7 Transmit diversity for OFDM.

SC modulation replaced by the tone index in OFDM. As an example, let us consider the Alamouti scheme, which extracts full spatial diversity in absence of channel knowledge at the transmitter with $N_T = 2$. The implementation of the Alamouti scheme requires that the channel remains constant over two consecutive symbol periods. Alamouti scheme applied in space and frequency domains is an example of Space–Frequency Block Code (SFBC) schemes. In SFBC schemes, the requirement of the channel being constant over consecutive symbols becomes a requirement over consecutive tones, i.e. $\mathbf{H}_k = \mathbf{H}_{k+1}$. The receiver detects the transmitted symbols from the signal received on the two tones using the Alamouti detection technique. The use of consecutive tones is not strictly necessary, any pair of tones can be used as long as the associated channels are equal. The technique can be generalized to extract spatial diversity over a larger number of antennas by using the ST techniques developed for SC modulation: in that case we need a block size $T \geq N_T$ and the channel must be identical over the T tones.

9.7.2 Spatial Multiplexing

SM-OFDM system model is shown in Figure 9.8. Analogously to SM for MIMO systems with SC modulation, the objective of SM in conjunction with MIMO-OFDM is to achieve the spatial rate $R = N_T$ by transmitting parallel streams [143]. Thus, $N N_T$ scalar data symbols are transmitted over one OFDM symbol, with N_T symbols being transmitted over each sub-carrier. Thus SM in MIMO-OFDM system reduces to SM over each tone. The receiver architecture

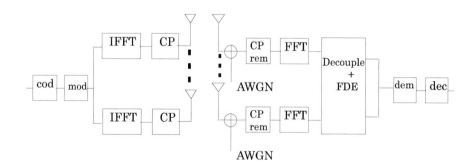

Fig. 9.8 Spatial multiplexing for OFDM.

for SM is identical to that for SC modulation. As in SC case we require that $N_R \geq N_T$ in order to support the symbol streams reliably.

9.7.3 Beam-Forming

In a BF system, the weights to be multiplied with the signals have to be carefully chosen. From antenna theory, an array with N_T antenna elements has different excitation currents according to the direction of the waves arriving or departing from each element of the array. Considering linear phase progression, the weights have a phase that increases the same amount from one element to the next. Usually, BF is treated in literature as a method for increasing array gain when receiving signals from a specific direction. Direction of Arrival (DoA) is the angle of the wave arriving to the antenna.

Since it is more feasible to have multiple antennas at the BS than at the MS, it makes sense to place the beamformer at the BS, i.e. at the transmitter side in case of DL [172]. This will bring additional complications if it is assumed that there is complete channel knowledge at the receiver, but not at the transmitter. In order to solve this problem, the Downlink (DL)-BF technique can extract the weights to be used in DL from the weights calculated in UL. This method can be used in Time Division Duplex (TDD) systems without losing performance. However, in Frequency Division Duplex (FDD) systems, as a consequence of the frequency dependent steering array response and uncorrelated fading, the UL weight reuse in DL degrades system's performance, since the frequency for UL is different from the one used in DL. A block diagram showing the implementation of DL-BF in an OFDM transmitter is presented in Figure 9.9.

9.7.4 Usability of Multi-antenna Techniques in OFDM Systems

Together with other important parameters, sub-carrier bandwidth, Δf, and OFDM symbol duration, T_s, are closely related to the choice of multi-antenna techniques in OFDM systems. In general, when Δf is large enough, so that T_s is very small compared to channel coherence time, T_c, then a number of algorithms that are dependent on the channel information at the transmitter are easily and effectively implementable. In this case, the feedback can be used for several OFDM symbols, thus, the total amount of feedback to be transported can be reduced. In the other case, when T_s is large (and conversely Δf is small), the Channel State Information (CSI) feedback will be outdated quite

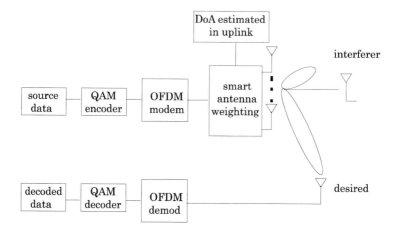

Fig. 9.9 OFDM system with beamforming.

fast and would require very frequent transmission of CSI between transmitter and receiver. This will cost in terms of spectral efficiency.

Considerations with receive diversity: In general, receive antenna diversity is practical, effective and hence, commonly-used technique for mitigating the effect of multi-path fading. By employing multiple antennas at the receiver and having a suitable combining scheme, significant diversity gain can be achieved, which can be used to maximize the coverage and/or link reliability [173]. The diversity gain comes without having to increase the transmit power and/or to extend the frequency spectrum. It also does not generally introduce delay like time diversity technique. Receive diversity is relatively simple to implement, as compared to transmit antenna diversity and other MIMO techniques. Nevertheless, additional antenna(s) also means increasing in size, complexity, and power consumption, and these are key limiting factors for receive diversity. As a result, in today's systems, receive diversity is mostly applied at the BS to compensate for low transmit power at MSs. All processing tasks are performed at BS side, therefore MS structure can be quite simple.

Being one of space diversity techniques, receive diversity requires that antenna separation be larger than the coherence distance of the channel, so that the spatial fading becomes uncorrelated. The coherence distance depends on the richness of scattering environment, i.e. how many multipath are available or whether LOS or a dominant reflection is present. The coherence distance of the channel is inversely proportional to the Root Mean Square (RMS) angle spread — the larger the angle spread, the shorter the coherence distance [139].

There are two methods for receive diversity combining for OFDM systems, namely Post-DFT and Pre-DFT combining. In Post-DFT combining schemes, the time-domain OFDM samples from multiple receive antennas are first DFT demodulated, and then combined on sub-carrier basis. As deep fades are distributed randomly among sub-carriers, combining at sub-carrier level enables the combination (or selection) of only the best-quality sub-carriers from different diversity branches. Therefore, Post-DFT Maximal Ratio Combining (MRC) scheme is optimal in terms of BER performance. However, Post-DFT requires separate DFT processor for each antenna branch, which increases the computational complexity and power consumption at the receiver.

On the other hand, Pre-DFT combining techniques combine the diversity branches before performing OFDM demodulation. This reduces the complexity of the receiver, but at the expense of performance degradation. For instance, if the channel is flat over the entire OFDM spectrum, the performance of Pre-DFT Maximal Average Ratio Combining (MARC) scheme is comparable to that of the Post-DFT MRC scheme. However, when the channel is frequency-selective fading, its performance is degraded and less diversity gain is achieved compared to Post-DFT MRC [174].

Considerations with transmit diversity: The implementation of multiple antennas at the transmitter side brings the possibility of exploiting redundancy of information by sending signals through different paths. Transmit diversity used alone falls under the group of MISO techniques and it is employed by using different ways to transmit the signal, separating them in time, frequency and/or space. However, applying diversity in time-domain only, e.g. repetition codes, considerably reduces the data rate [40]. Also, applying frequency-domain diversity alone, e.g. sending the same signal over different carriers, is not feasible due to the high cost of available spectrum [141]. Hence, space diversity, by applying more antennas is often a preferable option, given the decreasing cost of hardware. To overcome channel impairments, it is also beneficial to exploit diversity in more than one domain. The most common method employed to take advantage from transmit diversity is the STCs, i.e. Space–Time Block Code (STBC) and Space–Time Trellis Code (STTC). Since the MS has more restrictions, in order to lower its cost, to save on battery consumption

and to have small terminals, it may not be feasible to include multiple antennas at handsets. On the other hand, BS does not have such restrictions. It
should also be noted that usually all contemporary BSs have more than one
receive antennas for receive diversity purposes. These antennas can be used
for transmit diversity purpose in the DL commonly used in 2G BSs, whereas
transmit space diversity is proposed for The Third Generation Partnership
Project 2 (3GPP2) specifications and in IEEE 802.16a standards.

In the group of STBC, the so-called Alamouti scheme, presented in [30],
stands for being an optimal code to achieve both full data rate and full diversity
gain by employing two transmit antennas and having no CSI at transmitter.
STBC using more transmit antennas have been developed and an extended
performance analysis presented in [175]. However, with more than two antennas there are no codes achieving both full rate and full diversity simultaneously. It should also be pointed out that, with the increase of order of
STBC techniques, the latency of the transmission gets larger, e.g. with a four
Transmitter (Tx) antennas STBC there is the need to wait for four symbol durations before decoding the same number of symbols. Even not affecting data
rate, the block-by-block processing may introduce unacceptable delay in real
time services. Another group of transmit diversity techniques are included in
SFBC, exploiting frequency diversity [176]. This is employed across OFDM
tones. Since in SFBC the time domain is not used, there is no latency. To
make full use of diversity over the three domains (i.e time, frequency, and
space domains), more complex codes are required and they are referred to as
Space–Time–Frequency Block Code (STFBC) [177].

It is worth noting here that if no CSI is available at transmitter, the array gain
cannot be exploited at the transmitter. CSI is usually estimated at the receiver
side, hence, to achieve transmit array gain, feedback should be sent from
the MS. However, this option comprises a high amount of information to be
exchanged and may not be feasible to implement, especially when the wireless
channel is severely time-variant and when the usable data bandwidth is not
sufficient. Nevertheless, it can be used in very low mobility situations such as
indoor scenarios, where $T_s \ll T_c$. Two such systems are Institute of Electrical
and Electronics Engineers (IEEE) 802.11a and IEEE 802.16a. Another option
to have CSI at transmitter is using the CSI from the UL and apply it to DL.
If the channel remains static for two symbol durations, this method can be
easily implemented in TDD systems. On the other hand, in FDD systems the

CSI from UL cannot be directly re-used in DL, since the carrier frequency is different and usually not within the coherence bandwidth. In that case, a proper frequency calibration is required.

Choice of diversity technique: There is a great degree of freedom in combining OFDM and diversity techniques. Diversity gain can be obtained by employing either space, time or frequency diversity, or a combination of any of these three (i.e. space–time or space–frequency or time–frequency) or all three diversity techniques together (i.e. space–time–frequency) in an OFDM system. In general, obtaining space diversity would require additional antennas in the transmitter or receiver or at both. Obtaining time diversity would always introduce a time delay in service. Obtaining frequency diversity will always depend on the channel itself. It should also be noted that higher frequency diversity can be exploited in OFDM itself by using channel coding and interleaving. A careful choice of channel coding and interleaving can give a very good frequency diversity gain. In this section, the choice of diversity technique is considered for different scenarios, such as uplink vs downlink and indoor vs outdoor situations.

Uplink vs downlink scenario: Due to the limited dimensions of the mobile terminals, the complexity of processing has to be put in the BS, not in the MS. This trend seems to remain the same even in the future, since the economy of scale may allow more complex BS, thus the BS complexity may be the only trade space for achieving the requirements of next generation wireless systems. Considering the complexity issues, for Uplink (UL) (transmission from MS to BS), the preferable choice is receive diversity, while for DL (transmission from BS to MS) is transmit diversity. As shown in [30] both kinds of diversity can be combined, to achieve full diversity gain.

Outdoor vs indoor scenario: In a multi-path channel, each ray arriving to an antenna element at the receiver has a specific angle of incidence of the plane wave, which is referred to as DoA. Equivalently, each ray departing from a transmitter antenna has a Direction of Departure (DoD) [178]. The angle spread is a measure referring to the spread in DoA at the receive antenna arrays or the spread in DoD at the transmit antenna arrays. In an outdoor environment, the angle spread of the signals received by the MS is much lower than that of an indoor environment. This is mainly caused by the longer distance that separates BS and MS in outdoor situation, hence narrowing the angles of the multipaths of the received or transmitted signals. As a result, it is difficult to get the most out of angle diversity in outdoor scenario. Since BF performs better when desired signal arrives or departs from a single or narrow direction (i.e. having high spatial correlation) [172], this could be an interesting option to follow. On the other hand, in indoor channels, due to a broader angle spread, the spatial fading is uncorrelated, and therefore spatial diversity can be effectively exploited. Furthermore, BF may not be suitable for indoor scenario, since there is no narrow beam to increase SNR and separate from neighboring interferers [172].

Another important issue to diversity is the time variance of the channel. In order to implement Space-Time Block Code (STBC), the channel has to remain static over several symbol durations, depending on the code length, e.g. two symbols for Alamouti scheme. In addition, to achieve array gain in transmit diversity, it is required that CSI is available at the transmitter. In indoor environment, the channel does not change so fast as in outdoor scenario, hence it is easier to implement such schemes. For instance, in IEEE 802.11a a bandwidth of 20 MHz corresponds a symbol period of $4 \mu s$, while in Orthogonal Frequency Division Multiple Access-Fast Sub-Carrier Hopping (OFDMA-FSCH) (IEEE 802.16a Working Group) the symbol duration is 0.1 ms [18]. In WLAN environment with pedestrian mobility (i.e. 3 kmph velocity), the channel coherence time is 63.5 ms for 2.4 GHz carrier frequency [18, Appendix-A]. Thus, in ideal case, for $\frac{63.5 \text{ ms}}{4 \mu s} = 15,875$ OFDM symbols, we need to transmit the CSI only once from MS to BS. This is reasonably low signaling overhead compared to other cases when the user mobility is very high. For fast-variant channel (e.g. outdoor environment) or when CSI feedback to transmitter is costly, receive diversity might be a better choice.

Special care needs to be taken into account in case of indoor scenario: it provides very little delay spread and this means the available frequency diversity may be very small, thus the system may suffer from flat fading situation (or very little frequency selective situation). This situation is worsened by the fact that indoor channels may experience very little fading in time domain also. This, in turns, means that when a terminal enters in flat fading scenario, it will remain there for some time. At this stage, the frequency diversity exploitation via channel coding and interleaving may not be enough. Thus, some intelligent techniques are required to combat this situation. Rahman *et al.* [179] propose Cyclic Delay Diversity (CDD) as a probable solution for this.

Considerations with spatial multiplexing: Depending on the final goal, different techniques have to be adopted. If the goal is to improve the SNR, this can be obtained by the increase of the array gain, provided, e.g. by BF. If the objective is to achieve QoS, an improvement in link reliability is needed, obtained by the increase of diversity gain, and the right choice is employing diversity techniques. If the goal is the increase of throughput, the multiplexing gain has to be maximized, through SM techniques [180].

Indoor vs outdoor: If $T_s << T_c$, such as in indoor case, the CSI is available, so the non blind SM techniques are more suitable (e.g. MIMO with Singular Value Decomposition (SVD) or Adaptive Modulation and Coding (AMC)) than the blind ones. If T_s is comparable to T_c, as in the outdoor case (high mobility of the users) the feedback is expensive or impossible, and in this case it is better to choose the blind SM techniques, e.g. Bell labs LAyered Space Time (BLAST).

In indoor environment, there is a rich scattering (large Angle Spread (AS)) such that the antennas are uncorrelated. This implies an high rank of $\overline{\mathbf{H}}$ (where $\overline{\mathbf{H}}$ is the equivalent channel matrix between the MIMO transmitter and receiver of size $Q \times P$, with P and Q are the number of transmit and receive antennas, respectively) and since the SM performance depends on the rank, this case is suitable for high data rate applications. In outdoor the scattering is poor (small AS), so the antennas are correlated and this leads to a low rank of $\overline{\mathbf{H}}$: in this case the performance of SM is poor. In [181], it is shown that the presence of transmit correlation due to lack of scattering and/or insufficient transmit antenna spacing can have a detrimental impact on multi-antenna signaling techniques. In fact in case of antenna correlation, the diversity and multiplexing gains vanish, preserving only the receive array gain [182]. In general, antenna spacings of several wavelengths are required to achieve sufficient de-correlation. Antenna correlation causes rank loss, but the converse is not true.

SM is good at short distance from the BS (high SNR region), but the drawback here is the LOS situation (Ricean fading) that leads to rank decrease. Diversity techniques, such as STBC, perform better at large distance from the BS (low SNR). In Figure 9.10 this situation is visualized, where also a comparison with the SISO case is shown.

Polarized Antennas: Recently, the use of dual-polarized antennas has been proposed for SM as a cost-effective way of realizing significant multiplexing

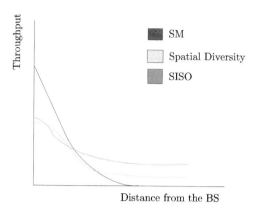

Fig. 9.10 Qualitative representation of throughput versus distance from BS for SISO, SM, and spatial diversity.

gain, where two spatially separated antennas are replaced by a single antenna structure with orthogonally polarized elements [139, 183]. Antennas with different polarizations at transmitter and receiver lead to a power imbalance between the elements of $\overline{\mathbf{H}}$, i.e. the power is not transmitted anymore all over the parallel pipes (the higher the power imbalance the higher the Cross Polarization Discrimination (XPD)). Low XPD hurts the information rate at low SNR and enhances rate at high SNR [139].

Multi-user: SM can also be applied in a scenario where U users are present [139]. In the UL case, U users transmit each with a single antenna, and the transmitted signals arrive at a BS equipped with U antennas. The BS can separate the incoming signals to support simultaneous use of the channel by the users. This technique can also be termed as Space Division Multiple Access (SDMA). In the DL case, the BS can transmit U signals with spatial filtering so that each user can decode his own signal. This technique can also be called beamforming in multi-user context. In conclusion, a SM-Multi User (MU) system can achieve a capacity increase proportional to the number of antennas at the BS and the number of users.

10

Transmit Diversity Vs Beamforming

In this chapter, we study the impact of spatial diversity and array gain obtained from transmit diversity and beamforming, respectively, in the resultant system performance for indoor and outdoor cellular systems. Following a brief overview of the multi-antenna techniques considered in this study, to be found in Section 10.2, we analyze the impact of the two above multi-antenna gain components in terms of downlink capacity and error probability in Section 10.3. The way to employ DL-BF in multi-user OFDM systems is explained in Section 10.4. Performance analysis and comparison of the techniques are explained in Section 10.5. Finally conclusions are drawn in Section 10.6.

10.1 Introduction

Wireless systems which operate at high data rate, providing higher multi-user capabilities, are always impaired by harsh wireless channel. Multi-antenna techniques can be used to overcome these unwanted situations, e.g. diversity techniques can be used to obtain reliable transmission systems or beamforming can be used to increase the signal strength towards a particular user, thus reducing interference to others. Traditionally spatial diversity is exploited involving multiple antennas in transmitter (Transmit Diversity) and/or receiver (Receive Diversity). Transmit diversity is a lucrative and reasonable choice for Downlink (DL), i.e. Base Station (BS)-to-Mobile Station (MS), especially for portable receivers where current drain and physical size are important constraints.

Orthogonal Frequency Division Multiplexing (OFDM) itself does not pose any built-in diversity, thus it is necessary to install some forms of diversity in an OFDM system for the purpose of achieving higher link quality and link availability without using any extra bandwidth. For example, channel coding and interleaving are used in IEEE 802.11a to obtain frequency diversity. Contrary to this, BF techniques can be used to achieve similar performance.

In this chapter, we compare the usability of transmit diversity and BF for DL of OFDM based cellular systems at indoor (micro and picocells) and outdoor urban macrocell scenarios. As seen in subsequent analysis in this chapter, the trade-off between transmit diversity and Transmit Beamforming (TxBF) is actually a trade-off between diversity gain and array gain. *Diversity Gain* is related to shape of the probability density function (pdf) of instantaneous Signal-to-Noise Ratio (SNR), while *Array Gain* is related to the improvement in the average SNR. In both approaches the configuration used is Multiple Input Single Output (MISO), since it is supposed that multiple antennas are employed at the BS transmitter and a single antenna at the MS receiver. The target is to define the conditions in which one of the two techniques is preferred to the other for DL cellular systems. We compare the schemes based on BER performance, on various angular spread values and corresponding channel correlation status.

In any OFDM-type multi-carrier system, DL beamforming can be implemented either before IDFT (frequency domain) or after IDFT (time domain) module in a BS transmitter. We denote the former scheme as Pre-IDFT DL beamforming, and the latter as Post-IDFT DL beamforming. In this chapter, we have compared Pre-IDFT DL beamforming with Space–Time Block coded and Space–Frequency block coded transmit diversity schemes for 4×1 downlink OFDM systems. The study is performed for indoor microcells and picocells and urban macro cells. The studies can also be adapted to multi-users scenarios, e.g. Orthogonal Frequency Division Multiplexing–Time Division Multiple Access (OFDM–TDMA) and Orthogonal Frequency Division Multiple Access (OFDMA) systems. Regardless of the multiple access scheme, it is found that beamforming always performs better in outdoor environment, where angular spread is lower, thus spatial correlation is higher. Similarly, indoor environment (high angular spread and low spatial correlation) suggests that transmit diversity schemes performs better than beamforming strategies.

10.2 A Brief Look at Diversity and Beamforming

10.2.1 Beamforming

In a BF system, the weights to be multiplied with the signals have to be carefully chosen. From antenna theory, an array with P antenna elements has different excitation currents according to the angle of direction of the waves arriving or departing from each element of the array. Considering linear phase progression, the weights have a phase that increases the same amount from one element to the next. For Uniform Spaced Linear Arrays (USLA), the array factor becomes:

$$\mathbf{F}(\theta) = \sum_{p=0}^{P-1} A_p e^{j2\pi (p-1)\frac{d}{\lambda} \sin\theta}, \quad (10.1)$$

where θ is the Direction of Arrival (DoA), d is inter-element distance, λ is the wavelength of the transmitted signal, and A_p is the instantaneous signal amplitude at pth transmit antenna element [139]. This assumption is termed as *Narrow-band Antenna Array* condition, where neighboring antenna elements see same signal amplitude and only a progressive rotations in corresponding signal phases.

Usually, BF is treated in literature as a method of increasing array gain when receiving signals from a specific direction. DoA is the angle of the wave arriving to the antenna. However, in this study it is intended to analyze the DL case, placing the beamformer at the transmitter side, i.e. at the BS, since it is more feasible to have multiple antennas at the BS than at the MS. This will bring additional complications if it is assumed that there is complete channel knowledge at the receiver, but not at the transmitter. In order to solve this problem, the DL-BF technique to be employed in this analysis will extract the weights to be used in DL from the weights calculated in UL. This method can be used in Time Division Duplex (TDD) systems without losing performance. However, in Frequency Division Duplex (FDD) systems, as a consequence of the frequency dependent steering array response and uncorrelated fading, the weights determined in UL cannot be directly used in DL, because the system performance will be degraded. So, for FDD systems, a proper frequency calibration technique is required, for determining the DL-BF weight vectors based on received signals in the UL.

A block diagram showing the implementation of DL-BF in an OFDM transmitter is presented in Figure 10.1(a). In order to concentrate most of the transmit energy toward the DoA, Minimum Mean Square Error (MMSE) criterion can be used for determining the BF weight criterion. There are a number of other BF algorithms, that can be used also. We consider a beamformer with weight vector, $\mathbf{w} = [w_1 \quad w_2 \quad \ldots \quad w_P]$, with $|\mathbf{w}| = 1$ (i.e. BF weights are chosen so that $\mathbf{w}$ is an unit vector) and P is the number of transmit antennas.

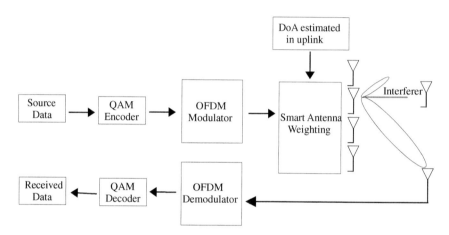

(a) OFDM system with beamforming operation.

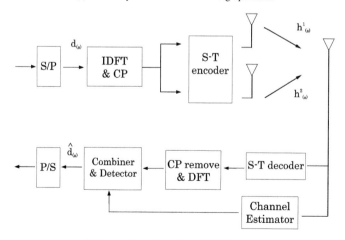

(b) Space–Time diversity in OFDM system.

Fig. 10.1 Beamforming and transmit diversity systems in OFDM system.

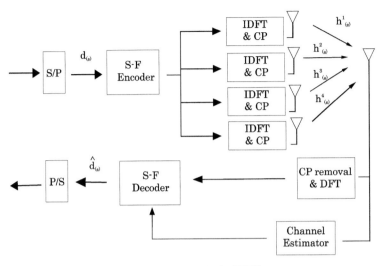

(c) Space–Frequency diversity in OFDM system.

Fig. 10.1 (Continued)

We denote the instantaneous SNR at the receiver after reception of the BF signals, γ as,

$$\gamma = \overline{\gamma}|\alpha|^2 = \frac{\overline{S}}{\sigma_n^2}\left|\sum_{p=1}^{P} w_p h_p\right|^2, \tag{10.2}$$

where $\overline{\gamma} = \frac{\overline{S}}{\sigma_n^2}$ is the average SNR with $\overline{S}$ as the average transmit power and σ_n^2 is the variance of AWGN. $\alpha = \sum_{p=1}^{P} w_p h_p$ is a Zero Mean Circularly Symmetric Complex Gaussian (ZMCSCG) random variable with a variance g of,

$$g = \mathbf{w}^H \mathbf{R} \mathbf{w}, \tag{10.3}$$

where $\mathbf{w}^H$ is the conjugate transpose of $\mathbf{w}$, and $\mathbf{R}$ is the covariance matrix of the channel transfer vector, $\mathbf{h} = [h_1 \quad h_2 \quad \dots \quad h_P]$. All these statements are true when the channel experienced between the transmit antenna elements and the receiver is frequency flat. Thus, we should be able to use this analysis in sub-carrier basis of an OFDM system. Depending on the fading correlation across different transmit antenna elements, we can find two extreme cases:

Independent fading: When independent fading across the transmit antennas is observed, then $\mathbf{R} = \mathcal{E}\{\mathbf{h}\mathbf{h}^H\} = \mathbf{I}$ and $g = 1$. This means that the transmit BF array provides no gain compared to any single antenna transmitter. So, $\mathcal{E}\{\gamma\} = \overline{\gamma}$.

Correlated fading: When the transmit antennas are completely correlated, we have $\mathbf{R} = \mathcal{E}\left\{\mathbf{h}\mathbf{h}^H\right\} \neq \mathbf{I}$. Indeed, $\mathbf{R} = [\mathbf{1}]_P$, where $[\mathbf{1}]_P$ is a $P \times P$ square matrix with all unity entry. So, $g = \mathbf{w}^H[\mathbf{1}]_P\mathbf{w} = P$. In this case, we can write that [184]:

$$\gamma = P\frac{\overline{S}}{\sigma_n^2}|\varsigma|^2, \tag{10.4}$$

where ς is zero mean and unit variance complex Gaussian random variable, i.e. γ has a chi-square distribution, i.e. χ_2^2 distribution. This also means that $\mathcal{E}\{\gamma\} = P\overline{\gamma}$. Thus, the pdf of the received SNR for Beamforming (BF) system at full correlation across the transmit antennas can be written as

$$\mathfrak{f}(\gamma) = \frac{1}{P\overline{\gamma}}\exp\left(-\frac{\gamma}{P\overline{\gamma}}\right). \tag{10.5}$$

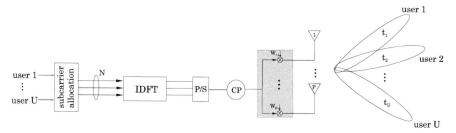

(a) Time-domain beamforming approach.

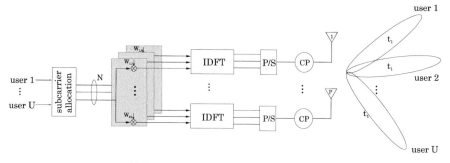

(b) Frequency-domain beamforming approach.

Fig. 10.3 Time and frequency-domain beamforming approach on an OFDMA system.

Looking at the above expressions, we can see that BF obtains complete array gain when the antennas are completely correlated, while spatial diversity techniques require independent fading across different spatial branches for obtaining complete diversity gain. For a better comparison of BF and SISO systems, we can envision that BF and SISO would require $\frac{\overline{S}}{\sigma_n^2}$ and $\frac{\overline{S}}{\sigma_n^2} + 10\log_{10} P$ amount of power, respectively, to obtain the same SNR level at the receiver.

10.2.2 Space–Time Block Coding (STBC)

Space–Time Block Coding (STBC) consists of applying STC on blocks of data symbols instead of individual symbols. Its implementation in an OFDM system is shown in Figure 10.1(b). A special case of STC is Alamouti's STBC [30], which is actually a full diversity and full rate STBC scheme for 2×1 system. In this process, where two transmitter antennas and one receiver antenna are used, two symbols are being transmitted through two transmitting antennas at every symbol duration. Tarokh *et al.* [175] proved that full rate and full diversity is only achievable for 2×1 transmit diversity system, when only linear receiver is considered. They proposed partial rate and full diversity codes for 4×1 space–time transmit diversity system. One such example of code is described here in Table 10.1.

By combining the received eight symbols and applying Maximum Likelihood (ML) detector, the four transmitted symbols can be obtained [175]. The abstract transceiver structure of this scheme in an OFDM system is shown in Figure 10.1(b).

Table 10.1 Tarokh's $\frac{1}{2}$ rate space–time block encoding scheme; s_n denotes the transmitted symbol at nth OFDM symbol duration.

OFDM Symbol	Ant 1	Ant 2	Ant 3	Ant 4
T_n	s_n	s_{n+1}	s_{n+2}	s_{n+3}
$T_n + 1$	$-s_{n+1}$	s_n	$-s_{n+3}$	s_{n+2}
$T_n + 2$	$-s_{n+2}$	s_{n+3}	s_n	$-s_{n+1}$
$T_n + 3$	$-s_{n+3}$	$-s_{n+2}$	s_{n+1}	s_n
$T_n + 4$	s_n^*	s_{n+1}^*	s_{n+2}^*	s_{n+3}^*
$T_n + 5$	$-s_{n+1}^*$	s_n^*	$-s_{n+3}^*$	s_{n+2}^*
$T_n + 6$	$-s_{n+2}^*$	s_{n+3}^*	s_n^*	$-s_{n+1}^*$
$T_n + 7$	$-s_{n+3}^*$	$-s_{n+2}^*$	s_{n+1}^*	s_n^*

Using the notations from the earlier sections, we can write for any open loop transmit diversity scheme that:

$$\gamma = \frac{\overline{S}\sum_{p=1}^{P}|h_p|^2}{P\sigma_n^2},\tag{10.6}$$

where $\overline{S}$ is the total transmitted from one sub-carrier and divided equally across all P transmit antennas. $\sum_{p=1}^{P}|h_p|^2$ has a χ_{2P}^2 distribution, thus, we can write the pdf of SNR for open loop transmit diversity scheme as

$$f(\gamma) = \frac{\gamma^{P-1}}{(P-1)!\left(\frac{\overline{\gamma}}{P}\right)^P}\exp\left(-\frac{P\gamma}{\overline{\gamma}}\right).\tag{10.7}$$

We can write that:

$$\mathcal{E}\{\gamma\} = \frac{\overline{S}}{\sigma_n^2} = \overline{\gamma}.\tag{10.8}$$

This is the case when independent fading is assumed across the antenna elements, thus, $\mathcal{E}\left\{\sum_{p=1}^{P}|h_p|^2\right\} = P$. So, it is clear that the independent spatial branches are a requirement for obtaining the transmit diversity benefits. Although there is no difference in the average SNR when a SISO and open-loop MISO is compared, the probability distribution of SNR for MISO system is very different corresponding to the number of transmit antennas available at the system. *Channel tightening* impact is seen when the number of transmit branches is increased. As the order of the distribution $2P$ increases, the pdf is concentrated more and more around the average SNR, $\overline{\gamma}$; thus the change in the shape of the SNR distribution provides us the diversity gain.

Space–frequency block coding (SFBC): STBC algorithms can be implemented in OFDM systems with a condition that the channel is time-invariant for the duration of the orthogonal transmission block (i.e. coherence time, $T_c \geq \frac{PT_s}{R_{td}}$, where T_s is total OFDM symbol duration including the cyclic prefix) and R_{td} is the STBC code rate. This can be a bottleneck for highly time-variant mobile environment. Besides, effectively the latency that the receiver incurs because of STBC algorithm is essentially equal to $\frac{PT_s}{R_{td}}$. This can also be unsatisfactory for some particular real-time services.

An alternative to Space–Time Block Coded OFDM (STBC-OFDM) system in this regard is Space–Frequency Block Coded Orthogonal Frequency

Division Multiplexing (SFBC-OFDM) system. In SFBC-OFDM system, the orthogonal code blocks are designed across OFDM sub-carriers. Thus, as long as neighboring sub-carriers have similar frequency response, there will be no problems with SFBC algorithms. It is worth noting here that SFBC-OFDM systems will suffer in performance in situations where $\Delta f \leq B_c \leq \frac{P\Delta f}{R_{td}}$. Here, Δf is the sub-carrier spacing of the OFDM systems. For cases when $\frac{P}{R_{td}}$ is large, special care needs to be taken in SFBC system design, so that degradation due to change of CSI inside the coding span in frequency does not occur. Moreover, the orthogonal coding is not done across OFDM symbols, rather in one OFDM symbol, so as long as the channel remains static for one OFDM symbol duration (i.e. T_s), the coding should provide optimum diversity gain to the system [176]. Its implementation in OFDM system is shown in Figure 10.1(c). The orthogonal space–frequency blocks can be defined in a similar fashion like Table 10.1. All the considerations related to array gain and diversity gain done in case of STBC also hold for SFBC system.

10.2.3 Receive Diversity System

As a comparison, we also introduce the receive diversity scheme, namely Single Input Multiple Output (SIMO) systems, in this section. In a SIMO system, the signal is transmitted from a single transmit antenna and received at multiple antennas. After that, they are combined using Maximal Ratio Combining (MRC). So, the instantaneous SNR can be written as

$$\gamma = \overline{\gamma} \sum_{q=1}^{Q} \left| h_q \right|^2 = \frac{\overline{S}}{\sigma_n^2} \sum_{q=1}^{Q} \left| h_q \right|^2, \tag{10.9}$$

where Q is the number of receive antennas. The pdf of instantaneous SNR at the output of the MRC combiner can be written as

$$f(\gamma) = \frac{\gamma^{Q-1}}{(Q-1)! \overline{\gamma}^Q} \exp\left(-\frac{\gamma}{\overline{\gamma}}\right). \tag{10.10}$$

We can see from Equation (10.9) that the instantaneous SNR is increased by the number of spatial branches in MRC, which is the direct result of *array gain* obtained in MRC receiver. The average SNR after the receiver combining can be written as

$$\mathcal{E}\{\gamma\} = Q\overline{\gamma}. \tag{10.11}$$

Thus, array gain can be defined as:

$$\text{Array gain} = \frac{\text{Average SNR at the output of the linear combiner}}{\text{Average SNR at the input, i.e. at the transmitter}}$$

$$= \frac{\mathcal{E}\{\gamma\}}{\bar{\gamma}}. \tag{10.12}$$

10.2.4 MIMO Diversity System

A particular interest regarding the array gain and diversity gain is MIMO diversity system, with P number of transmit branches while the combiner has Q branches. When open-loop transmit (e.g. Alamouti's STBC) is used at the transmitter and MRC combining is done at the receiver, then the resultant SNR, γ and the pdf of γ at the output of the combiner can be written as

$$\gamma = \frac{\bar{\gamma}}{P} \sum_{p=1}^{P} \sum_{q=1}^{Q} |h_{p,q}|^2 \tag{10.13}$$

$$f(\gamma) = \frac{\gamma^{PQ-1}}{(PQ-1)! \left(\frac{\bar{\gamma}}{P}\right)^{PQ}} \exp\left(-\frac{P\gamma}{\bar{\gamma}}\right). \tag{10.14}$$

Channel tightening due to the transmit diversity scheme, together with receive diversity and array gains are seen in MIMO diversity schemes.

10.2.5 SNR Statistics of Diversity and Beamforming Systems

We present the pdf of instantaneous SNRs in Figure 10.3 for all the four schemes that are described in the above sections. For brevity, we have only shown the cases with 2 and 8 antennas for BF, transmit diversity and MRC schemes, and 2×4 and 4×2 transmit-schemes. This figure highlights different aspects of diversity and array gain that can be obtained by exploiting the available spatial degrees of freedom in different ways. Looking at the pdfs, we can summarize the following:

1. We obtain complete array gain in an ideal MRC receiver, while we obtain no array gain in open loop transmit diversity. That is the reason that 2×1 and 8×1 transmit diversity has almost similar average SNR, while 8×1 transmit diversity has more "concentrated" pdf, coming from channel tightening due to 8-order of diversity.

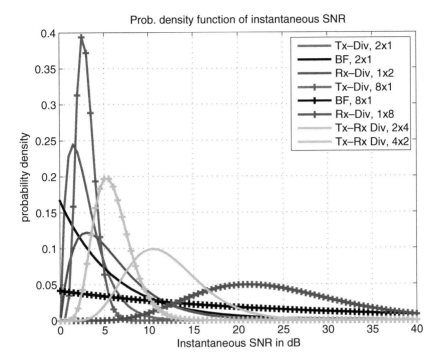

Fig. 10.3 Probability density function (pdf) of SNR for different combinations of diversity and beamforming systems.

2. Beamforming obtains full array gain when spatial branches are completely correlated. In this case, no diversity gain is obtained.
3. 4×2 and 2×4 transmit–receive diversity has same diversity order of 8. But, they have different array gain of 3 and 6 dB respectively. Thus, the pdf is right-shifted for 2×4 system. Similarly, 4×2 also experiences higher channel tightening compared to 2×4 system, thus, we see that pdf of 4×2 system is more concentrated compared to that of 2×4 system.

10.3 Downlink Capacity and Error Probability Analysis

10.3.1 Ergodic Capacity

For any specific system, when the received SNR can be characterized by γ, the well-known Shannon capacity formula for nonfading Gaussian channel can be used [44], as shown below:

$$C = \log_2(1 + \gamma). \tag{10.15}$$

The capacity in C is also referred as error-free deterministic spectral efficiency for any specific γ, or the data rate per unit bandwidth that can be sustained reliably over the link with instantaneous SNR of γ [139].

Since the wireless channel coefficients are random variables, so is the parameter γ, which means that the information rate associated with the capacity is also a random variable. That is why we use a term called *Ergodic Capacity* of the system for further analysis. The ergodic capacity, $\overline{C}$, of any wireless channel is the ensemble average of the deterministic capacity, C over the distribution of the elements of the random channel parameter. When we use asymptotically optimal transmit codebooks, then we can send the information at the channel at a rate of $\overline{C}$ with vanishing error, thus, ergodic capacity is equal to shannon capacity in this sense.

Based on some well-known works, we can write the ergodic capacity of receive diversity schemes as [147]:

$$\overline{C} = \mathcal{E}\{C\}$$
$$= \mathcal{E}\left\{ \log_2\left(1 + \overline{\gamma}\sum_{q=1}^{Q}|h_q|^2\right)\right\}$$
$$= \mathcal{E}\left\{ \log_2\left(1 + \frac{\overline{S}}{\sigma_n^2}\sum_{q=1}^{Q}|h_q|^2\right)\right\} \tag{10.16}$$

Similar to this, we can write the ergodic capacity of open loop transmit diversity scheme (i.e. no Channel State Information (CSI) present at the transmitter) as [142]:

$$\overline{C} = \mathcal{E}\{C\}$$
$$= \mathcal{E}\left\{ \log_2\left(1 + \frac{\overline{\gamma}}{P}\sum_{p=1}^{P}|h_p|^2\right)\right\}$$
$$= \mathcal{E}\left\{ \log_2\left(1 + \frac{\overline{S}}{P\sigma_n^2}\sum_{p=1}^{P}|h_p|^2\right)\right\}. \tag{10.17}$$

For beamforming, we can consider the system to be equivalent to a single antenna system with an increased average gain. So, the capacity of the beamformer can be written as [184]:

$$\overline{C} = \mathcal{E}\left\{\log_2(1 + P\overline{\gamma}|h|^2)\right\} = \mathcal{E}\left\{\log_2\left(1 + \frac{P\overline{S}}{\sigma_n^2}|\varsigma|^2\right)\right\}. \quad (10.18)$$

The notations in the above equations are same as the ones used before. Note that the capacity expression for BF case is valid when the transmit antenna elements are completely correlated, while they are assumed to completely independent for both the diversity methods. Figure 10.4 shows us the ergodic capacity of all three schemes when 2, 4, and 8 antennas are used either at transmitter, or at receiver. The ergodic capacity is an average parameter, thus, only the impact of array gain can be seen here. Receive diversity offers both diversity and array gain, thus, it always achieves higher capacity than other

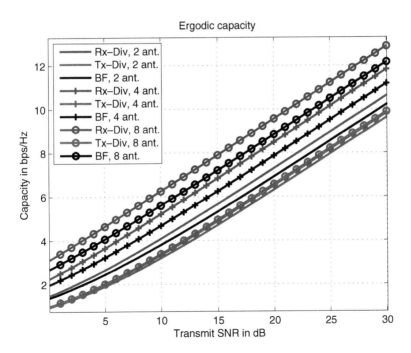

Fig. 10.4 Ergodic channel capacity for transmit diversity, receive diversity, and beamforming techniques with 2, 4, and 8 antennas.

two techniques for the same antenna configuration. Beamforming performs better than transmit diversity, because it achieves array gain.

10.3.2 Outage Capacity

One problem with Shannon capacity parameters is that, it assumes very ideal scenario, thus, it only provides an upper bound for any practical system. The underlying assumptions are, infinite block length, vanishing error probability, etc. Thus, outage capacity is used in several literature for determining a more realistic capacity parameter. Outage capacity can be defined as the capacity with outage probability p, so that the random capacity C always resides below the outage capacity with a probability of p.

In this section, we only concentrate on transmit diversity and beamforming for our studies, as these two techniques are the focus of our analysis. For receive diversity and transmit–receive diversity, the outage capacity expression can be derived in the similar fashion.

If we define, the pdf of C as $\mathfrak{f}(C)$, then outage capacity, $C_{\text{out}}(\mathfrak{p}_c)$ is related to corresponding outage probability $\mathfrak{p}_c$ as:

$$\mathfrak{p}_c = \int_{-\infty}^{C_{\text{out}}(\mathfrak{p}_c)} \mathfrak{f}(C)\, dC. \tag{10.19}$$

For brevity, we denote the following:

Tx-Div: $\quad C = \log_2\left(1 + \frac{\overline{\gamma}}{P}\phi\right); \quad$ where $\phi = \sum_{p=1}^{P} |h_p|^2$ (10.20)

Tx-BF: $\quad C = \log_2\left(1 + P\overline{\gamma}\psi\right); \quad$ where $\psi = |\varsigma|^2$ (10.21)

As mentioned earlier, the pdf of ϕ and ψ are known as χ^2_{2P} and χ^2_2, respectively.

Case study: transmit diversity: Using standard random variable transform formula as in [184], we can write the following transmit diversity:

$$\mathfrak{f}(C) = \frac{\chi^2_{2P}(\phi)}{\frac{dC}{d\phi}}, \tag{10.22}$$

where

$$\frac{dC}{d\phi} = \frac{1}{\left(1 + \frac{\overline{\gamma}}{P}\phi\right)\ln 2}\frac{\overline{\gamma}}{P} = \frac{\overline{\gamma}}{2^C P \ln 2} \tag{10.23}$$

using $\phi = (2^{\mathcal{C}} - 1)\frac{P}{\gamma}$. So, pdf of $\mathcal{C}$ can be written as

$$
\begin{aligned}
\mathfrak{f}(\mathcal{C}) &= \frac{2^{\mathcal{C}} P \ln 2}{\overline{\gamma}} \chi_{2P}^2 \left((2^{\mathcal{C}} - 1)\frac{P}{\overline{\gamma}} \right) \\
&= \frac{2^{\mathcal{C}} P \ln 2}{\overline{\gamma}} \frac{1}{(P - 1)!} \left[(2^{\mathcal{C}} - 1)\frac{P}{\overline{\gamma}} \right]^{P-1} \exp\left(-\frac{(2^{\mathcal{C}} - 1)P}{\overline{\gamma}} \right) \\
&= \frac{2^{\mathcal{C}} \ln 2}{(P - 1)!} \left(\frac{P}{\overline{\gamma}} \right)^P (2^{\mathcal{C}} - 1)^{P-1} \exp\left(-\frac{(2^{\mathcal{C}} - 1)P^2}{\overline{\gamma}} \right). \quad (10.24)
\end{aligned}
$$

The $\mathfrak{p}_c$ can actually be calculated from the CDF of the capacity. So, from here, we can find the $\mathfrak{p}_c$ by using a known result from [44, Section 2.1.4]:

$$
\mathfrak{p}_c = e^{\left(\frac{P}{\overline{\gamma}}\right)} \left[\sum_{p=0}^{\lfloor \frac{P}{2} \rfloor - 1} \frac{1}{p!} \left(\left(-\frac{P}{\overline{\gamma}} \right)^p - \left(2^{\mathcal{C}_{out}} - 1 \right)\frac{P}{\overline{\gamma}} \right)^p e^{(1 - 2^{\mathcal{C}_{out}})} \right]. \quad (10.25)
$$

Case study: beamforming: Using the same procedure, we can define the pdf of $\mathcal{C}$ for BF case as follows:

$$
\mathfrak{f}(\mathcal{C}) = \frac{\chi_2^2(\psi)}{\frac{d\mathcal{C}}{d\psi}} \quad \text{with} \quad \frac{d\mathcal{C}}{d\psi} = \frac{P\overline{\gamma}}{2^{\mathcal{C}} \ln 2}. \quad (10.26)
$$

Thus, for beamforming systems,

$$
\begin{aligned}
\mathfrak{f}(\mathcal{C}) &= \frac{2^{\mathcal{C}} \ln 2}{P\overline{\gamma}} \chi_2^2 \left(\frac{2^{\mathcal{C}} - 1}{P\overline{\gamma}} \right) \\
&= \frac{2^{\mathcal{C}} \ln 2}{P\overline{\gamma}} \exp\left(-\frac{2^{\mathcal{C}} - 1}{P\overline{\gamma}} \right). \quad (10.27)
\end{aligned}
$$

From here, we can calculate the outage probability corresponding to any outage capacity values as follows:

$$
\begin{aligned}
\mathfrak{p}_c &= \int_{-\infty}^{\mathcal{C}_{out}} \frac{2^{\mathcal{C}} \ln 2}{P\overline{\gamma}} \exp\left(\frac{2^{\mathcal{C}} - 1}{P\overline{\gamma}} \right) d\mathcal{C} \\
&= \exp\left(-\frac{1}{P\overline{\gamma}} \right) - \exp\left(-\frac{2^{\mathcal{C}_{out}} - 1}{P\overline{\gamma}} \right). \quad (10.28)
\end{aligned}
$$

Using Equations (10.25) and (10.28), we produce the outage capacity for specific outage probability and outage probability for specific outage capacity in Figures 10.5(a) and 10.5(b), respectively. In terms of outage capacity, we can see that the outage capacity is almost two times higher than the BF scheme for all antenna configurations. Note that this result is valid for independent fading for transmit diversity and fully correlated fading for beamforming. Thus, favorable conditions for both the schemes. For a fixed outage capacity of 3 bps/Hz, we obtain the outage probability in Figure 10.5(b). At higher outage probability, the array gain is more useful compared to diversity gain, thus, BF performs better. As the outage probability decreases, the advantage of BF diminishes and transmit diversity schemes starts to perform better. For example, at an outage probability of 1%, the 8 × 1 transmit diversity outperforms 8 × 1 BF system by around 7 dB.

10.3.3 Error Probability

In this section, we derive the average bit error probability expressions for transmit diversity and beamforming. The average BER can be computed for any system using the the following expression:

$$\bar{\mathbf{P}}_b = \int_0^\infty \mathbf{p}_b \mathfrak{f}(\gamma) d\gamma, \tag{10.29}$$

where $\mathbf{p}_b$ is the instantaneous BER for any M-QAM system in AWGN channel. $\mathbf{p}_b$ for M-QAM system can be written as [44]:

$$\mathbf{p}_b \approx 4\left(1 - \frac{1}{\sqrt{(M)}}\right) \mathbb{Q}\left(\sqrt{\frac{3\bar{\gamma}}{M-1}}\right), \tag{10.30}$$

where $\mathbb{Q}(x) = \frac{2}{\sqrt{\pi}} \int_x^\infty \exp\left(-t^2\right) dt$ is called complementary error function. It is difficult to obtain closed-form expressions when $\mathbb{Q}$-function is involved in calculation, so we take an approximation of $\mathbf{p}_b$ from [25]. Here it is shown that the instantaneous BER of an M-QAM system, with coherent detection and gray coding over Additive White Gaussian Noise (AWGN) channel, can be approximated within 1 dB tight and for BER $\leq 10^{-3}$ by, $\mathbf{p}_b \approx 0.2 \exp\left(-\frac{1.6\gamma}{M-1}\right)$.

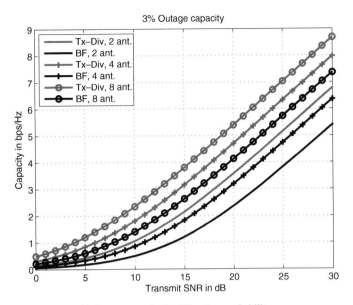

(a) Outage capacity for 3% outage probability.

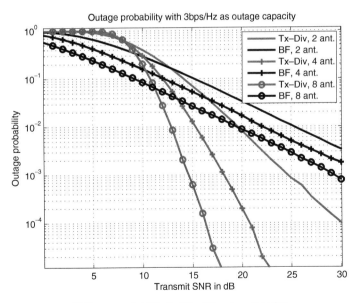

(b) Outage probability for 3 bps/Hz outage capacity.

Fig. 10.5 Outage capacity and outage probability for transmit diversity and beamforming with varying number of transmit antennas.

For a transmit diversity system, we can write the average BER as

$$\bar{\mathbf{P}}_b = \frac{0.2}{(P-1)!\left(\frac{\bar{\gamma}}{P}\right)^P} \int_0^\infty \gamma^{P-1} \exp\left[-\left(\frac{1.6}{M-1} + \frac{P}{\bar{\gamma}}\right)\gamma\right] d\gamma$$

$$= 0.2 \left(\frac{1}{1 + \frac{1.6\bar{\gamma}}{P(M-1)}}\right)^P \tag{10.31}$$

using the following identity: $\int_0^\infty x^k \exp(-\xi x) dx = \frac{k!}{\xi^{k+1}}$.

Similarly, we can obtain the average BER expression for beamforming as

$$\bar{\mathbf{P}}_b = \frac{0.2}{P\bar{\gamma}} \int_0^\infty \exp\left[-\left(\frac{1.6}{M-1} + \frac{1}{P\bar{\gamma}}\right)\gamma\right] d\gamma$$

$$= \frac{0.2}{1 + \frac{1.6P\bar{\gamma}}{(M-1)}}. \tag{10.32}$$

The average Bit Error Rates (BERs) of transmit diversity and beamforming systems are shown in Figure 10.6, by using Equations (10.31) and (10.32), respectively. The BERs for different number of transmit antennas are presented. Similar to the case of outage probability for 3 bps/Hz outage capacity

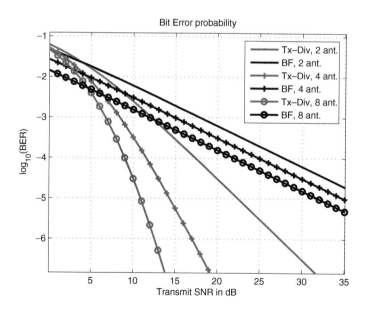

Fig. 10.6 Theoretical average bit error probabilities for transmit diversity and beamforming systems.

in Figure 10.5(b), transmit diversity obtains the diversity gain, while beam-forming only shows the array gain benefits. So, roughly for BER of 10^{-2}, the diversity gain is more dominant and transmit diversity performs better than beamforming. Below that range, the array gain is more important and beamforming outperforms transmit diversity systems.

10.4 Downlink Beamforming and Transmit Diversity in Multi-user OFDM Systems

10.4.1 Issues in OFDM–TDMA

In an OFDM–TDMA system, one complete OFDM symbol at a certain time is allocated to a user [185]. If there are U users in the system and if all users require same bit rate services, we can assign every $(lU + 1)$th OFDM symbol to Uth user ($l = 0, 1, \ldots$). The number of OFDM symbols per TDMA frame can be varied according to each user requests. Its main drawback is the increase of the latency when the number of users gets higher in the cell.

Keeping in mind that STBC algorithm requires more than one consecutive OFDM symbols for orthogonal transmission (to be exact, P number of consecutive OFDM symbols for $P \times 1$ STBC-OFDM-TDMA system), it is clear that STBC orthogonal blocks need to be implemented with data from different users, when TDMA slots consists of less than P consecutive OFDM symbols. This will mean that though an OFDM symbol slot is not allocated to any user, that particular user has to receive the data in that slot to decode STBC blocks. This is not desired, so we propose to use Space–Frequency Block Code (SFBC) algorithms in OFDM–TDMA systems for transmit diversity purposes. In this case, only one user at a time is involved in the transmission process and so the benefits of OFDM–TDMA are preserved.

It is fairly easy to implement beamforming algorithms in OFDM–TDMA systems. Because the data carried by one whole OFDM symbol is intended for one single user, it is possible to direct the transmitted signal at that symbol duration to the direction of respective user. So, the goals of beamforming are preserved, i.e. only the involved user's data is directed to concerned user and interference to other users is reduced. If the beamformer is placed before the IDFT, the problem is that the number of IDFTs must be equal to the number of transmitting antennas P, instead of only one as if the beamformer was placed after the IDFT. It is therefore better to allocate the beamformer after the IDFT, in fact in this case the complexity of the system is lower (one IDFT is needed instead of P).

10.4.2 DL-BF in OFDMA

In OFDMA system, users are separated across sub-carriers. BS allocates the available sub-carriers to users while transmitting in the DL. After that IDFT modulation is performed on frequency domain data. So, if the sub-carrier allocation is kept static for at least μP number of OFDM symbols (or even if it is dynamic allocation via sub-carrier hopping for each OFDM symbol), then the creation and transmission of orthogonal transmit block is done simply by implementing the algorithm across OFDM symbols. Here, we denote transmit diversity coding rate as μ. As usual, it will not change any OFDMA properties, only that it will increase the latency of detection process (which is typical to any Space-Time Block Code (STBC) algorithm). Similar to this, SFBC can also be used in OFDMA system, satisfying the condition that the users are allocated contiguous sub-carriers and the number of sub-carriers assigned to one user, $\eta = l\mu P$, where $l \in [1, \ldots, \lfloor \frac{N}{\mu P} \rfloor - 1]$.

After IDFT modulation, time-domain sampled data are transmitted toward the user. It is obvious that based on the CSI of each user, we can apply the BF weights either before IDFT modulation, or after IDFT modulation. These are called Post-IDFT Beamforming and Pre-IDFT Beamforming, respectively.

Post-IDFT downlink beamforming: A block diagram of an implementation of Post-IDFT BF technique in an OFDMA system is shown in Figure 10.2(a). After the IDFT operation in the transmitter, the signal is indeed wideband time-domain signal, so any wideband BF technique can be used. A multi-tap beamformer can be used for this purpose. The wideband OFDM signal is frequency selective, thus the weight vectors are required for each user which can be used to steer the signal toward the user. The BF problem can be expressed as [186],

$$\frac{\mathbf{w}_u^H \mathbf{R}_u \mathbf{w}_u}{\sum_{l=1;l\neq u}^{U} \mathbf{w}_l^H \mathbf{R}_u \mathbf{w}_l + \sigma_u^2} \geq \gamma_u, \quad u = 1, 2, \ldots, U. \tag{10.33}$$

Here, $\mathbf{R}$ denotes the covariance matrix as it is shown in Section 10.5. γ_u, $\mathbf{w}_u$, and σ_u^2 denote the required SNR, beamformer tap weights and additive white gaussian noise variance of uth user, respectively. Solving this problem with algorithm proposed by Schubert and Boche [186] will give us the beamformer weights $\mathbf{w}_u$ for all users.

For this purpose, the channel needs to be estimated in time domain. Once the beamformer weights are decided, then the received signal at the MS, $\tilde{y}_s$, can be written as

$$\tilde{y}_s = \mathbf{w}_u * \tilde{x}_s. \tag{10.34}$$

Pre-IDFT downlink beamforming: Before applying the IDFT, the BF can be implemented across OFDM sub-carriers, as shown in Figure 10.2(b). Sub-carriers poses narrowband frequencies and thus, any narrowband scheme can be used in this case. Similar to Equation (10.33), we can write,

$$\frac{w_{u,k}^{H}\mathbf{R}_{u,k}w_{u,k}}{\sum_{l=1;l\neq u}^{U} w_{l,k}^{H}\mathbf{R}_u w_{l,k} + \sigma_u^2} \geq \gamma_u^k,$$

$$u = 1, 2, \ldots, U \ \& \ k = 0, \ldots, N - 1. \tag{10.35}$$

Here, the number of sub-carriers is identified by N and γ_u^k defines required SNR for uth user's Nth sub-carrier.

10.4.3 DL-BF in Clustered OFDMA

Both the Pre- and Post-IDFT BF schemes clearly show that the user data is not separated before steering the beams to user's location, rather all the OFDM modulated signal is sent toward all users. In this way, it is made sure that the users receive the transmitted data with higher SNR, but they also receive other users' data at the same time. In a sense, it can be said that using these kind of BF schemes, it is not possible to obtain all the goals of BF.

In order to reduce the power inefficiency of BF in OFDMA, in the way that user individual highly directive beams are carrying data from all users in the cell, we take another approach which partially solve this issue. If using a number of IDFTs equal to the number of antennas, the BS can transmit through that same number of beams to groups of users. So, what we do is to create clusters of users who are co-located. This means grouping the users spatially, i.e. define clusters of users that are close to each other, each beam can be set to acquire a shape that covers the users in that set and protects them from inter-ference coming from other user sets in the cell. The process of grouping users into clusters, shall be undertaken by the BS using an algorithm that identifies DoA from the CSI and organizes the transmission radiation pattern according to an error minimization criterion like MMSE. This clustering beamforming approach is employed in our studies.

10.5 Performance Analysis and Comparison

10.5.1 Channel Model

Indoor (micro and picocells) and outdoor (macrocells) wireless channels for low and high user mobility are considered at 5 GHz carrier frequency. For indoor channel delay profiles, we use HiperLan/2 channel model A, corresponding to a typical office environment for Non Line Of Sight (NLOS) conditions and 50 ns average rms delay spread [187]. For outdoor delay profiles, we use typical urban 12-path channel model [188, Appendix-E].

10.5.2 Angular Spread and Spatial Correlation

Angular spread is a measure of the angular dispersiveness of DoAs. Angular spread is usually higher in indoor scenario, whereas it is smaller in urban macrocell scenario. Thus, indoor channel provides enough angular diversity so that transmit diversity schemes can be easily implemented. Angular spread can be measured as

$$AS^j = \sqrt{\frac{\sum_i \left(\phi_i^j - E[\phi^j]\right)^2 |A_i^j|^2}{\sum_i |A_i^j|^2}}, \tag{10.36}$$

where ϕ^j is the Angle of Arrival (AoA) of a certain mobile user j and A_i is the field amplitude, hence $|A_i|^2$ is the power of a signal coming from that scatterer i. In an indoor environment, the angular spread of the signals received by the MS is much higher than when the communication is established in an outdoor environment. This is mainly caused by the longer distance that separates BS and MS in outdoor situation, hence narrowing the angles of the received signals from multipath [189].

This characteristic makes it difficult to get the most out of angle diversity in outdoor channels, thus BF could be an interesting option to follow. On the other hand, in indoor channels, due to a broader angular spread, the multi-path signals have low correlation between them, therefore angle diversity can be effectively exploited. Furthermore, BF may not be suitable for indoor scenarios, since there is no narrow beam to increase SNR and separate from neighboring interferers.

The rays departing from BS are spread over a certain angular spread. The spatial covariance matrix is characterized on the geometry relevant to a certain Direction of Departure (DoD) profile which is represented by Equation (10.37) [190].

$$\mathbf{R} = \sum_{l=1}^{L} \mathbf{a}^H(\theta_l)\mathbf{a}(\theta_l) \tag{10.37}$$

$$\mathbf{a}(\theta_l) = [1, \exp(-j\alpha_l), ..., \exp(-j(P-1)\alpha_l)] \tag{10.38}$$

$$\alpha_l = \frac{2\pi}{\lambda} d \sin\theta_l, \tag{10.39}$$

where L is the number of rays emitted from the BS, $\mathbf{a}(\theta_l)$ is the directional vector dependent on azimuth direction θ_l of the lth ray and P is the number of transmit antennas.

The channel covariance matrix $\mathbf{R}$ can be decomposed into eigenvectors and eigenvalues; and the correlated channel vector for multiple transmit antennas and single receive antenna can now be written as [190]:

$$\mathbf{h}^R = \mathbf{Q}\Sigma^{1/2}\mathbf{h}, \tag{10.40}$$

where $\mathbf{h}$ and $\mathbf{h}^R$ are independent channel frequency response vector and correlated frequency response vector, respectively. $\mathbf{Q}$ stands for the matrix which is composed with eigenvectors and Σ contains eigenvalues in diagonal locations for covariance matrix.

Using the spatial correlation model described above, we investigate the impact of transmit antenna correlation on transmit diversity and beamforming schemes. We use a semi-analytical method to calculate the outage probability corresponding to 3 bps/Hz outage capacity. The channel coefficients are generated so that they are zero-mean unit-variance independent Gaussian random variables. The channel is taken to be un-correlated in time between two consecutive symbol durations. Then, using Equations (10.38) and (10.39), we can introduce the spatial correlation corresponding to different angle spread values. 10^7 number of channel capacity values are calculated using Equations (10.17)

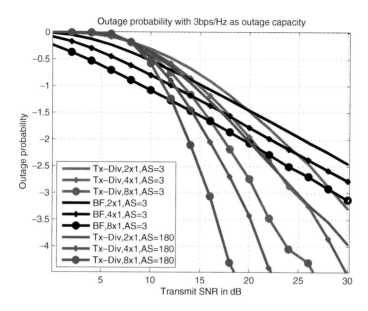

Fig. 10.7 Impact of antenna correlation corresponding to different angle spread on transmit diversity and beamforming.

and (10.18) for transmit diversity and beamforming, respectively. From these capacity values, we determine the outage probability for 3 bps/Hz indexoutage capacity, which is shown in Figure 10.7. This figure should be seen in collaboration with Figure 10.5(b), where independent fading for transmit diversity and complete correlation for BF is assumed. Comparing the result in Figure 10.7, we can see that for an angular spread of $3°$, the performance of transmit diversity suffers hugely. Only for outage probability less than 1%, the transmit diversity performs better compared to BF for 2×1 and 4×1 systems. When the angular spread is $180°$, then the diversity gains are more visible in transmit diversity schemes. In a nutshell, when spatial correlation is high due to very small angular spread, the beamforming scheme should perform very closely (or even better) at least for 2×1 and 4×1 systems.

10.5.3 Simulation Parameters

Indoor microcell and outdoor macrocell scenarios are considered for DL in our simulations. Simulations are performed for a single cell system to simplify

Table 10.2 Parameters for comparison between diversity and beamforming.

	Indoor	Outdoor
Max delay spread [189]	$\leq 0.5\,\mu s$	$5\,\mu s$
Max angular spread [189]	$360°$	$20°$
System bandwidth		20 MHz
Carrier frequency		5.0 GHz
OFDM sub-carriers, N	64	1024
Data sub-carriers	52	832
Null sub-carriers	12	192
Sub-carrier spacing, kHz	312.5	19.53125
CP length, N_g	16	200
OFDM Symbol Duration, $T_s = T_u + T_g$, μs	4	61.2
Useful data period, T_u, μs	3.2	51.2
CP period, T_g, μs	0.8	10
Sampling duration, T, ns	50	50
Symbol mapping	QAM	QAM
Channel coding	$\frac{1}{2}$-rate convolutional coding	
Data rate, Mbps	13	14.2
User velocity, kmph	3	50
Doppler spread, Hz	13.9	231.5
Coherence time, T_c, ms	30.5	1.8

the analysis. Obviously this means that no CCI is present in the system. The simulation parameters are listed in Table 10.2.

Four antennas at the transmitter side is chosen for BF system. By looking at the outage capacity results in previous sections, we have seen that two transmit antennas can provide at best 3 dB of array gain for BF, while this gain may not be enough for some scenarios. With four transmit antennas, the maximum achievable array gain will be 6 dB. More number of antennas are also possible to use for the simulations, but, in that case, we may not be able to ensure that the channel is constant for a duration of $\frac{PT_s}{R_{td}}$ seconds.

We use the $\frac{1}{2}$-rate Tarokh code described in Table 10.1, which is suitable for any 4×1 system, giving full diversity but losing in bandwidth efficiency. In both transmit diversity approaches, STBC and SFBC, the antenna spacing has to be such that low correlation between signals is achieved at MS, hence it is chosen to use $d = 5\lambda$, since at the BS there are no space constraints as at the MS. However, since the performances of STBC and SFBC do not differ much, we have only included STBC for this simulations. Regarding BF, in order to obtain a narrow beam, the antenna elements must be placed closely, hence the spacing chosen was $d = \lambda/4$.

10.5.4 BER Results and Discussions

Space–Time Block Coding (STBC) and BF applied to the multi-access schemes OFDM–TDMA and OFDMA are compared regarding their performance and implementation complexity. We assume that the synchronization requirement of the OFDM receiver is perfectly met and perfect CSI is present at the receiver. When comparing the systems using different techniques, it is assumed that they are using the same number of antennas, as well as the same transmit power level. The performance is measured in terms of raw BER.

Based on [189], angular spread is chosen for indoor channel as [120°, 240°, 360°]. On the other hand, for outdoor channel, the angular spread is made to vary within [5°, 10°, 20°].

For these different values of angular spread, BER curves for STBC-OFDM system are presented in Figure 10.8. The STBC is implemented with four antennas using Tarokh's orthogonal code with half-rate. By analyzing the results in that same figure, it can be seen that for indoor scenarios (i.e. higher angular spread), the BER performance of transmit diversity is generally better

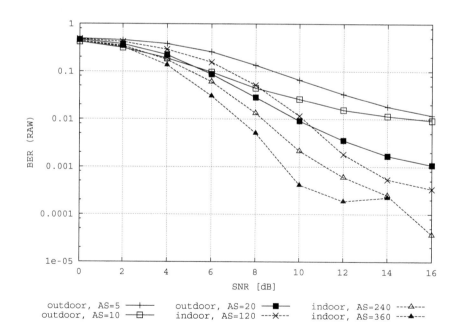

Fig. 10.8 BER performance for STBC/SFBC for different AS.

than for outdoor scenarios (i.e. lower angular spread). This means that the advantage that can be taken out of the use of STBC is more significant in indoor environments than in outdoor scenarios, due to higher angular spread that makes the arriving signals from different spatial location less correlated. Hence, more reliability is achieved by combining those different sources of data.

The simulations are performed in an analogous way for BF-OFDM system and the results are shown in Figure 10.9. Unlike the previous case, it can now be seen that for BF, a smaller angular spread results in better performance of the system.

In order to compare which method better fits to each environment, the results for outdoor and indoor cases are shown in Figures 10.10 and 10.11, respectively.

From the results obtained in this work, it can be highlighted that in outdoor scenarios, BF performs better than transmit diversity in OFDM systems. It is found that transmit diversity increases reliability of the received signals with the increase of angular spread. It has been noticed that in an outdoor

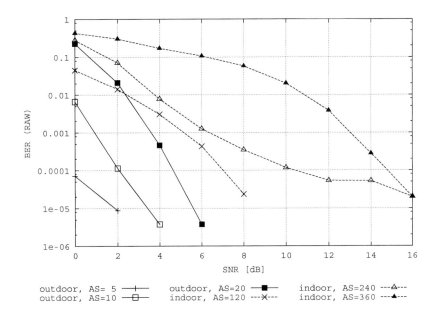

Fig. 10.9 BER performance for BF for different AS.

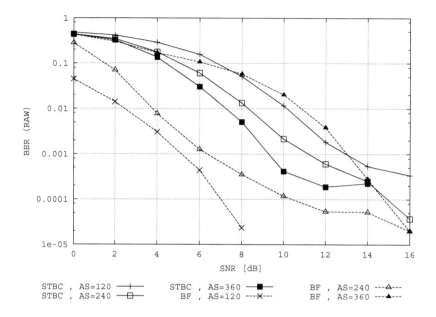

Fig. 10.10 BER comparison between BF and STBC for indoor channel.

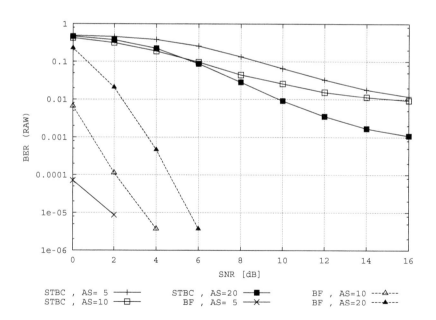

Fig. 10.11 BER comparison between BF and STBC for outdoor channel.

environment, targeting an error rate of 10^{-2} with $20°$ of angular spread (i.e. in outdoor scenario), BF has a gain of more than 7 dB compared to transmit diversity. Although the complexity of employing BF is much higher than the one for transmit diversity, it can be said that for outdoor environments, it is worthy.

For indoor environments, it is seen that transmit diversity does not always perform better than BF. In some particular cases where the user is located in LOS in indoor environment, then BF certainly performs better than transmit diversity. However, even when it seems that employing BF would easily improve system's performance, one has to bear in mind that the complexity of this method is much higher than the simple transmit diversity scheme. Since the gain of employing BF in indoor scenarios is not significantly higher as in outdoor, the strategy could pass for targeting a cheaper cost by employing transmit diversity.

Although we have not investigated the immediate outdoor scenario, the user mobility is higher than indoor in this case and the feedback channels may not be quite efficient to provide the Channel Quality Index (CQI) at the beamformer and/or closed-loop transmit diversity system, so that closed-loop schemes can be implemented. So, it is our suggestion that open loop system should be used for immediate outdoor channels.

Although it is suggested that BF is a better option compared to transmit diversity in outdoor scenario, BF requires some form of closed-loop information, and the fact that higher user speed in outdoor scenario can somewhat preclude the possibility of closed-loop operation. This provides a conflicting scenario, which may become a hindrance in using closed-loop BF even in outdoor scenario. Thus, BF methods that utilizes the long-term channel statistics may become more useful in outdoor scenario due to system complexity constraints, rather than the ones that require "short"-term channel information.

10.5.5 Pilot Design Issue

For a satisfactory system performance, proper design of pilot channels are very important. In case of transmit diversity and beamforming systems, the total available power is shared among all transmit antennas. This means that a pilot in these systems will have $\log_{10}(P)$ dB less power, on the average.

This is not advantageous for transmit diversity schemes, as the receiver needs to estimate all P number of channels based on these pilots which has an average transmit power reduced by $\log_{10}(P)$ dB. For beamforming schemes, the pilots also obtain the array gain, provided that proper beamforming technique is employed and thus, the received power at the pilots is $\log_{10}(P)$ dB higher compared to the case of transmit diversity scenario. Thus, we can see that beamforming has an advantage in terms of pilot design compared to transmit diversity schemes.

10.6 Chapter Summary

Space–Time Block Code (STBC) [30] is an open loop transmit diversity scheme where the diversity is achieved at the receiver without the knowledge of the channel at the transmitter. A complementary to this kind of transmit diversity scheme is BF [191]. Transmit Diversity (TxDiv) and BF are two techniques that describes the trade-off between diversity and array gain, respectively.

When the wireless channels between the transmit and the receive antennas are correlated to each other, then transmit diversity scheme is not expected to perform well, i.e. if independent fading among the antenna signals cannot be achieved, BF is preferred over transmit diversity. In BF we exploit the fact that the antenna elements are close together so that appreciable coherence between the antenna signals is present.

In this chapter, we have studied the transmit diversity and beamforming strategies from diversity and array gain perspective. Special attention is given in outage probability for certain outage capacity for the systems. From this point onwards, we have performed numerical evaluations to compare the usability of diversity and array gain in indoor and outdoor scenario.

Regardless of the multiple access scheme, it is found that BF always performs better in outdoor environment, where angular spread is lower, thus spatial correlation is higher. Similarly, indoor environment (high angular spread and low spatial correlation) suggests that transmit diversity schemes are more suitable to be employed than BF strategies.

In a practical cellular system, the total system performance depends on a number of issues that provides different forms of diversity, such as spatial diversity, multi-user diversity, retransmission diversity, macro diversity, etc. It is understood that, when different forms of diversity benefits are available, e.g. when macro diversity is achieved using handover between neighboring cells, then the benefits out of spatial degrees in transmit diversity scheme diminishes and BF may become more preferable, as array gain becomes more important. In the end, the system design depends on available angular spread and number of antennas available at the terminals.

11

Exploiting Cyclic Delay Diversity
in OFDM System

Cyclic Delay Diversity (CDD) is an interesting way to transform the available spatial diversity into extended frequency diversity in an Orthogonal Frequency Division Multiplexing (OFDM) system. Due to its simplicity in its implementation and its conformance with any established standard, recently CDD has obtained a lot of consideration for broadband OFDM system.

In this chapter, we study the exploitation of CDD properties in an OFDM system. The basic OFDM system model is given in Section 11.2. We briefly explain the well-known Post-DFT MRC scheme in Section 11.3. Section 11.4 presents the basic mechanisms of CDD technique, and the system capacity of CDD based transmit diversity systems. Our proposed Pre-DFT Maximal Average Ratio Combining (MARC) scheme is described in Section 11.5. We analyze the optimum Signal-to-Noise Ratio (SNR) for maximum cyclic shift and optimum gain factors in this section. We also present a simple case of a dual receive antenna system. Performance results and discussions are placed in Section 11.5.4. This section also describes the comparison of computational complexity between Post-DFT MRC and Pre-DFT MARC techniques and the effect of time-variance in the combining scheme.

Exploiting CDD in Spatial Multiplexing (SM) systems are described in Section 11.6. We have outlined the system architecture in Section 11.6.1. The ergodic capacity analysis of Cyclic Delay Assisted Spatially Multiplexed Orthogonal Frequency Division Multiplexing (CDA-SM-OFDM) system is described in Section 11.6.2. Finally, the chapter summary in presented in Section 11.7.

11.1 Introduction

Orthogonal Frequency Division Multiplexing (OFDM) has been successfully used in Wireless Local Area Network (WLAN), such as IEEE 802.11a,

European HiperLAN/2 or Japanese Multimedia Mobile Access Communication (MMAC) standards as high-data rate PHY layer transmission scheme for local area coverage. All three of these standards are almost similar in their PHY layers. Higher data rate along with higher Quality-of-Service (QoS) are one of the prime demands for next generations of wireless communications systems. The omnipresent WLANs standards are capable of providing data rates up to 54 Mbps for local area coverage. The IEEE 802.11a WLAN standard specifies channel coding ($\frac{1}{2}$ rate convolutional coding with a constraint length of 7) and frequency interleaving to exploit the frequency diversity of the wideband channel. Efficiency can only be achieved if the channel is sufficiently frequency-selective, corresponding to long channel delay spreads. In a flat fading situation (or in relatively lesser frequency-selective fading situation which we often encounter in indoor wireless scenario), all or most sub-carriers are attenuated simultaneously leading to long error bursts. In this case, frequency interleaving does not provide enough diversity to significantly improve the decoding performance as reported in [192].

Traditionally space domain is exploited at the receiver to obtain multipath diversity, so schemes like Maximal Ratio Combining (MRC), Equal Gain Combining (EGC), or Selection Combining (SCo) are used to obtain a better link quality. MRC is more complex compared to SCo and EGC, but yields the highest SNR [44, 58]. In case of MRC, the signals at the output of the receivers are linearly combined so as to maximize the instantaneous SNR. The linear combining in MRC is achieved by combining the co-phased signals, which requires that the Channel State Information (CSI) is known at the receiver. The SNR of the combined signal is equal to the sum of the SNRs of all the branch signals [44].

For an MRC-OFDM system as shown in Figure 11.1, the combining operations are performed at sub-carrier level after the Discrete Fourier Transform (DFT) operation, thus we denote the process as Post-DFT MRC or sub-carrier combining receiver [174]. The received OFDM signals at different antenna branches are first transformed via Q separate DFTs, when Q is the number of receive branches. Their outputs are assigned to N diversity combiners where N refers to number of OFDM sub-carriers. Note that, similar to MRC, all of the above spatial diversity schemes in an OFDM system requires multiple DFT blocks in the receiver.

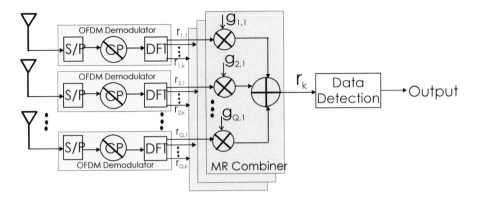

Fig. 11.1 Multiple antenna receiver diversity with MRC at sub-carrier level.

Space–Time Coding (STC), such as Alamouti's Space–Time Block Code (STBC) [30] can be implemented in an OFDM system that takes advantage of flat fading channel and thus achieve full diversity [176, 193]. Using Alamouti's scheme, the source symbols are transmitted in such a way that they can be combined in the receiver using an MRC like method.

Another way of achieving transmit diversity is using a simple Delay Diversity (DD) scheme [194]. Using DD, the original signal is transmitted via the first antenna and a (several) linearly time-delayed version(s) of original signal is (are) transmitted via one (or more) additional antenna elements. The limiting factor for such a diversity system is that the introduced delay should always be shorter than the Cyclic Prefix (CP) to make sure that ISI is avoided.

To overcome this limiting problem, CDD has been proposed in [28, 195, 196]. In this case, the signal is not truly delayed between respective antennas but cyclically shifted and thus, there are no restrictions for the delay times. The receiver structure in the DD and CDD schemes are similar to each other. It is worth noting here that, similar to Alamouti scheme, it is possible to apply CDD as a transmit diversity technique without knowing the CSI at the transmitter.

In this chapter, CDD concept is exploited and implemented in OFDM receiver [197], to obtain receiver diversity like MRC. All the signal processing needed is performed in *time domain* when CDD is used to exploit multiple receive antennas, so the duplication of the DFT operation for each receiving antenna branch is not a requirement any more, thus the receiver has lower computational cost compared to Post-DFT MRC (Figure 11.1). At the receiver, the antenna branch signals can be used for estimating the channel responses for each individual receiver antenna in order to optimize the diversity combining based on the instantaneous channel behavior. This allows for an optimized diversity combining using cyclic delays in received data samples and complex gain factors [197].

We have studied a scheme named MARC, which is basically application of CDD in an OFDM receiver. In effect, we obtain partial array gain by employing the weight estimation method explained in this chapter. A detailed discussion on the scheme is presented, where the optimum weighting factors and cyclic shifts are derived for the multiple antenna case, based on the estimated CSI in the receiver. We present a comparative study between MRC receiver diversity and CDD in the IEEE 802.11a and/or HiperLAN/2 WLAN context. The traditional use of MRC is studied along with usage of CDD in both OFDM transmitter and receiver. Special attention is given to the application of CDD at a dual antenna receiver, where the optimum weighting factors and cyclic shifts are derived for the two antenna case, based on the known CSI at the receiver side. Thus, this receive-CDD technique performs better compared to transmit (or receive) CDD techniques studied in [195]. Simulations based on the WLAN standard are performed to compare the different diversity schemes in terms of BER with and without error correction coding.

The Pre-DFT combining simplifies the receiver operations, but it also costs the system performance, because the operations are done on wideband, thus, proper receiver combining gain cannot be archived for highly frequency-selective channels. Thus, an alternative is proposed in [174], where Pre-DFT combining is used together with sub-carrier combining after the DFT operation. In this way, the loss due to wideband combining is gained back using sub-carrier level combining, but it also costs in spectral efficiency, because symbols need to be re-transmitted across many sub-carriers, so that sub-carrier level combining can be done. Compared to this work, we do not re-transmit same symbols across many sub-carriers to avoid the loss in spectral efficiency, rather we employ the CDD property at the receiver, so that transformed frequency diversity via CDD principle can be exploited for recovering the loss incurred from wideband Pre-DFT combining.

11.2 OFDM System Model

The OFDM signal consists of N orthogonal sub-carriers modulated by N parallel data streams. Denoting the frequency and complex source symbol of the kth sub-carrier as f_k and d_k, respectively, the baseband representation of an OFDM symbol is

$$x(t) = \frac{1}{\sqrt{N}} \sum_{k=-\frac{N}{2}}^{\frac{N}{2}-1} d_k e^{j2\pi f_k t}, \quad -\frac{N}{2}T \leq t \leq \left(\frac{N}{2} - 1\right)T. \quad (11.1)$$

d_k is typically taken from a Phase Shift Keying (PSK) or Quadrature Amplitude Modulation (QAM) symbol constellation, and NT is the duration of the OFDM symbol. The sub-carrier frequencies are equally spaced at $f_k = k/NT$. The OFDM signal in Equation (11.1) can be derived by using a single (inverse) DFT operation rather than using a bank of oscillators. Thus the discrete time representation of an OFDM symbol is

$$x(t) = \frac{1}{\sqrt{N}} \sum_{k=-\frac{N}{2}}^{\frac{N}{2}-1} d_k W_N^{kn}, \quad -\frac{N}{2} \leq n \leq \frac{N}{2} - 1, \quad (11.2)$$

where $W_N = e^{j2\pi/N}$. Assuming that the channel impulse response is shorter than the guard interval and that perfect synchronization is achieved, the received signal constellations after OFDM demodulation (which includes the removal of the guard interval and a DFT operation) can be related to the data symbols d_k by [192]

$$r_k = d_k h_k + n_k, \quad (11.3)$$

where h_k is the (complex-valued) channel transfer function at frequency f_k and n_k denotes an Additive White Gaussian Noise (AWGN) sample with variance $\sigma^2 = \mathcal{E}\{n_k^2\}$. The BER performance depends on the SNR at a certain sub-carrier, which is given by

$$\gamma_k = \frac{\mathcal{E}\{|d_k|^2\}|h_k|^2}{\sigma^2} \propto |h_k|^2. \quad (11.4)$$

Note that the SNR is proportional to the squared magnitude of transfer function; therefore, maximizing h_k will yield improved performance.

11.3 Post-DFT Maximum Ratio Combining

In a traditional wireless system, MRC is widely used as a receiver diversity combining technique where multiple receive antennas are used along with a single transmit antenna. In MRC, the signals at the output of the receivers are linearly combined so as to maximize the instantaneous SNR, array gain, and diversity gain [44]. The signals are co-phased and combined together. The assumption here is that the CSI must be known perfectly at the receiver.

For an MRC-OFDM system as shown in Figure 11.1, the combining operations are performed at sub-carrier level after the DFT operation, thus we denote the process as Post-DFT MRC-OFDM system. Recognizing the fact that combining is performed at sub-carrier level, the Post-DFT MRC-OFDM receiver can also be termed as sub-carrier combining receiver [174]. The received OFDM signals at different antenna branches are first transformed via separate DFT processors and the values of kth data sub-carrier in an OFDM symbol ($r_{1,k}$ up to $r_{Q,k}$) are introduced to diversity combiner, given Q receive antennas are present. The combiner combines the signals based on SCo, EGC or MRC on sub-carrier basis. We will focus on MRC in the following discussions.

Ignoring the noise present in the reception process, the received signal in Qth antenna for kth data sub-carrier can be expressed as $r_{q,k}$. In the linear combiner, the sub-carrier data in each branch is multiplied by

$$g_{q,k} = \alpha_{q,k} e^{-j\theta_{q,k}} \tag{11.5}$$

This multiplication is performed such that each of the signal branches are co-phased (i.e. all branches have zero phase). Here, $g_{q,k}$ is a complex number with $\theta_{q,k}$ is the corresponding phase of received signal branch, and $\alpha_{q,k}$ is the weighting factor defined as

$$\sum_{q=1}^{Q} |\alpha_{q,k}|^2 = 1. \tag{11.6}$$

Then co-phased branches are added together. The resultant signal envelope is

$$r_k = \sum_{q=1}^{Q} r_{q,k} g_{q,k}, \tag{11.7}$$

The above-mentioned Post-DFT combining diversity scheme is optimum in terms of its Bit Error Rate (BER) performance [174]. As deep fades are often uniformly distributed among sub-carriers, combining at sub-carrier level enables the selection and combination of only the best-quality sub-carriers from uncorrelated antenna branches. For Post-DFT combining schemes, the OFDM signal is treated in the same way as in the single carrier system.

On the other hand, the computational complexity of Post-DFT combining schemes increases with the number of receive antennas and number of sub-carriers used. As shown in Figure 11.1, the Post-DFT combining scheme requires Q independent OFDM demodulators, one for each receive antenna. In many cases, the number of receive antennas is limited to 2, since the largest diversity gain is obtained with $Q = 2$, and diminishing returns are realized with increasing of Q [198]. In addition, the number of diversity combining modules is equal to the number of sub-carriers of interest at the receiver. For instance, in the Downlink (DL), the number of diversity combiners is equal to the number of sub-carriers assigned to that particular user. In the Uplink (UL), the Base Station (BS) needs N diversity combining module, one for each sub-carrier. These requirements will result in a very complex and power-hungry receiver, as 4G wireless system is expected to operate with large number of sub-carriers [199]. In Section 11.5, a less complex diversity combining technique is introduced, which is important for Mobile Station (MS) whose low hardware and algorithm complexity as well as low power consumption are the key requirements.

11.4 Benefitting from Cyclic Delay Property in OFDM System

In this section, we describe the effect of CDD on an OFDM system. CDD is a simple, efficient, and effective way to achieve diversity in such a flat fading scenario. The technique is originally proposed as a transmit diversity [28, 195, 196]. As mentioned earlier, when we shift the OFDM signals cyclically and add them up in the receiver linearly, we actually insert some virtual echoes on the channel response. This effect increases the channel frequency-selectivity, thus higher order frequency diversity can be achieved, which is effectively exploited by a Coded Orthogonal Frequency Division Multiplexing (COFDM) system.

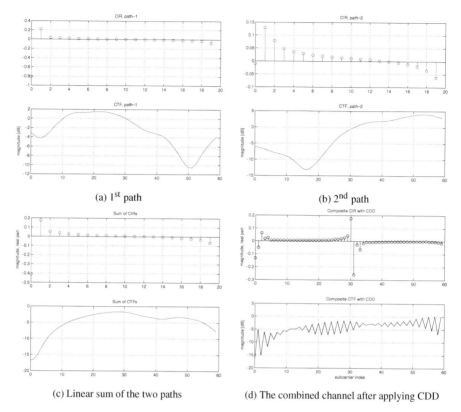

(a) 1st path

(b) 2nd path

(c) Linear sum of the two paths

(d) The combined channel after applying CDD

Fig. 11.2 CIR and CTF of two separate paths, their linear sum and the combined channel after applying cyclic delay principle.

Figures 11.2(a)–11.2(d) explain the effect of employing CDD seen at the receiver with two diversity paths. Here, we assume that we have 2 transmit antennas (i.e. $P = 2$). The second path is assumed to be cyclically shifted by $\frac{N}{2}$ number of samples, where N denotes total number of OFDM sub-carriers in the system. The 1st and 2nd SISO paths show a frequency flat (or less frequency selective) wideband scenario in Figures 11.2(a) and 11.2(b), respectively. When we linearly add the paths, we do not obtain much diversity, because linear addition of two short channel impulse responses creates another short channel impulse response, thus the frequency selectivity remains same. On the other case, when cyclic delays are applied in the transmitter branches, CDD is achieved and the resulting channel response seen from the receiver is equivalent to frequency selective channel.

11.4.1 System Model with Cyclic Delay Diversity

Figure 11.3 shows the transmission scheme using cyclic delays at the transmitter antenna branches. At first, user data is Forward Error Correction (FEC) coded and baseband modulated from any PSK or QAM constellation points. Then the modulated symbols are used at the IFFT inputs and $\tilde{x}_s n$, $\forall n$ is produced at the Inverse Fast Fourier Transform (IFFT) output. The time-domain samples are parallel-to-serial converted. At this stage, an antenna specific cyclic delay, Ω_p, is inserted. So the transmit symbol from pth transmit antenna at sth OFDM symbol time window is written as

$$\tilde{x}_{s,p}n = \tilde{x}_s\big((n - \Omega_p)\big)_N, \tag{11.8}$$

where $\big((n - \Omega_p)\big)_N$ means $(n - \Omega_p)\bmod N$. The delay Ω_p is shown here in samples, it's equivalent to $T\Omega_p$ seconds in time scale. After this step, CP is inserted at all the antenna branches separately before transmission.

These cyclically delayed branches together convert a MISO channel into a SISO channel. Similarly, a MIMO channel will become SIMO channel when CDD is applied at the transmitter [200]. This is because, the addition of all cyclically delayed transmit branches at each receive antenna will equivalently "see" a more frequency selective SISO. The equivalent Channel Impulse Response (CIR) seen at the qth receive antenna can be written as

$$\mathbf{g}_{q,\mathrm{equ}} = \frac{1}{\sqrt{P}} \sum_{p=1}^{P} \mathbf{g}_{q,p}\big((n - \Omega_p)\big)_N. \tag{11.9}$$

Here, $\mathbf{g}_{q,\mathrm{equ}}$ is a $(L + \max(\Omega)) \times 1$ vector, whose lth element is $g_{q,\mathrm{equ}}l$, i.e. the Channel Impulse Response (CIR) of lth tap for the corresponding chan-

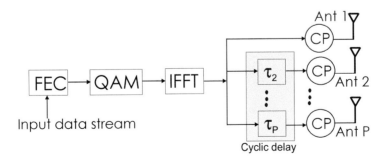

Fig. 11.3 Transmitter model for CDD based MISO-OFDM transmission scheme.

nel. The normalization factor $\frac{1}{\sqrt{P}}$ is used to keep the transmission power constant. $\max(\Omega)$ is the maximum amount of cyclic delay inserted in the transmit branches.

The Channel Transfer Function (CTF) of the equivalent Channel Impulse Response (CIR) can be obtained as

$$\mathbf{h}_{q,\text{equ}} = \mathbb{F}\{\mathbf{g}_{q,\text{equ}}\}, \tag{11.10}$$

where $\mathbb{F}$ shows the DFT operation. Equivalently,

$$
\begin{aligned}
h_{q,\text{equ}}[k] &= \sum_{l=0}^{L+\max(\Omega)-1} g_{q,\text{equ}}[l]\, e^{-j2\pi\frac{k}{N}l} \qquad k = 0, 1, \ldots, N-1 \\
&= \sum_{l=0}^{L+\max(\Omega)-1} \frac{1}{\sqrt{P}} \sum_{p=1}^{P} g_{q,p}\big[\big((l - \Omega_p)\big)_N\big] e^{-j2\pi\frac{k}{N}l} \\
&= \frac{1}{\sqrt{P}} \sum_{p=1}^{P} \sum_{l=0}^{L+\max(\Omega)-1} g_{q,p}\big[\big((l - \Omega_p)\big)_N\big] e^{-j2\pi\frac{k}{N}l} \\
&= \frac{1}{\sqrt{P}} \sum_{p=1}^{P} h_{q,p}[k] W_N^{-k\Omega_p}.
\end{aligned}
\tag{11.11}
$$

As mentioned above, Figure 11.2(d) shows the corresponding time and frequency response of equivalent SISO channel as described in Equations (11.9) and (11.11). It is worth noting that the increased frequency diversity at the receiver is originated from available spatial diversity from the transmitter, thus it can be said that CDD utilizes the available spatial diversity for better frequency de-correlation for the OFDM system. When the frequency selectivity is increased, then it is understood that the average BER can be improved when CDD can be used in co-ordination with FEC coding in near frequency flat (or less frequency selective) scenario.

Choice of delay parameter in Tx-CDD: As we noted that no CSI is available at the transmitter, we have to insert the cyclic delays in a blind manner. As it is shown in [26] that, if Ω is chosen to be equal to or smaller than L, then the delay diversity cannot convert the available spatial diversity into frequency diversity, because the delayed channel impulse responses will "fall" on each other in the equivalent CIR. Thus, a good choice of cyclic delay can be [201]

$$\Omega_p = \frac{N(p-1)}{P} = \frac{N}{P} + \Omega_{p-1}. \tag{11.12}$$

Diversity order in CDD: The orthogonal STBC (and corresponding SFBC) that are designed in [30, 175] obtain full diversity, which can also be seen from Equation (10.6). This is ensured by utilizing linear receivers that adds the channel coefficients only constructively (i.e. MRC like receiver). Thus, the full diversity is already obtained before the constellation de-mapper. As opposed to this, when we look at Equation (11.11), we can see that the channel is added up incoherently for CDD systems. It can be easily shown that $h_{q,\text{equ}}[k]$ has a normal distribution with zero mean and unity variance, i.e. $h_{q,\text{equ}}[k] \sim c\mathcal{N}(0, 1)$. This is the reason that we do not see any gain in terms of BER between SISO and CDD system in Figure 11.10 for uncoded case. Thus, the available frequency diversity can only be picked up when the Forward Error Correction (FEC) coding and de-interleaving is performed. In this way, in collaboration with FEC, CDD systems obtain full diversity [195, 196, 201]. For original CDD proposals [28, 195], the delay parameter is inserted blindly either at transmitter or at receiver. In both of these cases, the diversity gain is full, i.e. the diversity order is always P and Q for the cases, when blind CDD is employed in transmitter and receiver, respectively.

Note that, for any arbitrary number of antennas, it is possible to obtain full rate and full diversity in CDD with linear receivers, while original Space-Time Block Codes (STBCs) cannot obtain full rate and full diversity simultaneously with linear receivers when $P > 2$ [175].

11.4.2 Capacity of CDD Based OFDM System

In the following we assume that for each channel use, we obtain an independent realization of channel response vectors for all channels between transmit and receive antennas, thus, no spatial correlation is present. It is also assumed that the channel remains constant for one OFDM channel use. When we use CDD, then the equivalent CTF of P number of CDD branches can be treated like any SISO channel CTF. As we use a normalization constant in Equation (11.11), the total transmitted power is constant in all CDD branches compared to a single SISO channel. We can write the capacity for any kth sub-carrier of the equivalent CDD channel as

$$C_k^{\text{CDD}} = \log_2\left[1 + \overline{\gamma} h_{q,\text{equ}}[k] h_{q,\text{equ}}^*[k]\right], \qquad (11.13)$$

where Υ, is the average received SNR, and $h_{q,\text{equ}}^*[k]$ means the complex conjugate of $h_{q,\text{equ}}[k]$.

The total capacity of the broadband channel can be written as

$$
\begin{aligned}
\mathcal{C}^{\mathrm{CDD}} &= \frac{1}{N} \sum_{k=0}^{N-1} \mathcal{C}_k^{\mathrm{CDD}} \\
&= \frac{1}{N} \sum_{k=0}^{N-1} \log_2 \left[1 + \overline{\gamma} h_{q,\mathrm{equ}}[k] h_{q,\mathrm{equ}}^*[k] \right] \\
&= \frac{1}{N} \sum_{k=0}^{N-1} \log_2 \left[1 + \overline{\gamma} \left(\frac{1}{\sqrt{P}} \sum_{p=1}^{P} h_{q,p}[k] W_N^{-k\Omega_p} \right) \right. \\
&\qquad\qquad \left. \times \left(\frac{1}{\sqrt{P}} \sum_{p=1}^{P} h_{q,p}[k] W_N^{-k\Omega_p} \right)^* \right] \\
&= \frac{1}{N} \sum_{k=0}^{N-1} \log_2 \left[1 + \frac{\overline{\gamma}}{P} \sum_{p=1}^{P} \sum_{p'=1}^{P} h_{q,p}[k] h_{q,p'}[k] W_N^{-k\left(\Omega_p - \Omega_{p'}\right)} \right].
\end{aligned}
$$

$$(11.14)$$

The capacity expression in Equation (11.14) is in fact the average instantaneous capacity across OFDM sub-carriers. This average is the instantaneous capacity of the whole wideband system. To obtain the ergodic capacity, we average this instantaneous wideband capacity over long time.

Figure 11.4(a) shows the CDF considering capacity of Single Input Single Output (SISO) and various CDD systems. Corresponding outage capacity is seen in Figure 11.4(b). These curves are taken against the following parameters: $N = 64$, $\tau_{\mathrm{rms}} = 100\,\mathrm{ns}$, $T_s = 4.0\,\mu\mathrm{s}$ and $T_g = 0.8\,\mu\mathrm{s}$. In the ergodic capacity results, the diversity effect cannot be seen, as the time average introduces the time diversity. This diversity effect is added to all the spatial diversity schemes, thus the impact of spatial diversity on ergodic capacity cannot be seen. From the outage capacity results in Figure 11.4(b), it is seen that outage capacity is improved when higher number of CDD branches are used. It should be noted here that the diversity effect is captured in the CDD system capacity without increasing the system complexity too much, i.e. no extra processing is required in the receiver and a minimal processing is needed at the transmitter. Compared to other transmit diversity techniques, such as Alamouti's STBC, the CDD is much simpler and easier to implement.

11.5 Pre-DFT Maximum Average Ratio Combining

Introducing CDD in an OFDM system (either in the transmitter or in the receiver) amounts to increasing the frequency-selectivity of a relatively flat fading channel seen from the receiver side [192]. When CDD is introduced at the receiver, the diversity combining is performed prior to the DFT operation [192, Section 8.3], as shown in Figure 11.5. At the receiver, the antenna branch signals can be used to estimate the channel responses for each individual receiver antenna in order to optimize the diversity combining based on the instantaneous channel behavior. This allows for an optimized diversity

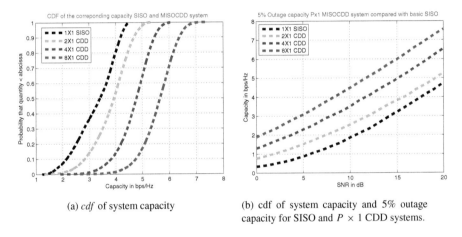

(a) *cdf* of system capacity

(b) cdf of system capacity and 5% outage capacity for SISO and $P \times 1$ CDD systems.

Fig. 11.4 CDF of system capacity and 5% outage capacity for SISO and $P \times 1$ CDD systems.

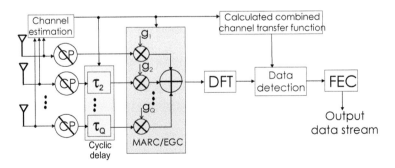

Fig. 11.5 OFDM receiver with Pre-DFT Combining CDD. The instantaneous channel is estimated from the received signals to determine the optimum cyclic shifts (and gain factors, if MARC combining is performed).

combining using cyclic delays, $\tau_n^{(q)}$ in received data samples and complex gain factors, $g_q = a_q e^{j\phi_q}$, where $a_q = |g_q|$, $\phi_q = \angle g_q$ and $\sum_{q=1}^{Q} |g_q|^2 = 1$, for $q = 1, 2, \ldots, Q$.

We denote this combining technique in the OFDM receiver as Pre-DFT Maximum Average (signal-to-noise) Ratio Combing (Pre-DFT MARC). If we select equal values for the gain magnitudes (i.e. $|g_q| = \sqrt{\frac{1}{Q}}$), the combining technique is named as Pre-DFT EGC.

11.5.1 Optimum SNR for the Combined Signal

We denote the discrete time and discrete frequency index as l and k, respectively; and define, $r_q(l)$, $r_{q,\text{CDD}}(l)$, and $r_{\text{comb}}(l)$ as received signal in time-domain at q th receive antenna, signal after applying CDD at q th diversity branch and combined signal after the combining, respectively. All of these vectors are defined for one OFDM symbol, so they have a dimension of $N, 1$. Denoting the complex valued time-invariant CIR of the q th diversity branch as $\mathbf{c}_q(l)$, we can write that [44]

$$r_q(l) = \left[d(l) \overset{N}{\circledast} c_q(l) \right],\qquad(11.15)$$

where $\overset{N}{\circledast}$ denotes N-point circular convolution, and by definition, $0 \leqslant l \leqslant N - 1$. The length of $c_q(l)$ is actually the length of the CIR for q th diversity branch which is obviously less than N, so the length is extended to N by padding zeros at the end of the vector.

The signals at each diversity branch after applying CDD can be written as

$$r_{q,\text{CDD}}(l) = \left[d(l) \overset{N}{\circledast} c_q((l - \tau_n^{(q)}))_N \right],\qquad(11.16)$$

where $c_q((l))_N = c_q((l) \bmod N)$. The CDD signal, $r_{q,\text{CDD}}(l)$ is multiplied with the gain factor g_q to obtain the combined signal,

$$r_{\text{comb}}(l) = \frac{1}{\sqrt{Q}} \sum_{q=1}^{Q} g_q \left[d(l) \overset{N}{\circledast} c_q((l - \tau_n^{(q)}))_N \right].\qquad(11.17)$$

The data part of the convolution in Equation (11.17) is the same for all diversity branches, so we can write the effective channel impulse response of the combined channel as

$$c_{\text{comb}} = \sum_{q=1}^{Q} g_q c_q ((l - \tau_n^{(q)}))_N. \tag{11.18}$$

We assume, independent additive white gaussian noise with equal powers are present at the branches, $\sigma_1 = \sigma_2 = \cdots = \sigma_Q = \sigma$. After diversity combining, the noise powers are scaled by the squared magnitude of the gain factors and summed up. Since we choose $\sum_{q=1}^{Q} g_q^2 = 1$, the resulting noise level after diversity combining is constant and equal to the noise power of each antenna branch. Therefore, we can derive a measure that is proportional to the SNR after diversity combining as $SNR \propto w0$, where wl is

$$wl = c_{\text{comb}}l \overset{N}{\circledast} c_{\text{comb}}^* N - l$$

$$= \sum_{q=1}^{Q} g_q c_q ((l - \tau_n^{(q)}))_N \overset{N}{\circledast} \sum_{n=1}^{Q} g_n^* c_n^* ((\tau_n^{(n)} - l))_N$$

$$= \sum_{q=1}^{Q} |g_q|^2 c_q l \overset{N}{\circledast} c_q^* N - l$$

$$+ \sum_{q=1}^{Q} \sum_{n=q+1}^{Q} \left\{ g_q g_n^* c_q ((l - \tau_n^{(q)}))_N \overset{N}{\circledast} c_n^* ((\tau_n^{(n)} - l))_N \right.$$

$$\left. + g_n g_q^* c_n ((l - \tau_n^{(n)}))_N \overset{N}{\circledast} c_q^* ((\tau_n^{(q)} - l))_N \right\} \tag{11.19}$$

$$= \sum_{q=1}^{Q} |g_q|^2 c_q l \overset{N}{\circledast} c_q^* N - l$$

$$+ 2 \sum_{q=1}^{Q} \sum_{n=q+1}^{Q} \Re \left\{ g_q g_n^* c_q ((l - \tau_n^{(q)}))_N \overset{N}{\circledast} c_n^* ((\tau_n^{(n)} - l))_N \right\}, \tag{11.20}$$

where * denotes the conjugate complex.

It is evident that the first part of Equation (11.19) of this expression is independent of the cyclic delays and the phases of the gain factors. Thus, the SNR can be optimized with respect to these parameters by maximizing the second part of Equation (11.20). Unfortunately it is not possible to optimize these parameters independently, for the following reasons.

Between each pair of signals q and n, $n \neq q$, the cyclic delay leading to maximum SNR is given by the index of the maximum value in the respective summation term of Equation (11.20),

$$(\tau_n^{(q)} - \tau_n^{(n)})_{\text{opt}} = \arg\max_l \left| c_q((l - \tau_n^{(q)}))_N \overset{N}{\circledast} c_n^*((\tau_n^{(n)} - l))_N \right|. \quad (11.21)$$

This optimum SNR would be reached by selecting the phase terms according to

$$\angle(g_q g_n^*) = -\angle \left\{ \max_l \left[c_q((l - \tau_n^{(q)}))_N \overset{N}{\circledast} c_n^*((\tau_n^{(n)} - l))_N \right] \right\}. \quad (11.22)$$

It becomes visible at this point that an independent optimization of the parameters $\angle(g_q)$ and $\tau_n^{(q)}$ is *not* possible if $Q > 2$, e.g. if we optimize the delays and phase-rotations for the antenna pairs 1-2 and 1-3, the corresponding parameters of pair 2-3 will be determined implicitly. We suggest to use the $Q - 1$ largest terms obtained by Equation (11.21) for optimizing the cyclic delays. The gain factors will then be optimized using the approach described in the next section (i.e. Section 11.5.2).

11.5.2 Optimum Diversity Weights

As we have derived an optimum way to determine the cyclic delays for respective antennas, now the next step should be to determine the diversity branch weight factors. A method to derive optimum diversity weight factors for multiple antenna Pre-DFT processing OFDM receiver is presented in [174]. We adopt a similar weight estimation scheme for Pre-DFT MARC with CDD receiver diversity scheme.

The SNR after sub-channel diversity combining can be written as:

$$\text{SNR} = \Gamma \mathbf{g}^H \mathbf{C} \mathbf{g}, \quad (11.23)$$

where Γ is the average SNR per diversity branch, $\mathbf{C}$ is the covariance matrix of the CIRs of all the diversity branches, and $\mathbf{g}$ is the weight vector, defined as $\mathbf{g} = \left[g_1, g_2, \ldots, g_Q \right]^T$. Eigen analysis can be performed on the Hermitian matrix $\mathbf{C}$ according to $\mathbf{C} = \mathbf{Z}\Lambda\mathbf{Z}^H$, where $\Lambda = \text{diag}\left(\lambda_1, \lambda_2, \ldots, \lambda_Q\right)^T$ is the diagonal matrix whose diagonal elements consist of eigenvalues λ_q of $\mathbf{C}$ and $\mathbf{Z}$ is the unitary matrix whose columns are the eigenvectors corresponding to λ_q.

It is found that the optimum diversity weight vector $\mathbf{g}_{\text{opt}}$ is the eigenvector, which corresponds to maximum eigenvalue from diagonal matrix Λ.

In the method described above, the covariance matrix $\mathbf{C}$ is derived from the CIRs of all diversity branches. The time-domain CIR estimators that are found in literature impose a high computational complexity, thus utilizing the correlation among the signals of the diversity branches directly in time domain instead of explicitly estimating the CIRs or the CTFs of all the diversity branches yields a satisfactory trade-off solution.

We denote the sampled received signal vector at l th sampling instant as $\mathbf{r}^{(l)} = \left[r_1^{(l)}, r_2^{(l)}, \ldots, r_Q^{(l)} \right]^T$. The correlation matrix of the received signal vector at any sampling instant is given by

$$\mathbf{R} = \frac{1}{2} E\{\mathbf{r}^{(l)} \mathbf{r}^{(l)H}\} = \left[\rho_{p,q} \right], \tag{11.24}$$

where $\rho_{p,q}$ is the p, q th element of $\mathbf{R}$ matrix, that represents the correlation among received signal sample for l th instant between p th and q th receive diversity branches, $\rho_{p,q} = \frac{1}{2} E r_p^{(l)} r_q^{(l)H}$. Denoting σ_x^2 as the variance of the transmitted signal and σ_n^2 as the variance of the noise component, we find from Equation (11.24) that

$$\mathbf{R} = \sigma_x^2 \mathbf{C} + \sigma_n^2 \mathbf{I}. \tag{11.25}$$

The equation above shows that the eigenvectors of $\mathbf{R}$ and $\mathbf{C}$ are the same, hence we can estimate the optimum weight factors based on the correlation matrix [174]. As it is shown in Figure 11.6, the received signals corresponding to all diversity branches for any sampling instant are put together in a vector ($\mathbf{r}^{(l)}$) and the autocorrelation of that vector is calculated according to Equation (11.24). After that the optimum weights for all the receive antenna branches are determined using Eigen analysis as described in Equation (11.25).

11.5.3 System Analysis with Dual Antenna Receiver

In the following analysis, we will concentrate on two branch diversity. Analysis on two-branch receiver is simpler and easier to understand. The channel transfer functions of the two branches are denoted as $\mathbf{h}_1$ and $\mathbf{h}_2$, where

$$\mathbf{h}_q = \left[\begin{array}{cccc} h_{q,0} & h_{q,1} & \cdots & h_{q,N-1} \end{array} \right]^T. \tag{11.26}$$

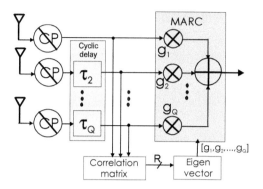

Fig. 11.6 Diversity weight estimation method for Pre-DFT MARC with CDD receiver diversity scheme; for clarity, channel estimation procedure and COFDM part of the receiver is not shown in the figure.

$h_{q,k}$ are samples of the Channel Transfer Function (CTF). Using the above vector notation, we can obtain a measure that is proportional to the average SNR over all sub-carriers as

$$\text{SNR}_i = \mathbf{h}_i^H \mathbf{h}_i, \tag{11.27}$$

where $(.)^H$ denotes the complex conjugate of a transposed vector $(.)$. For dual antenna system (i.e. $M = 2$), we denote the complex gain factors as $g_1 = \sqrt{a}$ and $g_2 = \sqrt{1-a}\, e^{j\phi}, 0 \le a \le 1$. $\tau_n^{(2)} = n^{(2)}$ (in more general term, $\tau_n^{(q)} = n^{(q)}$ in data samples) is the inserted cyclic delay in branch two (in samples). The combined channel frequency response as a function of inserted cyclic delay, $n^{(2)}$, and diversity weight factor, g, can be written as

$$\mathbf{h} = \sqrt{g}\,\mathbf{h}_1 + \sqrt{1-g}\,e^{j\phi}\mathbf{W}_N^{n^{(2)}}\mathbf{h}_2, \tag{11.28}$$

where $\mathbf{W}_N^{n^{(q)}} = \text{diag}(W_N^{0^{(q)}}, W_N^{n^{(q)}}, W_N^{2n^{(q)}}, \ldots, W_N^{(N-1)n^{(q)}})$, $W_N = e^{j2\pi/N}$, represents the cyclic delay in the qth branch.

Following Equations (11.27) and (11.28), we can obtain an expression that is proportional to the average SNR over all sub-carriers in terms of a and $n^{(2)}$ as

$$\gamma(a, n) = a\mathbf{h}_1^H \mathbf{h}_1 + (1-a)\mathbf{h}_2^H \mathbf{h}_2 + 2\Re\left(e^{j\phi}\sqrt{a}\sqrt{1-a}\,\mathbf{h}_1^H \mathbf{W}_N^{n^{(2)}} \mathbf{h}_2\right) \tag{11.29}$$

The phase, ϕ, for maximum SNR is found as

$$\phi_{\max} = -\angle(\mathbf{h}_1^H \mathbf{W}_N^{n^{(2)}} \mathbf{h}_2), \tag{11.30}$$

which maximizes the additive third term in Equation (11.29). Substituting this phase, the optimum SNR can be calculated as

$$\gamma(a, n) = a\mathbf{h}_1^H\mathbf{h}_1 + (1-a)\mathbf{h}_2^H\mathbf{h}_2 + 2\sqrt{a}\sqrt{1-a}|\mathbf{h}_1^H\mathbf{W}_N^{n^{(2)}}\mathbf{h}_2|. \quad (11.31)$$

Optimum weight factors for maximum SNR when $Q = 2$: We will achieve two sets of g for any given n, $g_{q|n}$ for $q \in \{1, 2\}$ by setting the first derivative of Equation (11.31) to zero with respect to g. After some manipulations, we obtain:

$$g_{q|n} = \frac{1}{2} \pm \frac{A}{2}\sqrt{\frac{1}{4B^2 + A^2}} \quad \text{for } q \in \{1, 2\}, \quad (11.32)$$

where $A = \mathbf{h}_1^H\mathbf{h}_1 - \mathbf{h}_2^H\mathbf{h}_2$ and $B = |\mathbf{h}_1^H\mathbf{W}_N^{n^{(2)}}\mathbf{h}_2|$. It is found that the maximum SNR is obtained using the $+$ sign. A simple observation from Equation (11.32) shows that when both channels have equal average SNR, then $A = 0$ and the two weight factors are equal, being $g_{1|n} = g_{2|n} = 0.5$.

If the channel transfer functions $\mathbf{h}_1$ and $\mathbf{h}_2$ are orthogonal to each other for some specific cyclic delay n, then $B = 0$ leading to $g_{1|n} = 0$, $g_{2|n} = 1$ or $g_{1|n} = 1$, $g_{2|n} = 0$, unless $A = 0$. This means that for these particular channel realizations, nothing can be gained on the average over all sub-carriers compared to a single channel system. In another words, only one diversity branch can be exploited effectively in this case.

For one set of channel realizations, the average SNR as a function of the shift parameter n is shown in Figure 11.7. Pre-DFT MARC yields two solutions of weighting factors as indicated by Equation (11.32), corresponding to the minimum and maximum average SNR. With respect to n, the maximum is reached with a cyclic shift of 2, for this particular set of channels. There is a gain of about 0.9 dB compared with the best solution without CDD (i.e., when $n^{(2)} = 0$). The maximum SNR is always achieved for a very small or very large value of n as analyzed below.

Optimum cyclic shift for maximum SNR when $Q = 2$: The maximum SNR with respect to n can be found from Equation (11.31) by determining the maximum of B with respect to n (because A does not depend on n). $B(n)$ is in fact the IFFT of the product of two CTFs. So, to determine the delay n for maximum average SNR efficiently, we can compute

$$\mathbf{c} = \|\mathbb{F}_N^H\{\mathbf{h}_1^* \cdot \mathbf{h}_2\}\|, \quad (11.33)$$

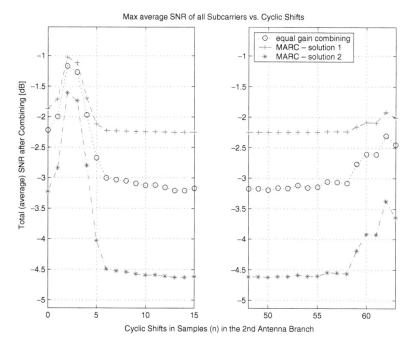

Fig. 11.7 Average SNR with two-branch Pre-DFT MARC and Pre-DFT EGC with cyclic delays as a function of the delay parameter *n* (in samples).

where the IFFT is zero padded to N points and . denotes element-wise multiplication of two vectors. The index to the maximum of the resulting vector **c** equals the cyclic delay n_{max} that provides maximum average SNR. Figure 11.8 presents the histogram of the optimum cyclic shifts for maximum average SNR, calculated by using Equation (11.33). The figure confirms that the optimum delays are confined to small values of n between 0–4 and 60–63. Note that the number of sub-carriers in the OFDM system in 64. This is due to the fact that the IFFT of the product of two CTFs is equivalent to the cyclic convolution of the respective channel impulse responses (where the impulse response corresponding to the conjugated transfer function is time-reversed). For relatively flat fading indoor wireless channel scenario, the channel impulse responses span over a few samples each, so their convolution is roughly twice the length of one impulse response, located around both sides of 0. Note that, for the channel parameters used, in more than 50% of the simulated channel realizations some gain was obtained by introducing the cyclic delay. For the

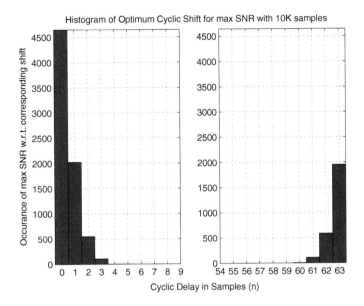

Fig. 11.8 Histogram of cyclic shifts obtained for maximum average SNR with 10000 samples.

readers information, without cyclic delays, Pre-DFT diversity combining was studied in [174].

11.5.4 Numerical Analysis and Discussions

Channel model: A second-order stochastic channel model (WSSUS model) suitable for Rayleigh and Ricean fading distributions was used in this work. The frequency-selectivity is described by the spaced-frequency correlation function and by the delay power spectrum (DPS) [192]. In the simulations, realizations of channel transfer functions are generated directly, based on well-defined channel parameters, such as the normalized (or average) received power P_0, the Ricean parameter K and the RMS delay spread τ_{rms}. Indoor WLAN channels with $\tau_{rms} = 5\,\text{ns}$ to $50\,\text{ns}$ are generated. This corresponds to 0.1 to 1 sample considering a sampling frequency of 20 MHz as used in IEEE 802.11a. Rayleigh fading scenarios (i.e. $K = 0$) and Ricean fading scenarios with $K = 4$ were considered.

Simulation parameters: Simulations are performed with parameters stipulated by the IEEE 802.11a WLAN standard: number of OFDM sub-carriers,

$N = 64$, length of Cyclic Prefix (CP), $N_{CP} = 16$ samples, OFDM symbol duration, $T_S = 4\,\mu s$ (consists of useful data period of $3.2\,\mu s$ and CP duration of $0.8\,\mu s$), QPSK symbol mapping with half-rate convolutional coding (corresponds to 12 Mbps raw bit rate at the receiver), system bandwidth of 20 MHz and operating at the 5 GHz band. The simulations presented in this book only considered the dual antenna case (i.e. $M = 2$).

Analysis of channel responses after combining: The analysis shows that the amount of cyclic shift and the combiner weight factors can be determined effectively in order to achieve optimum SNR in all cases for a Pre-DFT *Receiver* CDD system, because the CSI can be estimated. Figure 11.9 shows the magnitude response of the combined channels (equivalent to the SNR per sub-carrier) for several receiver diversity schemes, along with the channel responses of the branch channels. It is seen that the Post-DFT MRC scheme shows better SNR characteristics, though the responses for Pre-DFT MARC and Pre-DFT EGC are also very close.

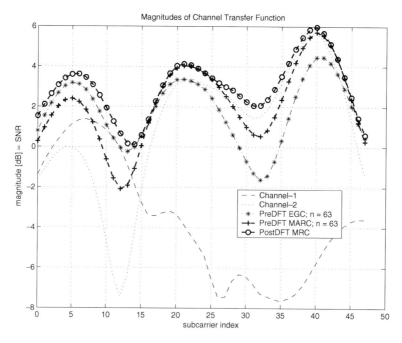

Fig. 11.9 Magnitudes of channel transfer functions before and after diversity combining. Pre-DFT MARC, Post-DFT MRC, and Pre-DFT EGC are shown in the figure.

Performance results and discussions: BER simulations have been performed for dual antenna receiver diversity using Post-DFT MRC, Pre-DFT MARC, and Pre-DFT EGC. For comparison, pure CDD at the transmitter (Tx-CDD) with fixed cyclic delay of 16 samples [196] and Pre-DFT MARC without cyclic delay (which is equivalent to the technique described in [174]) are also simulated.

Figures 11.10 and 11.11 show uncoded BER results, which were calculated in a semi-analytical way as follows. For the various receiver concepts compared, the SNR values on the OFDM sub-carriers were simulated. Based on these simulated channels, the BERs were determined analytically, using the Q-function, and averaged. The E_b/N_0 shown is the ratio of the average symbol energy per sub-carrier to the noise power density. Coherently detected QPSK with perfect channel estimation is assumed in this analysis.

On the Rayleigh channel, Tx-CDD does not give any performance advantage in terms of uncoded BER compared with a single antenna receiver, because the channel gains are added up incoherently just like the noise. On

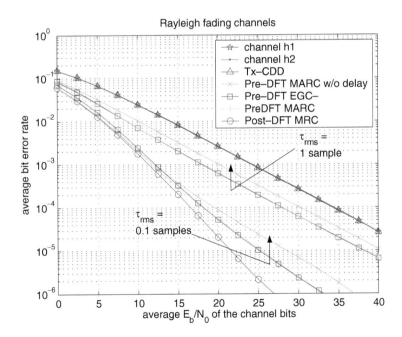

Fig. 11.10 Uncoded BER with and without application of diversity, Rayleigh fading channels with various τ_{rms}.

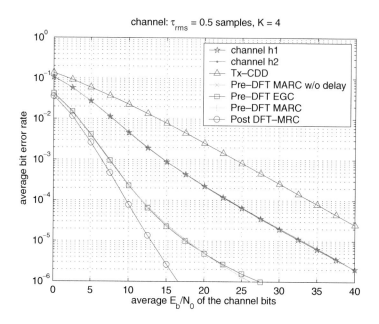

Fig. 11.11 Uncoded BER with and without application of diversity, Ricean channel with $K = 4$.

the Ricean channel, the performance with Tx-CDD is even worse, because the combined channel has deeper fades than the component channels. This is one of the main drawbacks of Tx-CDD. The best performance is achieved with Post-DFT MRC at the cost of high computational complexity. The Pre-DFT receiver diversity schemes lie in between those results. It is evident that more can be gained over "flatter" channels (lower τ_{rms} and/or higher K), which is not surprising since in these cases the Pre-DFT combining schemes can add up the channel transfer functions constructively over a wider frequency range. Under the same condition, we observe less performance difference among the Pre-DFT schemes exploiting the CSI. Pre-DFT MARC and Pre-DFT EGC show very similar performance although the average SNR over the sub-carriers is significantly higher using MARC. The gain over Pre-DFT MARC without delay can be significant, but it reduces on channels with a very short channel impulse response.

In Figure 11.12, performance results are given in terms of bit error rate for the coded OFDM system. The source data is FEC-coded with a $\frac{1}{2}$ rate convolutional coder, whose constraint length is 5, i.e., the effect of each information

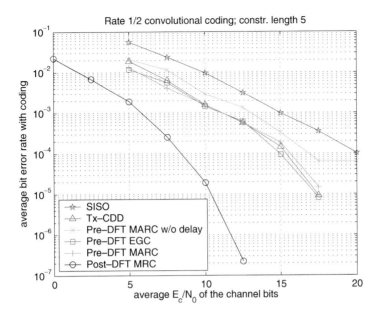

Fig. 11.12 Performance results in terms of coded BER; Rayleigh channel, $\tau_{rms} = 1$ sample; rate $\frac{1}{2}$ convolutional coding with constraint length 5.

bit is spread over roughly 10 FEC coded bits. A block interleaver (with interleaver depth $= 4$) is used to exploit the available frequency diversity. After interleaving the FEC-related bits are spread over 40 sub-carriers in an OFDM symbol (which consists of 48 data sub-carriers). Note that the constraint length was reduced compared with the WLAN standard in order to speed up the computer simulations. Although this affects the absolute results, we expect the general trends and conclusions to be equivalent. BER performance shows that the largest gain is achieved with the "traditional" Post-DFT MRC technique, amounting to almost 7 dB at $BER = 10^{-4}$. About 3–4 dB gain are observed from the Pre-DFT diversity combining techniques, whose performance is remarkably similar. Only MARC without delay is slightly weaker. In particular, after coding, the "optimized" techniques applied at the receiver, which can use CSI, perform not much better than the "unsupervised" combining technique using a fixed cyclic delay.

It is evident from Figures 11.10–11.12 that the scheme works better in situation where the RMS delay spread is quite small, which means that, in a typical indoor WLAN environment, Pre-DFT MARC will perform well.

It is a prolific advantage that Pre-DFT MARC scheme works better even if the diversity branches are correlated to each other due to LOS scenario. This brings another benefit, i.e. usually the size constraint in MS causes insufficient spatial separation between antennas. When the antennas are closely placed to each other, then the diversity branches will experience sufficient correlation which will destroy the benefits that the diversity schemes (such as Post-DFT MRC or EGC or SCo) bring. In those cases, Pre-DFT MARC scheme can be used to combine the signals efficiently from correlated diversity branches.

Efficient implementation: We compared the complexity of the schemes in terms of number of multiplications required. Considering that we have a channel which is time-invariant for considerable amount of time, so that N_{pkt} number of OFDM symbols can be put in one OFDM packet, then the number of multiplications required for one OFDM symbol are

$$X_{\text{pre}} = \frac{Q(Q-1)N_o}{2} + N_{pkt}\left(\frac{N}{2}\log_2 N + QN\right) \quad (11.34)$$

$$X_{\text{post}} = N_{pkt}\left(\frac{QN}{2}\log_2 N + QN\right), \quad (11.35)$$

where $\frac{N}{2}\log_2 N$ multiplications are required for FFT module per OFDM symbol [11]. The 1st term of Equation (11.34) corresponds to the calculation of covariance matrix shown in Figure 11.6. N_o is the number of time-domain data samples that need to be acquired to obtain the correlation matrix. The computational complexity associated with Eigen analysis for gain factors is not taken into account, as it is only required once for complete OFDM packet [174].

Figures 11.13(a) and 11.13(b) show the relative processing cost between Pre-DFT MARC and Post-DFT MRC in comparison to N and N_{pkt}, respectively, for different values of Q. In both figures, we can see that the processing cost is drastically reduced in the presented scheme when N or N_{pkt} or Q increases.

Effect of channel time-variance: Referring to Figure 11.5, Pre-DFT MARC scheme requires channel estimates in time domain for delay insertion and gain factor estimation; and in frequency domain for data detection. If the channel is severely time variant, then the channels need to be estimated very frequently, and thus the savings in complexity due to time-domain combining will be lost in excessive channel estimation burden. Table 11.1 summarizes the

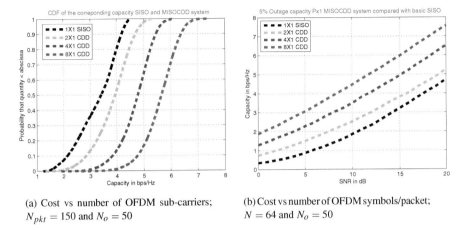

(a) Cost vs number of OFDM sub-carriers; $N_{pkt} = 150$ and $N_O = 50$

(b) Cost vs number of OFDM symbols/packet; $N = 64$ and $N_O = 50$

Fig. 11.13 Relative processing cost for Pre-DFT MARC and Post-DFT MRC in comparison to number of OFDM sub-carriers and number of OFDM symbols/packet.

Table 11.1 Comparison of channel coherence time with OFDM packet duration for IEEE 802.11a WLAN Standard.

v, km/h	f_d, Hz	T_c, ms	max N_{pkt}
3	54	7.83	1957.5
10	180	2.35	587.5
20	360	1.175	293.75
50	900	0.47	117.5
100	1800	0.235	58.75
250	4500	0.094	23.5

relationship between channel coherence time with respect to user velocity in the context of IEEE 802.11a WLAN system. If the coherence time is defined as the time over which the time correlation function is 0.5, then the coherence time is approximately, $T_c = \sqrt{\frac{9}{16\pi f_d^2}} = \frac{0.423}{f_d}$ [40, Section 4.4.3], when f_d is maximum doppler shift. In the last column of Table 11.1, we have shown the number of OFDM symbols that can be transmitted during T_c at corresponding user velocity. In this case, we have taken that one OFDM symbol duration is $4\,\mu s$. The standard specifies a test case of 1000 octets in one OFDM packet, this corresponds to $\frac{1000 \times 8}{24} = 333.3$ OFDM symbols in a packet with BPSK modulation and $\frac{1}{2}$ rate convolutional coding at the best-case scenario. We can see from the table that a velocity up to 20 km/h will have a coherence time that equals to 293.75 OFDM symbols. Twenty kilometer per hour is quite a high velocity (if not impractical) for a WLAN user. When the velocity is increased (in the order of tens of km/h), then obviously the complexity will go higher. In general, when the channel is less time-variant, then the coherence time is larger and so the packet duration can be made arbitrarily larger, so the channel estimation frequency will be smaller.

11.6 Cyclic Delay Assisted Spatial Multiplexing

Spatial Multiplexing (SM) is a promising and powerful technique to dramatically increase the system capacity in terms of throughput. The independent spatial channels in rich scattering environment can be exploited to send multiple signals at the same time and frequency, thus it is also spectrally very efficient. Most of the available SM techniques are effective in frequency flat scenario [139, 189]. Thus, in frequency selective environment, amalgamation of SM and OFDM techniques can provide a high data rate system where the spectral efficiency can also be high. All the algorithms can now be implemented on OFDM sub-carrier level, since OFDM converts a wideband frequency selective channel into a number of narrowband sub-carriers [27, 143].

Original SM techniques are proposed for increasing spatial data rate [142, 146]. These algorithms do not require any CSI at the transmitter, but require complete CSI at the receiver. The performance of such systems are always under investigation, because, for a sufficient Frame Error Rate (FER) performance, we need to have ML detection schemes, but the ML schemes are always very complex to implement. Thus, VBLAST is proposed, as a sub-optimal solution, and as a trade-off between performance and cost [29]. Recently there are some approaches of incorporating the VBLAST technique with some well known STC techniques, so that both diversity and multiplexing benefits can be obtained in one transmission structure [202, 203]. We call such a system as JDM system. Arguably, the performance of JDM system would be better compared to original SM system (as in [29]). In [202], the Spatially-Multiplexed Orthogonal Frequency Division Multiplexing (SM-OFDM) system uses two independent STC for two sets of transmit antennas. Thus, an original 2×2 SM-OFDM system is now extended to 4×2 STC aided SM-OFDM system. In the receiver, the independent STC are decoded first using pre-whitening, followed by maximum likelihood detection. Again, this increases the receiver complexity tremendously, though the system performance gets much better.

In this work, we discuss using CDD as the way to introduce the Space-Time Coding (STC) on Spatial Multiplexed Orthogonal Frequency Division Multiplexing (SM-OFDM) system. We denote the combined system as CDA-SM-OFDM system. Introducing CDD in SM system only requires changes in the transmitter and the receiver remains intact. Thus, the complexity increment is very minimal, but an improvement in performance is expected. We have compared the proposed CDA-SM-OFDM system with original SM schemes and have shown significant improvement in error rate.

11.6.1   Transmission Structure

Figure 11.14 shows the basic transmitter and receiver structure, when cyclic delays are used at the transmission side of a SM-OFDM system. The source data is FEC coded and baseband modulated. We denote this

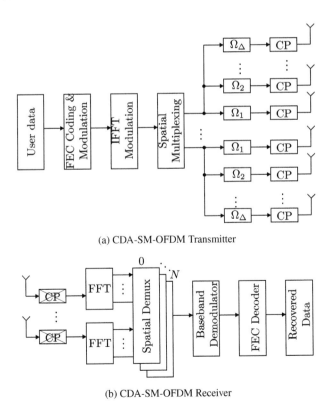

(a) CDA-SM-OFDM Transmitter

(b) CDA-SM-OFDM Receiver

Fig. 11.14 Transceiver architecture for CDA-SM-OFDM system.

baseband modulated symbols as m_k. Then IFFT modulation is performed and time-domain samples, xn, are placed for appropriate STC. For example, the sequence of xn is de-multiplexed into $\mathbf{x}_1, \ldots, \mathbf{x}_P$ signal vectors. After that $\mathbf{x}_p$ for each p is cyclically delayed. In Figure 11.14, the blocks noted by $\Omega_1, \Omega_2, \ldots, \Omega_\Delta$ perform the cyclic delay operation. The amount of delays for δth delayed branch for pth SM branch is given by Equation (11.39). $\mathbf{x}_p$ and its cyclically delayed branches are transmitted via Δ number of antennas, so we denote the transmitted symbol vectors as, $\mathbf{x}_p^{(\delta)} = \mathbf{x}\big((n - \Omega_\delta)\big)_N$. This means that total number of transmit antennas in the transmitter is $P\Delta$. For any p, all corresponding Δ number of cyclic delay branches are named together as pth group of transmit antennas. The receiver is similar to the original Vertical-Bell Labs LAyered Space-Time Architecture (VBLAST) receiver as explained in [29].

We denote the received signal vectors at qth receive antenna as $\mathbf{y}_q$ and the corresponding OFDM demodulated symbol vector as $\mathbf{z}_q$. For the following description, we use the notations, $g_{q,p}^{(\delta)}$ and $h_{q,p}^{(\delta)}$, as the CIR and CTF between δth cyclic delay branch of pth transmit antenna group and qth receive antenna, respectively. $\overset{N}{\circledast}$ denotes cyclic convolution operation of N size. The received signals at qth receive antenna can be written as

$$
\begin{aligned}
yn_q &= \frac{1}{\sqrt{\Delta}} \left[g_{q,1}^{(1)} \overset{N}{\circledast} xn_1 + g_{q,1}^{(2)} \overset{N}{\circledast} x\big((n - \Omega_2)\big)_{N1} + \cdots \right. \\
&\quad \left. + g_{q,1}^{(\Delta)} \overset{N}{\circledast} x\big((n - \Omega_\Delta)\big)_{N1} \right] + \cdots \\
&\quad + \frac{1}{\sqrt{\Delta}} \left[g_{q,P}^{(1)} \overset{N}{\circledast} xn_P + g_{q,P}^{(2)} \overset{N}{\circledast} x\big((n - \Omega_2)\big)_{NP} + \cdots \right. \\
&\quad \left. + g_{q,P}^{(\Delta)} \overset{N}{\circledast} x\big((n - \Omega_\Delta)\big)_{NP} \right] + vn_q
\end{aligned}
\tag{11.36}
$$

$$
\begin{aligned}
&= \frac{1}{\sqrt{\Delta}} \left[g_{q,1}^{(1)} + g_{q,1}^{(2)}\big((n - \Omega_2)\big)_N + \cdots + g_{q,1}^{(\Delta)}\big((n - \Omega_\Delta)\big)_N \right] \\
&\quad \overset{N}{\circledast} xn_1 + \cdots + \frac{1}{\sqrt{\Delta}} \left[g_{q,P}^{(1)} + g_{q,P}^{(2)}\big((n - \Omega_2)\big)_N + \cdots \right. \\
&\quad \left. + g_{q,P}^{(\Delta)}\big((n - \Omega_\Delta)\big)_N \right] \overset{N}{\circledast} xn_P + vn_q
\end{aligned}
\tag{11.37}
$$

$$= \frac{1}{\sqrt{\Delta}} \left[\sum_{\delta=1}^{\Delta} g_{q,1}^{(\delta)} \big((n - \Omega_\delta) \big)_N \right] \overset{N}{\circledast} xn_1 + \cdots$$

$$+ \frac{1}{\sqrt{\Delta}} \left[\sum_{\delta=1}^{\Delta} g_{q,P}^{(\delta)} \big((n - \Omega_\delta) \big)_N \right] \overset{N}{\circledast} xn_P + vn_q. \tag{11.38}$$

Equation (11.38) basically gives us the equivalent CIR of the wireless channel between qth receive antenna and all Pth transmit multiplexing branches, where $p \in \{1, \ldots, P\}$. For the above formulation, we took the CDD branches to be equal for all transmit antenna groups (i.e. $= \Delta$) for simplicity of expressions. In practice, there should not be any problem in choosing different number of antennas for each transmit antenna group. The amount of cyclic delays for Δ number of antennas in any transmit group can be chosen based on the analysis in [204] as

$$\Omega_\delta = \frac{N(\delta - 1)}{\Delta} = \frac{N}{\Delta} + \Omega_{\delta-1}. \tag{11.39}$$

Now we can obtain the equivalent frequency response of the system from Equation (11.38). We define the following relationships:

(1) $\mathbb{F}\{\mathbf{y}_q\} = \mathbf{z}_q$ and $\mathbb{F}\{yn_q\} = z_k^q$,
(2) $\mathbb{F}\{\mathbf{x}_p\} = \mathbf{m}_p$ and $\mathbb{F}\{xn_p\} = m_k^p$,
(3) $\mathbb{F}\{\mathbf{v}_q\} = \mathbf{n}_q$ and $\mathbb{F}\{vn_q\} = n_k^q$,

where z_k^q, m_k^p, and n_k^q are elements of $\mathbf{z}_q$, $\mathbf{m}_p$, and $\mathbf{n}_q$, respectively. Using above definitions and Equation (11.38), the DFT demodulated symbols at kth sub-carrier (i.e. in frequency-domain) of qth receive antenna can be written as

$$z_k^q = \frac{1}{\sqrt{\Delta}} \left[\sum_{\delta=1}^{\Delta} h_{q,1}^{(\delta)}[k] W_N^{-k\Omega_\delta} \right] m_k^1 + \cdots + \frac{1}{\sqrt{\Delta}} \left[\sum_{\delta=1}^{\Delta} h_{q,P}^{(\delta)}[k] W_N^{-k\Omega_\delta} \right] m_k^P + n_k^q. \tag{11.40}$$

So, the simplified system response for CDD assisted SM-OFDM system can be written as

$$\begin{bmatrix} z_k^1 \\ \vdots \\ z_k^Q \end{bmatrix} = \mathbf{H}^{\mathrm{CDD}}[k] \begin{bmatrix} m_k^1 \\ \vdots \\ m_k^P \end{bmatrix} + \begin{bmatrix} n_k^1 \\ \vdots \\ n_k^P \end{bmatrix}, \tag{11.41}$$

where

$$\mathbf{H}^{CDD}[k] = \frac{1}{\sqrt{\Delta}} \begin{bmatrix} \sum_{\delta=1}^{\Delta} h_{1,1}^{(\delta)}[k]W_N^{-k\Omega_\delta} & \cdots & \sum_{\delta=1}^{\Delta} h_{1,P}^{(\delta)}[k]W_N^{-k\Omega_\delta} \\ \vdots & \ddots & \vdots \\ \sum_{\delta=1}^{\Delta} h_{Q,1}^{(\delta)}[k]W_N^{-k\Omega_\delta} & \cdots & \sum_{\delta=1}^{\Delta} h_{Q,P}^{(\delta)}[k]W_N^{-k\Omega_\delta} \end{bmatrix}. \qquad (11.42)$$

In the receiver, Ordered Successive Interference Cancellation (OSIC) approach as devised in [29] can be used in straight-forward manner. The key idea of the nonlinear OSIC receiver is to decode the symbol streams successively and extract them away layer by layer. At the beginning of each stage, the stream with the highest SNR for peeling is extracted. Once one particular layer is decoded, then the effect of the detected layer is subtracted from the received signal and the next branch in terms of SNR strength is chosen. This is continued until the last layer is decoded.

Ordered Successive Interference Cancellation (OSIC) is performed on a sub-carrier by sub-carrier basis. On each sub-carrier, the detection scheme appears to be very similar to VBLAST detection, as derived in [29]. OSIC improves the detection quality compared to detection without ordering and is shown to be the optimal Successive Interference Cancellation (SIC) approach in terms of error rate performance [139].

11.6.2   System Capacity of CDA-SM-OFDM

One can easily notice the difference between the effective of channel response matrix kth sub-carrier for CDA-SM-OFDM system (as in Figure 11.14) and SM-OFDM system [29] by looking at Equations (11.41) and (11.44), respectively. The channel matrix for SM-OFDM scheme, $\mathbf{H}[k]$, is shown below:

$$\mathbf{H}^{CDD}[k] = \frac{1}{\sqrt{\Delta}} \begin{bmatrix} \sum_{\delta=1}^{\Delta} h_{1,1}^{(\delta)}[k]W_N^{-k\Omega_\delta} & \cdots & \sum_{\delta=1}^{\Delta} h_{1,P}^{(\delta)}[k]W_N^{-k\Omega_\delta} \\ \vdots & \ddots & \vdots \\ \sum_{\delta=1}^{\Delta} h_{Q,1}^{(\delta)}[k]W_N^{-k\Omega_\delta} & \cdots & \sum_{\delta=1}^{\Delta} h_{Q,P}^{(\delta)}[k]W_N^{-k\Omega_\delta} \end{bmatrix} \qquad (11.43)$$

$$\mathbf{H}[k] = \begin{bmatrix} h_{1,1}[k] & h_{1,2}[k] & \cdots & h_{1,P}[k] \\ h_{2,1}[k] & h_{2,2}[k] & \cdots & h_{2,P}[k] \\ \vdots & \vdots & \ddots & \vdots \\ h_{Q,1}[k] & h_{Q,2}[k] & \cdots & h_{Q,P}[k] \end{bmatrix}. \tag{11.44}$$

Here, we recall the well-known channel capacity equations for any SM system when CSI is unknown at the transmitter. This is given by [139, Section 4.3]

$$C_k = \log_2\left[\det\left(\mathbf{I}_Q + \frac{\overline{\gamma}}{P}\mathbf{H}[k]\mathbf{H}^H[k]\right)\right], \tag{11.45}$$

where $\mathbf{H}^H[k]$ denotes the complex conjugate transpose (or Hermitian transpose) of $\mathbf{H}[k]$. Equivalently, we can write that[1]

$$C_k = \sum_{i=1}^{r}\log_2\left(1 + \frac{\overline{\gamma}}{P}\lambda_i\right), \tag{11.46}$$

where r is the rank of the channel and λ_i are the positive eigenvalues of $\mathbf{H}[k]\mathbf{H}^H[k]$. Thus, the total Mimo system capacity is actually the sum of r SISO channels (or spatial modes) with a power gain of λ_i. In the absence of CSI at the transmitter, we cannot exploit the different channel gain for different spatial modes, and thus, the power division by P appears in Equations (11.45) and (11.46). Similar system capacity expressions can be obtained for CDA-SM-OFDM system based on Equations (11.45) and (11.46) as

$$C_k = \log_2\left[\det\left(\mathbf{I}_Q + \frac{\overline{\gamma}}{P\Delta}\mathbf{H}^{\mathrm{CDD}}[k]\mathbf{H}^{\mathrm{CDD},H}[k]\right)\right] \tag{11.47}$$

$$= \sum_{i=1}^{r}\log_2\left(1 + \frac{\overline{\gamma}}{P\Delta}\lambda_i^{\mathrm{CDD}}\right), \tag{11.48}$$

where $\mathbf{H}^{\mathrm{CDD},H}[k]$ is Hermitian transpose of $\mathbf{H}^{\mathrm{CDD}}[k]$, and λ_i^{CDD} are the positive eigenvalues of $\mathbf{H}^{\mathrm{CDD}}[k]\mathbf{H}^{\mathrm{CDD},H}[k]$.

For a better understanding of the system capacity, we have examined the statistical characteristics of the eigenvalues of both channel structures. We

---

[1] For derivation of this expression, interested readers are requested to read the above mentioned reference (i.e. [139, Section 4.3]).

define the *normalized eigenvalue spread* as the difference between the maximum and the minimum eigenvalues, normalized by mean eigenvalue. When the eigenvalue spread is higher, then it is understood that one or more of the spatial modes are far weaker than others, or one or more spatial modes are far stronger than others. So when the spread goes very high, we may end up with available spatial modes less than $\min(P, Q)$. This means that the effective capacity gain of the Multiple Input Multiple Output (MIMO) channel is reduced. One other property that can be checked very closely with normalized eigenvalue spread is *mean eigenvalue*. This is defined as mean of $\lambda_i; \forall i$. Though this parameter may be very misleading about the system capacity, it gives some valuable information when studied along with eigenvalue spread information.

Figures 11.15(a) and 11.15(b) show, the CDF of normalized eigenvalue spread and mean eigenvalue for the case when $P = 2$, $Q = 2$, and $\Delta = 2$. Figure 11.15(a) shows that the normalized eigenvalue spread for 4×2 CDA-SM-OFDM system is in between 2×2 and 4×2 SM-OFDM systems. For mean eigenvalue for the two 4×2 system are much higher compared to 2×2 system, as shown in Figure 11.15(b). This is because of the extra spatial diversity effect that is available in 4×2 system. Note that 2 receive antennas means only 2-spatial branches can be transmitted for all the systems, thus, the increase in capacity for 4×2 system is logarithmic, rather than linear. We know that when the eigenvalue spread is the same for two channel matrices,

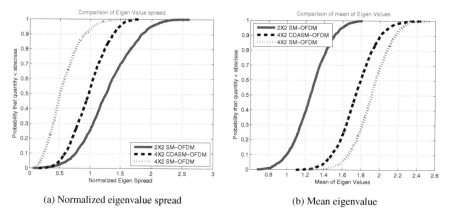

(a) Normalized eigenvalue spread   (b) Mean eigenvalue

Fig. 11.15 Comparison of normalized eigenvalue spread and mean eigenvalue for 4×2 CDA-SM-OFDM system with 2×2 and 4×2 SM-OFDM system.

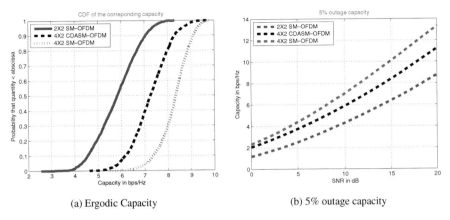

(a) Ergodic Capacity   (b) 5% outage capacity

Fig. 11.16 Comparison of system capacity and 5% outage capacity for 4×2 CDA-SM-OFDM system with 2×2 and 4×2 SM-OFDM system.

if the mean eigenvalue is higher for one of them, the capacity will be higher for that channel too. Thus, the system capacity is increased in 4×2 system compared to 2×2 system as shown in Figure 11.16(a). As expected, the capacity curves show that the system capacity for 4×2 CDA-SM-OFDM system is in between the capacity curve for 2×2 and 4×2 SM-OFDM systems.

The above comparison refers to the ergodic capacity of the respective systems. Now, from these ergodic capacity curves, we can obtain the outage capacity values. We define the *outage capacity* as the capacity in b/s/Hz obtained below 10% of the times. If we look at Figure 11.16(a), 10% of the times, the system capacity will be below 4.5 b/s/Hz approximately for 2×2 SM-OFDM system. For the same capacity level, the outage will be less than 1% for 4×2 CDA-SM-OFDM system and almost 0% for 4×2 ideal SM-OFDM system. Thus, the outage probability is also greatly reduced in CDA-SM-OFDM system compared to simple SM-OFDM system. Figure 11.16(b) shows us the outage capacity vs Pre-SNR for the systems, where we can see that outage capacity is increased when higher diversity order is available and when the Pre-SNR is increased.

It should be noted here that capacity for 4×2 and 2×2 SM-OFDM systems can be achieved when the best possible STCs (or Space–Frequency Codings (SFCs)) and channel coding are available in corresponding antenna configurations. Thus, the capacity curves in Figure 11.16(a) shows an upper limit of capacity for any 4×2 SM-OFDM system. In contrast to this, 4×2 CDA-SM-OFDM achieves a capacity closer to this upper limit with a simple STC, i.e. CDD. Introducing CDD does not require any change in the receiver, changes are required only in the transmitter. In this case, two more transmit antenna branches are required and the delay element need to be inserted in all the transmit branches.

11.7 Chapter Summary

Cyclic Delay Diversity (CDD) is shown to be a good technique for exploiting the available spatial diversity either at transmitter, or at receiver, or at both locations simultaneously. The technique can be implemented at the transmitter without altering the receiver technique. Similarly, when applied at receiver, significant performance gain can be achieved by exploiting the known channel information at the receiver. We have shown that the proposed receive CDD scheme performs considerably well in some scenarios considering the trade-off between performance and complexity. Because of the low complexity the presented CDD approach will allow low cost wireless modules targeting the mass market. For less frequency-selective channel, which is very similar to typical indoor or immediate outdoor wireless channel, Pre-DFT MARC with CDD combining is a lucrative option for cost effective, efficient and reliable diversity reception. In situations where diversity branches are correlated to each other because of the presence of LOS conditions, Pre-DFT MARC scheme works very well and efficiently combines the diversity branches to increase the transmission quality. It has also been understood that the scheme works well in severely time-variant situations.

We have only studied a single-tap weight estimation for Pre-DFT MARC scheme. The goal was to limit the system complexity, but still obtain a reasonable performance. For better performance, we can use multi-tap weight estimation, thus the scheme can become more robust to highly frequency-selective channels. Also, more taps in weight estimator can ease the receiver equalization process. But, one needs to keep in mind that, multi-tap time-domain weight estimation may require high equalization complexities and may jeopardize the simple single-tap equalization benefit that OFDM offers on every flat sub-carriers. Also, the proposed receiver Pre-DFT MARC principle can be implemented at the transmitter with similar system performance, but, it would require the channel information to be transported at the transmitter, thus system overhead would increase.

In collaboration to receive CDD proposed in this chapter, we can also use transmit-CDD with fixed delays simultaneously, so that the diversity principle can be used at both locations of the transmission. The improvement of the system performance in this case needs to be investigated.

When CDD is used in collaboration with any standard SM system, it increases the complexity marginally, but gives much better channel capacity values compared to basic SM-OFDM systems. CDD-based STC on SM-OFDM is much less complex compared to any other well-established STC techniques, such as Alamouti's 2×1 block codes. It is understood that CDA-SM-OFDM can be easily incorporated with any standard SM-OFDM system and can be directly used for DL using the proposed structure in Figure 11.14. When applied at the receiver, the scheme can also be used for UL.

However, it needs to be mentioned that channel estimation modules need to be taken care of, when CDD is used in the system. The technique essentially destroys the frequency correlation across neighboring sub-carriers, while channel estimation schemes sometimes exploit the correlation between the channel coefficients of adjacent sub-carriers in order to reduce the noise floor of the channel estimates [192]. Thus, eliminating this correlation would cause failure to some channel estimation schemes, and would require a re-design when CDD is used. Several solutions regarding channel estimation in CDD-based systems are proposed [205, 206] and can be modified to be used with the proposed receive CDD technique in this chapter. In [207], it is shown that the required additional pilot symbol overhead for channel estimation in CDD systems is the same compared to other STBC systems with same number of transmit antennas.

Future work: The impact of spatial correlation across the antennas at both transmitter and receiver on the proposed scheme can be studied in future. Incorporation of such a scheme in a multi-user scenario can also be investigated. It is also understood that the rate control across different multiplexing streams will yield throughput improvement for CDA-SM-OFDM system. For Bit-Interleaved Coded Modulation (BICM) systems, proper interleaver design together with efficient FEC coding can be studied for CDD based systems.

12

Joint Diversity and Multiplexing Schemes for MIMO-OFDM Systems

12.1 Introduction

Multiple antennas can be used at both ends of a MIMO wireless transmission system to exploit the benefits of spatial dimension. Two benefits can be obtained, namely Space Diversity (SD) and Spatial Multiplexing (SM). Space–Time Coding (STC) and Maximal Ratio Combining (MRC) can be used at the transmitter side and/or the receiver side, respectively, to exploit the maximum spatial diversity available in the channel. This can be used to increase the system reliability [139]. Similarly, parallel spatial channels between multiple antennas can be exploited for increasing the transmission rate without increasing required bandwidth. In rich scattering environment, the independent spatial channels can be exploited to send multiple signals at the same time and frequency, resulting in higher spectral efficiency [27].

Space Diversity (SD) schemes provide higher Signal-to-Noise Ratio (SNR) at the receiver, whereas Spatial Multiplexing (SM) schemes provide higher rate [31]. Transmit SD schemes have multiple antennas at the transmitter, and may have only one or more than one receive antennas, whereas SM schemes must have multiple antennas at both sides. A subset of SD schemes uses some sort of STBC (or SFBC) coding at the transmitter (i.e. $\Delta \times 1$ system, where Δ is the number of antennas at the transmitter side; conversely we define Q as the number of receive antennas). Depending on different Δ, we can classify the SD systems into three different categories: full diversity, full rate orthogonal [30]; partial diversity, full rate quasi-orthogonal [208] and full diversity, rate loss orthogonal [175] schemes for $\Delta = 2$, $\Delta > 2$ and $\Delta > 2$, respectively. These SD schemes do not require any channel information at the transmitter, thus, they can be implemented at fairly low system complexity and signaling overhead.

Spatial Multiplexing (SM) schemes are twofold, in fact they can either be Channel State Information (CSI) assisted or nonCSI based, i.e. so called blind SM schemes. Typical CSI assisted systems are Singular Value Decomposition (SVD) based SM, eigen-mode Multiple Input Multiple Output (MIMO), etc. These schemes provide higher system capacity, but require a feedback path, thus they are more complex and difficult to implement. Blind SM methods, such as BLAST, are easier to implement compared to SVD based SM schemes.

Blind SM schemes can have different kinds of receivers:

1. Linear Zero Forcing (ZF) or MMSE receiver: simple receiver, but poor performance [27].
2. Nonlinear ML receiver: optimal and most complex receiver [139, 143].
3. Nonlinear OSIC receiver with ZF or MMSE nulling: sub-optimal receiver [29].

Most of the above mentioned MIMO techniques are effective in frequency flat scenarios [169, 170]. All these narrowband algorithms can be implemented on Orthogonal Frequency Division Multiplexing (OFDM) sub-carrier level, because OFDM converts a wideband frequency selective channel into a number of narrowband sub-carriers. In another words, OFDM turns a frequency-selective MIMO fading channel into a set of parallel frequency-flat MIMO fading channels. So, in wideband scenarios, OFDM can be combined with MIMO systems, for both diversity and multiplexing purposes [154]. In frequency selective environments, amalgamation of SM and OFDM techniques can be a potential source of high spectral efficiency at reasonable complexity, because MIMO-OFDM drastically simplifies equalization in frequency-selective environments.

None of the schemes mentioned above can become an absolute choice for better system performance in all scenarios. To be exact, a user located very close to access point can exploit the multiplexing benefits, while a user at farther location may benefit more by using MIMO diversity schemes. That's why MIMO switching schemes, such as [32], are well studied and examined. In switching schemes, either SD or SM schemes are optimally chosen based on some specific selection criterion, so that optimum benefits from both type of schemes can be obtained in the system. This kind of combination is actually performed in time domain. In recent years, combining SM and SD schemes in space domain has gained a lot of interest, because these schemes are beneficial in effect that both diversity and multiplexing benefits are available at the same time. In this work, we study such MIMO structures which simultaneously exploit diversity and multiplexing gains at the same signaling scheme.

In a cellular wireless system, the STBC-OFDM [193] and SFBC-OFDM [176] can be used to increase the resultant SNR at the receiver, thus, increasing the coverage area. In contrast to this, as SM requires high receive SNR for reliable detection [32], it is evident that users at farther locations from BS cannot use SM techniques to enhance the spectral efficiency. Thus, it is required to combine both of these two techniques in one structure so that both the diversity and multiplexing benefits can be achieved at farther locations from transmission source. We call such schemes as Joint Diversity and Multiplexing (JDM) schemes.

To be specific, blind SD schemes enhance the receive SNR, thus they are good for users at locations farther from the BS; while blind SM schemes require higher SNR to efficiently decode the multiplexing branches, thus only users located near to the BS can get the services [31]. We can combine SM and SD, in one structure in the following ways (by P we mean the number of SM branches):

1. $P, Q =$ arbitrary; $\Delta = 2$; Alamouti scheme at every SM branch, thus, $2P \times Q$ Spatially-Multiplexed Orthogonal Space–Frequency Block Coding (SM-OSFBC) systems. Here, we denote Δ as the number of antennas per SM branch where Space–Frequency Block Code (SFBC) is applied;

2. $P, Q =$ arbitrary; $\Delta > 2$; Orthogonal partial rate SFBC scheme at every SM branch, thus, $P\Delta \times Q$ SM-OSFBC systems. These schemes lose SM benefits because of partial rate SFBC, so they are not considered in this work;

3. $P, Q =$ arbitrary; $\Delta > 2$; Quasi-orthogonal full rate SFBC scheme at every SM branch, thus, $P\Delta \times Q$ Spatially-Multiplexed Quasi-orthogonal Space–Frequency Block Coding (SM-QSFBC) systems.

When coding across S–T is envisioned, we can have Spatially-Multiplexed Orthogonal Space–Time Block Coding (SM-OSTBC) or

Spatially-Multiplexed Quasi-orthogonal Space–Time Block Coding (SM-QSTBC) systems.

Recently there are some approaches of incorporating the VBLAST technique with some well known STC techniques. One such work is described in [202], where a combination of SD and SM for MIMO-OFDM system is proposed. Arguably, the performance of such a JDM system would be better than SD only and SM only schemes. In [202], the SM-OFDM system uses two independent STC for two sets of transmit antennas. Thus, an original 2×2 SM-OFDM system is now extended to 4×2 STC aided SM-OFDM system. In the receiver, the independent STC are decoded first using pre-whitening, followed by maximum likelihood detection. Again, this increases the receiver complexity quite a lot, though the system performance gets much better. In later work, Alamouti's STBC is combined with SM for OFDM system in [209], and a linear receiver is designed for such a combination. Nonlinear receiver for similar SM-OSTBC scheme is mentioned in [210], though the receiver structure is not detailed. The main concentration of [210] is on switching between MIMO techniques, i.e. between SD only, SM only, and SM-OSTBC schemes. Choosing antenna pairs in SM-OSTBC scheme based on correlation criterion is studied in [211], while again receiver structures are not discussed at all. AMC on SM-OSTBC is studied in [212]. Detection schemes for SM-OSTBC system are analyzed in [213], where linear ZF and QR decomposition based receiver is discussed.

Following these trends, we have discussed the combination of SFBC with SM and obtained a linear receiver similar to [209] in this work. One advantage in using SFBC instead of STBC is that, in SFBC, the coding is done across the sub-carriers inside one OFDM symbol duration, while STBC applies the coding across a number of OFDM symbols equal to number of transmit antennas, thus, an inherent processing delay is unavoidable in STBC [214]. We have analyzed linear receiver for SM-OSFBC system with ZF and MMSE criterion: this linear receiver is very similar to the receiver structure for SM-OSTBC scheme in [209]. We have derived OSIC receiver with ZF and MMSE criterion, and their FER performance is compared with linear receivers via simulations. For SM-QSFBC scheme, OSIC receiver with ZF and MMSE is derived, analyzed, and simulated. To the best of our knowledge, SM-QSFBC scheme has not been studied and analyzed before this work. It is worth noting here that the OSIC approach that we have taken in this work is similar to original BLAST algorithm in [29]. With the understanding of different receiver structures, we have investigated the impact of spatial correlation at both sides of transmission and the impact of LOS component in wireless channel.

Organization of the chapter: The rest of this chapter is organized as follows. Section 12.2 explains the system model for SM-OSFBC and SM-QSFBC systems. We have explained LS-based linear receiver for SM-OSFBC scheme and Ordered Successive Interference Cancellation (OSIC)-based nonlinear receiver for both the systems. Numerous simulations are done to evaluate the systems and their performances under certain realistic wireless scenarios. The results of those simulations and the discussions can be found in Section 13.4.2. The numerical evaluations include comparison in terms of Frame Error Rate (FER) and outage spectral efficiency in realistic wireless channel scenario, such as spatial correlation at both ends of the transmission link and presence of Line of Sight (LOS). Conclusions are drawn in Section 12.5.

12.2 System Model

12.2.1 SM-OSFBC Transmission Scheme

In this section, we will explain the transmission structure of the JDM scheme based on combining SM and Orthogonal Space–Frequency Block Code (OSFBC). Following this, we propose a linear two-stages receiver, which is an extension of Least Squares (LS) receiver in [209], where the linear reception technique is used for SM-OSTBC system based on ZF and Minimum Mean Square Error (MMSE) criteria, on sub-carrier by sub-carrier basis. After this, we derive a nonlinear Successive Interference Cancellation (SIC) receiver where the detection is based on both ZF and MMSE nulling.

Transmitter: We denote the number of available transmit SM branches and receive antennas as P and Q, respectively. We have N number of sub-carriers in the system. We denote p, q, and k as indices for transmit SM branch, receive antenna, and sub-carrier group, respectively. For every pth SM branch, we implement an orthogonal block coding across the sub-carriers.

Figure 12.1 explains the basic transceiver architecture, while referring to a linear receiver. At first source bits are FEC coded. The coded bit stream is baseband modulated using an appropriate constellation diagram, such as Binary Phase Shift Keying (BPSK), Quadrature Amplitude Modulation (QAM) etc.

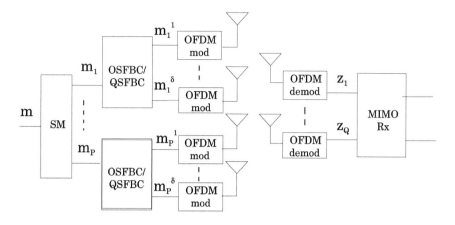

Fig. 12.1 Simplified system model for SM-OSFBC/SM-QSFBC transmission scheme.

We denote these baseband modulated symbols as $\mathbf{m}$. The sequence of $\mathbf{m}$ is demultiplexed into $\mathbf{m}_1, \ldots, \mathbf{m}_P$ vectors. $\mathbf{m}_p$ is transmitted via pth spatial channel, which can be written as, $\mathbf{m}_p = [\ m_{p,1} \quad m_{p,2} \quad \ldots \quad m_{p,N-1} \quad m_{p,N}\ ]^T$.

For every pth SM branch, we implement a block coding across the sub-carriers, thus SFBC is included in the system. For pth SM branch, we have Δ_p number of antennas where SFBC can be implemented. When $\Delta_p = \Delta$, $\forall p$, then we have $\Delta \cdot P$ number of transmit antennas at the transmission side. If we use well-known Alamouti coding [30] across the sub-carriers, then $\Delta = 2$.

For pth SM branch, $\mathbf{m}_p$ is coded into two vectors, $\mathbf{m}_p^{(\delta)}$; $\delta = 1, 2$. Thus, the output of the SFBC encoder block of the pth SM branch will be

$$\mathbf{m}_p^{(1)} = [\ m_{p,1} \quad -m_{p,2}^* \quad \ldots \quad m_{p,N-1} \quad -m_{p,N}^*\ ]^T \qquad (12.1)$$

$$\mathbf{m}_p^{(2)} = [\ m_{p,2} \quad m_{p,1}^* \quad \ldots \quad m_{p,N} \quad m_{p,N-1}^*\ ]^T. \qquad (12.2)$$

Following this, we define

$$\mathbf{m}_{p,o} = [\ m_{p,1} \quad m_{p,3} \quad \ldots \quad m_{p,N-3} \quad m_{p,N-1}\ ]^T \qquad (12.3)$$

$$\mathbf{m}_{p,e} = [\ m_{p,2} \quad m_{p,4} \quad \ldots \quad m_{p,N-2} \quad m_{p,N}\ ]^T. \qquad (12.4)$$

Using these equations, we can write that

$$\mathbf{m}_{p,o}^{(1)} = \mathbf{m}_{p,o}; \quad \mathbf{m}_{p,e}^{(1)} = -\mathbf{m}_{p,e}^* \qquad (12.5)$$

$$\mathbf{m}_{p,o}^{(2)} - \mathbf{m}_{p,e}; \quad \mathbf{m}_{p,e}^{(2)} - \mathbf{m}_{p,o}^*. \qquad (12.6)$$

After SM and SFBC operations, IFFT modulation is performed and CP is added before transmission via the respective transmit antenna. Transmitted time domain samples, $\mathbf{x}_p^{(\delta)}$, can be related to $\mathbf{m}_p^{(\delta)}$ as, $\mathbf{x}_p^{(\delta)} = \mathbb{F}^H\{\mathbf{m}_p^{(\delta)}\}$, where $\mathbb{F}^H$ is the IFFT matrix.

Two-stage linear receiver: In [215], a two stage interference cancelation receiver scheme for Space-Time Block Code (STBC) is presented. This receiver treats one of the branches as the interfering source for the other one. This receiver is used to derive a linear reception technique for SM-OSTBC system in [209]. In this work, we adopt a similar receiver structure for our SM-OSFBC system.

After DFT demodulation, we can express the frequency-domain sub-carrier signal as

$$\mathbf{z}_q = \sum_{\delta=1}^{2}\sum_{p=1}^{P}\mathbf{h}_{q,p}^{(\delta)} \odot \mathbf{m}_p^{(\delta)} + \mathbf{n}_q, \tag{12.7}$$

where $\mathbf{h}_{q,p}^{(\delta)}$ denotes the CTF vector of the wireless channel between qth receive antenna and δth SFBC antenna of pth SM branch, $\mathbf{n}_q$ is the frequency-domain sample of the received noise at qth received branch, and $\odot$ means element-wise multiplication.

We can divide $\mathbf{z}_q$ in odd and even components $\forall q$; $q = 1, 2$. Using Equations (12.3) and (12.4), we can express

$$\mathbf{z}_{q,o} = \mathbf{h}_{q,1,o}^{(1)} \odot \mathbf{m}_{1,o} + \mathbf{h}_{q,1,o}^{(2)} \odot \mathbf{m}_{1,e} + \cdots + \cdots$$
$$\mathbf{h}_{q,P,o}^{(1)} \odot \mathbf{m}_{P,o} + \mathbf{h}_{q,P,o}^{(2)} \odot \mathbf{m}_{P,e} + \mathbf{n}_{q,o} \tag{12.8}$$
$$\mathbf{z}_{q,e} = -\mathbf{h}_{q,1,e}^{(1)} \odot \mathbf{m}_{1,e}^* + \mathbf{h}_{q,1,e}^{(2)} \odot \mathbf{m}_{1,o}^* + \cdots - \cdots$$
$$\mathbf{h}_{q,P,e}^{(1)} \odot \mathbf{m}_{P,e}^* + \mathbf{h}_{q,P,e}^{(2)} \odot \mathbf{m}_{P,o}^* + \mathbf{n}_{q,e}. \tag{12.9}$$

From this point onwards, we use $P = 2$ and $Q = 2$. It must be fairly simple to extend this analysis to higher number of transmit and receive antennas. Let us introduce the following $Q \times 1$ column vectors, $\mathbf{h}_{p,o}^{(\delta)}, \mathbf{h}_{p,e}^{(\delta)}, \mathbf{z}_o, \mathbf{z}_e$, whose qth components are, respectively, the odd and even elements of CTF between δth transmit antenna of the pth SM branch and qth receive antenna, and the odd and even elements of received frequency domain signal at qth receive antenna. Furthermore, we denote the following $Q \times 1$ column vectors, $\mathbf{n}_o, \mathbf{n}_e$, whose

qth components are, respectively, the odd and even elements of frequency-domain noise at qth receive antenna.

Using these notations above, we can write Equation (12.10): in this equation, we have a sub-carrier notation, where it is understood that such a system model is intended $\forall k \in [1, \ldots, \frac{N}{2}]$.

We denote coherence bandwidth and sub-carrier spacing as B_c and Δf, respectively. We define severely frequency-selective scenario when coherence bandwidth is smaller than a pair of sub-carrier bandwidth, i.e. $B_c < 2\Delta f$. In this case, we use a tool called companion matrix explained in Appendix.

$$
\mathbf{z}_k = \begin{bmatrix} z_{1,o} \\ z_{2,o} \\ z_{1,e}^* \\ z_{2,e}^* \end{bmatrix}
$$

$$
= \begin{bmatrix} h_{11,o}^{(1)} & h_{11,o}^{(2)} & h_{12,o}^{(1)} & h_{12,o}^{(2)} \\ h_{21,o}^{(1)} & h_{21,o}^{(2)} & h_{22,o}^{(1)} & h_{22,o}^{(2)} \\ h_{11,e}^{(2)*} & -h_{11,e}^{(1)*} & h_{12,e}^{(2)*} & -h_{12,e}^{(1)*} \\ h_{21,e}^{(2)*} & -h_{21,e}^{(1)*} & h_{22,e}^{(2)*} & -h_{22,e}^{(1)*} \end{bmatrix} \begin{bmatrix} m_{1,o} \\ m_{1,e} \\ m_{2,o} \\ m_{2,e} \end{bmatrix} + \begin{bmatrix} n_{1,o} \\ n_{2,o} \\ n_{1,e}^* \\ n_{2,e}^* \end{bmatrix}
$$

$$
= \begin{bmatrix} \mathbf{z}_o \\ \mathbf{z}_e^* \end{bmatrix}_k
$$

$$
= \begin{bmatrix} \mathbf{h}_{1,o}^{(1)} & \mathbf{h}_{1,o}^{(2)} & \mathbf{h}_{2,o}^{(1)} & \mathbf{h}_{2,o}^{(2)} \\ \mathbf{h}_{1,e}^{(2)*} & -\mathbf{h}_{1,e}^{(1)*} & \mathbf{h}_{2,e}^{(2)*} & -\mathbf{h}_{2,e}^{(1)*} \end{bmatrix}_k \begin{bmatrix} m_{1,o} \\ m_{1,e} \\ m_{2,o} \\ m_{2,e} \end{bmatrix}_k + \begin{bmatrix} \mathbf{n}_o \\ \mathbf{n}_e^* \end{bmatrix}_k \qquad (12.10)
$$

Let us omit from now on the sub-carrier index k. We can represent Equation (12.10) as

$$
\mathbf{z} = \mathbf{Hm} + \mathbf{n} = \begin{bmatrix} \mathbf{H}_i \mid \mathbf{H}_j \end{bmatrix} \mathbf{m} + \mathbf{n} \qquad (12.11)
$$

with

$$
\mathbf{H}_i = \begin{bmatrix} \mathbf{h}_{1,o}^{(1)} & \mathbf{h}_{1,o}^{(2)} \\ \mathbf{h}_{1,e}^{(2)*} & -\mathbf{h}_{1,e}^{(1)*} \end{bmatrix}, \quad \mathbf{H}_j = \begin{bmatrix} \mathbf{h}_{2,o}^{(1)} & \mathbf{h}_{2,o}^{(2)} \\ \mathbf{h}_{2,e}^{(?)*} & -\mathbf{h}_{2,e}^{(1)*} \end{bmatrix} \qquad (12.12)
$$

We denote the companion matrices of $\mathbf{H}_i$ and $\mathbf{H}_j$ as $\widetilde{\mathbf{H}}_i$ and $\widetilde{\mathbf{H}}_j$, respectively. We define a new matrix, $\widetilde{\mathbf{H}} = \begin{bmatrix} \widetilde{\mathbf{H}}_i^T & \widetilde{\mathbf{H}}_j^T \end{bmatrix}^T$, with

$$\widetilde{\mathbf{H}}_i = \begin{bmatrix} \mathbf{h}_{1,e}^{(1)H} & \mathbf{h}_{1,o}^{(2)T} \\ \mathbf{h}_{1,e}^{(2)H} & -\mathbf{h}_{1,o}^{(1)T} \end{bmatrix}, \ \widetilde{\mathbf{H}}_j = \begin{bmatrix} \mathbf{h}_{2,e}^{(1)H} & \mathbf{h}_{2,o}^{(2)T} \\ \mathbf{h}_{2,e}^{(2)H} & -\mathbf{h}_{2,o}^{(1)T} \end{bmatrix}. \tag{12.13}$$

Now, at the beginning of the receiver, we can filter the received signal $\mathbf{z}$ with $\widetilde{\mathbf{H}}$ as it is shown below:

$$\mathbf{z}' = \widetilde{\mathbf{H}}\mathbf{z} = \begin{bmatrix} \widetilde{\mathbf{H}}_i \\ \widetilde{\mathbf{H}}_j \end{bmatrix} \begin{bmatrix} \mathbf{H}_i \mid \mathbf{H}_j \end{bmatrix} \mathbf{m} + \widetilde{\mathbf{H}}\mathbf{n}. \tag{12.14}$$

The part, $\widetilde{\mathbf{H}}\mathbf{H}$, can be extended as in Equation (12.15), where $\alpha_1' = \mathbf{h}_{1,e}^{(1)H}\mathbf{h}_{1,o}^{(1)} + \mathbf{h}_{1,o}^{(2)T}\mathbf{h}_{1,e}^{(2)*}$ and $\alpha_2' = \mathbf{h}_{2,e}^{(1)H}\mathbf{h}_{2,o}^{(1)} + \mathbf{h}_{2,o}^{(2)T}\mathbf{h}_{2,e}^{(2)*}$. We can see that $\mathbf{G}_{12}$ and $-\mathbf{G}_{21}$ form an orthogonal pair as defined in Appendix.

$$\widetilde{\mathbf{H}}\mathbf{H} = \begin{bmatrix} \mathbf{h}_{1,e}^{(1)H}\mathbf{h}_{1,o}^{(1)} + \mathbf{h}_{1,o}^{(2)T}\mathbf{h}_{1,e}^{(2)*} & 0 \\ 0 & \mathbf{h}_{1,e}^{(1)H}\mathbf{h}_{1,o}^{(1)} + \mathbf{h}_{1,o}^{(2)T}\mathbf{h}_{1,e}^{(2)*} \\ \mathbf{h}_{2,e}^{(1)H}\mathbf{h}_{1,o}^{(1)} + \mathbf{h}_{2,o}^{(2)T}\mathbf{h}_{1,e}^{(2)*} & \mathbf{h}_{2,e}^{(1)H}\mathbf{h}_{1,o}^{(2)} - \mathbf{h}_{2,o}^{(2)T}\mathbf{h}_{1,e}^{(1)*} \\ \mathbf{h}_{2,e}^{(2)H}\mathbf{h}_{1,o}^{(1)} - \mathbf{h}_{2,o}^{(1)T}\mathbf{h}_{1,e}^{(2)*} & \mathbf{h}_{2,e}^{(2)H}\mathbf{h}_{1,o}^{(2)} + \mathbf{h}_{2,o}^{(1)T}\mathbf{h}_{1,e}^{(1)*} \end{bmatrix}$$

$$\begin{bmatrix} \mathbf{h}_{1,e}^{(1)H}\mathbf{h}_{2,o}^{(1)} + \mathbf{h}_{1,o}^{(2)T}\mathbf{h}_{2,e}^{(2)*} & \mathbf{h}_{1,e}^{(1)H}\mathbf{h}_{2,o}^{(2)} - \mathbf{h}_{1,o}^{(2)T}\mathbf{h}_{2,e}^{(1)*} \\ \mathbf{h}_{1,e}^{(2)H}\mathbf{h}_{2,o}^{(1)} - \mathbf{h}_{1,o}^{(1)T}\mathbf{h}_{2,e}^{(2)*} & \mathbf{h}_{1,e}^{(2)H}\mathbf{h}_{2,o}^{(2)} + \mathbf{h}_{1,o}^{(1)T}\mathbf{h}_{2,e}^{(1)*} \\ \mathbf{h}_{2,e}^{(1)H}\mathbf{h}_{2,o}^{(1)} + \mathbf{h}_{2,o}^{(2)T}\mathbf{h}_{2,e}^{(2)*} & 0 \\ 0 & \mathbf{h}_{2,e}^{(1)H}\mathbf{h}_{2,o}^{(1)} + \mathbf{h}_{2,o}^{(2)T}\mathbf{h}_{2,e}^{(2)*} \end{bmatrix}$$

$$= \begin{bmatrix} \alpha_1'\mathbf{I}_2 & \mathbf{G}_{12} \\ \mathbf{G}_{21} & \alpha_2'\mathbf{I}_2 \end{bmatrix} \tag{12.15}$$

Now, Equation (12.14) can be written as

$$\mathbf{z}' = \begin{bmatrix} \alpha_1'\mathbf{I}_2 & \mathbf{G}_{12} \\ \mathbf{G}_{21} & \alpha_2'\mathbf{I}_2 \end{bmatrix} \mathbf{m} + \widetilde{\mathbf{H}}\mathbf{n}. \tag{12.16}$$

We define an LS receiver $\mathbf{W}$ as

$$\mathbf{W} = \frac{1}{\gamma} \begin{bmatrix} \alpha_2'\mathbf{I}_2 & -\mathbf{G}_{12} \\ -\mathbf{G}_{21} & \alpha_1'\mathbf{I}_2 \end{bmatrix}, \tag{12.17}$$

where $\gamma = \alpha_1' \alpha_2' - [\mathbf{G}_{12}(1,1)\mathbf{G}_{12}(2,2) - \mathbf{G}_{12}(1,2)\mathbf{G}_{12}(2,1)]$. Thus, the estimated symbol vector can be written as

$$\widehat{\mathbf{m}} = \mathbf{W}\mathbf{z}' = \mathbf{m} + \mathbf{W}\widetilde{\mathbf{H}}\mathbf{n}. \tag{12.18}$$

In relation to severely frequency-selective scenario, we define moderately frequency-selective scenario when $B_c > 2\Delta f$, and in that case we can easily say that neighboring sub-carriers have identical channel frequency response. In this case $\mathbf{h}_{p,o}^{(\delta)} = \mathbf{h}_{p,e}^{(\delta)} = \mathbf{h}_{p,oe}^{(\delta)}$. So we can write Equation (12.10) as

$$\mathbf{z} = \begin{bmatrix} \mathbf{z}_o \\ \mathbf{z}_e^* \end{bmatrix} = \begin{bmatrix} \mathbf{H}_1 | \mathbf{H}_2 \end{bmatrix}\mathbf{m} + \mathbf{n}, \tag{12.19}$$

where

$$\mathbf{H}_1 = \begin{bmatrix} \mathbf{h}_{1,oe}^{(1)} & \mathbf{h}_{1,oe}^{(2)} \\ \mathbf{h}_{1,oe}^{(2)*} & -\mathbf{h}_{1,oe}^{(1)*} \end{bmatrix}, \quad \mathbf{H}_2 = \begin{bmatrix} \mathbf{h}_{2,oe}^{(1)} & \mathbf{h}_{2,oe}^{(2)} \\ \mathbf{h}_{2,oe}^{(2)*} & -\mathbf{h}_{2,oe}^{(1)*} \end{bmatrix} \tag{12.20}$$

are two orthogonal matrices, so we can write that

$$\mathbf{H}_1^H \mathbf{H}_1 = \left(|\mathbf{h}_{1,oe}^{(1)}|^2 + |\mathbf{h}_{1,oe}^{(2)}|^2 \right)\mathbf{I}_2 = \alpha_1 \mathbf{I}_2 \tag{12.21}$$

$$\mathbf{H}_2^H \mathbf{H}_2 = \left(|\mathbf{h}_{2,oe}^{(1)}|^2 + |\mathbf{h}_{2,oe}^{(2)}|^2 \right)\mathbf{I}_2 = \alpha_2 \mathbf{I}_2, \tag{12.22}$$

where $\alpha_1 = |\mathbf{h}_{1,oe}^{(1)}|^2 + |\mathbf{h}_{1,oe}^{(2)}|^2$ and $\alpha_2 = |\mathbf{h}_{2,oe}^{(1)}|^2 + |\mathbf{h}_{2,oe}^{(2)}|^2$. Similarly,

$$\mathbf{H}_1^H \mathbf{H}_2 \left(\mathbf{H}_1^H \mathbf{H}_2 \right)^H = \mathbf{H}_1^H \mathbf{H}_2 \mathbf{H}_2^H \mathbf{H}_1 = (a+b)\mathbf{I}_2, \tag{12.23}$$

where $a = |\mathbf{h}_{1,oe}^{(1)H} \mathbf{h}_{2,oe}^{(1)} + \mathbf{h}_{1,oe}^{(2)T} \mathbf{h}_{2,oe}^{(2)*}|^2$ and $b = |\mathbf{h}_{1,oe}^{(1)H} \mathbf{h}_{2,oe}^{(2)} - \mathbf{h}_{1,oe}^{(2)T} \mathbf{h}_{2,oe}^{(1)*}|^2$.

Now we can write the estimated symbol vector by using LS estimator, i.e. by multiplying the received symbol vector with Moore–Penrose pseudo-inverse matrix for $\mathbf{H}$,

$$\widehat{\mathbf{m}} = \left(\mathbf{H}^H \mathbf{H} \right)^{-1} \mathbf{H}^H \mathbf{z}$$

$$= \frac{1}{\alpha_1 \alpha_2 - (a+b)} \begin{bmatrix} \mathbf{H}_2^H \mathbf{H}_2 & -\mathbf{H}_1^H \mathbf{H}_2 \\ -\mathbf{H}_2^H \mathbf{H}_1 & \mathbf{H}_1^H \mathbf{H}_1 \end{bmatrix} \begin{bmatrix} \mathbf{H}_1^H \\ \mathbf{H}_2^H \end{bmatrix} \mathbf{z}.$$

Using some definitions described above, we can express the former equation as follows:

$$\widehat{\mathbf{m}} = \frac{1}{\alpha_1 \alpha_2 - (a+b)} \begin{bmatrix} \alpha_2 \mathbf{I}_2 & -\mathbf{H}_1^H \mathbf{H}_2 \\ -\mathbf{H}_2^H \mathbf{H}_1 & \alpha_1 \mathbf{I}_2 \end{bmatrix} \begin{bmatrix} \mathbf{H}_1^H \\ \mathbf{H}_2^H \end{bmatrix} \mathbf{z}. \tag{12.24}$$

The MMSE receiver can be implemented in the same simple way. Defining ρ_n as the average SNR at each of the receive antennas, and the new constants $\beta_1 = \alpha_1 + \frac{\Delta P}{\rho_n}$, $\beta_2 = \alpha_2 + \frac{\Delta P}{\rho_n}$ then we can rewrite Equation (12.24) as

$$\widehat{\mathbf{m}} = \frac{1}{\beta_1\beta_2 - (a+b)} \begin{bmatrix} \beta_2\mathbf{I}_2 & -\mathbf{H}_1^H\mathbf{H}_2 \\ -\mathbf{H}_2^H\mathbf{H}_1 & \beta_1\mathbf{I}_2 \end{bmatrix} \begin{bmatrix} \mathbf{H}_1^H \\ \mathbf{H}_2^H \end{bmatrix} \mathbf{z}. \qquad (12.25)$$

OSIC-based nonlinear receiver: The basic transceiver architecture has been shown in Figure 12.1. In the case of OSIC-based nonlinear receiver, the detection part is shown in Figure 12.2. The key idea of the nonlinear receiver is to decode the symbol streams successively and extract them away layer by layer. At the beginning of each stage, the stream with highest SNR is chosen using ZF or MMSE detection for peeling. Once one particular layer is decoded, then the effect of the detected layer is subtracted from the received signal. From the remaining signal, the next branch in terms of SNR strength is chosen. This is continued until the last layer is decoded. This kind of nonlinear detection is called OSIC [29, 139]. OSIC is performed on sub-carrier by sub-carrier basis. On each sub-carrier, the detection scheme appears to be very similar to Vertical-Bell Labs LAyered Space-Time Architecture (VBLAST) detection, as derived in [29]. OSIC improves the detection quality compared to detection without ordering and is shown to be optimal for SIC approach [139].

We consider $P = 2$, $\Delta = 2$, and $Q = 2$. Considering the frequency domain, it can be written as

$$\mathbf{z}_k = \mathbf{H}_k\mathbf{m}_k + \mathbf{n}_k, \qquad (12.26)$$

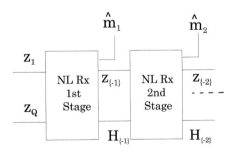

Fig. 12.2 Simplified data flow for OSIC receiver.

where

$$\mathbf{H}_k = \begin{bmatrix} h_{11,o}^{(1)} & h_{11,o}^{(2)} & h_{12,o}^{(1)} & h_{12,o}^{(2)} \\ h_{11,e}^{(2)*} & -h_{11,e}^{(1)*} & h_{12,e}^{(2)*} & -h_{12,e}^{(1)*} \\ h_{21,o}^{(1)} & h_{21,o}^{(2)} & h_{22,o}^{(1)} & h_{22,o}^{(2)} \\ h_{21,e}^{(2)*} & -h_{21,e}^{(1)*} & h_{22,e}^{(2)*} & -h_{22,e}^{(1)*} \end{bmatrix}_k \qquad (12.27)$$

and $\mathbf{z}_k = [z_{1,o}, z_{1,e}^*, z_{2,o}, z_{2,e}^*]_k^T$; $\mathbf{m}_k = [m_{1,o}, m_{1,e}, m_{2,o}, m_{2,e}]_k^T$; $\mathbf{n}_k = [n_{1,o}, n_{1,e}^*, n_{2,o}, n_{2,e}^*]_k^T$ with $k \in \left[1, \ldots, \frac{N}{2}\right]$.

The OSIC receiver consists of two steps: nulling and cancelation. While implementing the nulling operation, we consider two options, ZF and MMSE.

For ZF, we calculate the Moore–Penrose pseudo-inverse of the Channel Transfer Function (CTF)

$$\mathbf{G}_k = [\mathbf{g}_1 \, \mathbf{g}_2 \, \mathbf{g}_3 \, \mathbf{g}_4]_k = \sqrt{\frac{\Delta P}{P_T}} \left\{ \mathbf{H}_k^H \mathbf{H}_k \right\}^{-1} \mathbf{H}_k^H$$

$$= \sqrt{\frac{4}{P_T}} \left\{ \mathbf{H}_k^H \mathbf{H}_k \right\}^{-1} \mathbf{H}_k^H, \qquad (12.28)$$

where P_T is the total transmit power from all transmit antennas, while for the MMSE we consider a kind of pseudo-inverse that considers also the noise term

$$\mathbf{G}_k = \sqrt{\frac{4}{P_T}} \left\{ \mathbf{H}_k^H \mathbf{H}_k + \frac{4\sigma^2}{P_T} \mathbf{I}_4 \right\}^{-1} \mathbf{H}_k^H, \qquad (12.29)$$

where $[\mathbf{g}_i]_k$ is the ith column of $\mathbf{G}_k$, σ^2 is the noise variance at each receive antenna and $\mathbf{I}_4$ is the 4×4 identity matrix, and for both ZF and MMSE we have considered that $\Delta = 2$ and $P = 2$. At this stage, we calculate the metric:

$$l = \arg\min_l \{d_1, d_2\}, \qquad (12.30)$$

where $d_1 = \mathbf{c}(1) + \mathbf{c}(3)$, $d_2 = \mathbf{c}(2) + \mathbf{c}(4)$. For ZF the vector, $\mathbf{c} = [\mathbf{c}(1) \, \mathbf{c}(2) \, \mathbf{c}(3) \, \mathbf{c}(4)] = \mathrm{diag}\left(\{\mathbf{H}_k^H \mathbf{H}_k\}^{-1}\right)$ and for MMSE the vector, $\mathbf{c} = [\mathbf{c}(1) \, \mathbf{c}(2) \, \mathbf{c}(3) \, \mathbf{c}(4)] = \mathrm{diag}\left(\{\mathbf{H}_k^H \mathbf{H}_k + \frac{4\sigma^2}{P_T} \mathbf{I}_4\}^{-1}\right)$.

If $l = \arg\min_l \{d_1, d_2\} = 1$ then we decode first the stream associated to $p = 1$ by doing

$$\widehat{\mathbf{m}}_{1,k} = \begin{bmatrix} \widehat{m}_{1,o} \\ \widehat{m}_{1,e} \end{bmatrix}_k - \begin{bmatrix} \mathbf{g}_1^H \\ \mathbf{g}_2^H \end{bmatrix}_k \mathbf{z}_k \qquad (12.31)$$

and the second stream by doing the following steps (in Equation (12.32) the cancelation operation is performed)

$$\mathbf{z}'_k = \mathbf{z}_k - [\mathbf{h}_1 \ \mathbf{h}_2]_k \begin{bmatrix} \widehat{m}_{1,o} \\ \widehat{m}_{1,e} \end{bmatrix}_k \tag{12.32}$$

$$\mathbf{H}'_k = [\mathbf{h}_3 \ \mathbf{h}_4]_k \tag{12.33}$$

$$\mathbf{H}'^H_k \mathbf{H}'_k = \frac{1}{|h^{(1)}_{12,o}|^2 + |h^{(2)}_{12,o}|^2 + |h^{(1)}_{22,o}|^2 + |h^{(2)}_{22,o}|^2} \mathbf{I}_2$$

$$= \frac{1}{\alpha} \mathbf{I}_2 \tag{12.34}$$

$$\widehat{\mathbf{m}}_{2,k} = \begin{bmatrix} \widehat{m}_{2,o} \\ \widehat{m}_{2,e} \end{bmatrix}_k = \frac{1}{\alpha} \mathbf{H}'^H_k \mathbf{z}'_k, \tag{12.35}$$

where $[\mathbf{h}_i]_k$ is the ith column of $\mathbf{H}_k$ and $\mathbf{I}_2$ is the 2×2 identity matrix. In (12.34) we assume that neighboring sub-carriers have identical channel frequency response. Exactly the dual procedure of (12.31–12.35) has to be applied if $l = \arg \min_l\{d_1, d_2\} = 2$, to decode the second SM branch at the beginning and then the first one.

12.2.2 SM-QSFBC-OFDM Transmission Scheme

Transmitter: Similar to SM-OSFBC scheme, Figure 12.1 explains the basic transmitter architecture for proposed SM-QSFBC system, with the receiver depicted in Figure 12.2. For $\Delta > 2$, such as $\Delta = 4$, we use quasi-orthogonal codes as in [208] for the studied SM-QSFBC system, so that SM-rate of P is maintained. For pth SM branch, $\mathbf{m}_p$ is coded into Δ number of vectors, $\mathbf{m}_p^{(\delta)}$; $\delta = 1, \ldots, \Delta$. After SM and Quasi-orthogonal Space-Frequency Block Code (QSFBC) operations, IFFT modulation is performed and Cyclic Prefix (CP) is added before transmission via the respective transmit antenna. N sub-carriers are divided into $N/4$ groups of 4 sub-carriers and on each group QSFBC is performed. Note that $P \leq Q$.

OSIC receiver: The OSIC receiver for SM-QSFBC scheme is similar to the OSIC receiver for SM-OSFBC scheme described in Section 12.2.1. We can write the equivalent system model as follows:

$$\mathbf{z}_k = \mathbf{H}_k \mathbf{m}_k + \mathbf{n}_k \quad k \in \left[1, \ldots, \frac{N}{4}\right], \tag{12.36}$$

where $\mathbf{z}_k$ and $\mathbf{n}_k$ are the concatenations of the received signals and noise, respectively, for the kth group of sub-carriers. Here $\mathbf{z}_k = \left[\mathbf{z}_k^{(1)} \ldots \mathbf{z}_k^{(Q)}\right]^T$, with $\mathbf{z}_k^{(q)} = \left[z_{4k-3}^{(q)} \, z_{4k-2}^{(q)*} \, z_{4k-1}^{(q)*} \, z_{4k}^{(q)}\right]^T$. Similarly, $\mathbf{m}_k = \left[\mathbf{m}_k^{(1)} \ldots \mathbf{m}_k^{(P)}\right]^T$, with $\mathbf{m}_k^{(p)} = \left[m_{4k-3}^{(p)} \, m_{4k-2}^{(p)} \, m_{4k-1}^{(p)} \, m_{4k}^{(p)}\right]$. $\mathbf{m}_x^{(y)}$ and $\mathbf{z}_x^{(y)}$ denote transmitted and received samples respectively, for xth sub-carrier group on yth SM branch and receive antenna, respectively. $\mathbf{H}_k$ is shown in (12.37) as

$$\mathbf{H}_k = \begin{bmatrix} [\mathbf{H}_{11}]_k & \cdots & [\mathbf{H}_{1P}]_k \\ \vdots & \ddots & \vdots \\ [\mathbf{H}_{Q1}]_k & \cdots & [\mathbf{H}_{QP}]_k \end{bmatrix}, \tag{12.37}$$

where $\left[\mathbf{H}_{ij}\right]_k$ with $i \in \{1, \ldots, Q\}$ and $j \in \{1, \ldots, P\}$ is

$$[\mathbf{H}_{ij}]_k = \begin{bmatrix} h_{ij,4k-3}^{(1)} & h_{ij,4k-3}^{(2)} & h_{ij,4k-3}^{(3)} & h_{ij,4k-3}^{(4)} \\ h_{ij,4k-2}^{(2)*} & -h_{ij,4k-2}^{(1)*} & h_{ij,4k-2}^{(4)*} & -h_{ij,4k-2}^{(3)*} \\ h_{ij,4k-1}^{(3)*} & h_{ij,4k-1}^{(4)*} & -h_{ij,4k-1}^{(1)*} & -h_{ij,4k-1}^{(2)*} \\ h_{ij,4k}^{(4)} & -h_{ij,4k}^{(3)} & -h_{ij,4k}^{(2)} & h_{ij,4k}^{(1)} \end{bmatrix}. \tag{12.38}$$

For ZF, we calculate the Moore–Penrose pseudo-inverse, $\mathbf{G}_k$, of the equivalent CTF, $\mathbf{H}_k$,

$$\mathbf{G}_k = [\mathbf{g}_1 \ldots \mathbf{g}_{4Q}]_k = \sqrt{\frac{\Delta P}{P_T}} \{\mathbf{H}_k^H \mathbf{H}_k\}^{-1} \mathbf{H}_k^H$$

$$= \sqrt{\frac{8}{P_T}} \{\mathbf{H}_k^H \mathbf{H}_k\}^{-1} \mathbf{H}_k^H, \tag{12.39}$$

where P_T is total transmit power from all transmit antennas.

For the MMSE nulling operator, the pseudo-inverse is defined considering the noise variance [139], $\mathbf{G}_k = \sqrt{\frac{4P}{P_T}} \{\mathbf{H}^H \mathbf{H} + \frac{4P}{\rho} \mathbf{I}_{4P}\}^{-1} \mathbf{H}^H$. Note that in our case, the number of transmit antennas is $\Delta P = 4P$. The SNR is $\rho = \frac{P_T}{\sigma^2}$, with σ^2 noise variance at each receive antenna. Using this definition of ρ, we finally obtain the MMSE nulling operator as

$$\mathbf{G}_k = [\mathbf{g}_1 \ldots \mathbf{g}_{4Q}]_k = \sqrt{\frac{\Delta P}{P_T}} \left\{\mathbf{H}_k^H \mathbf{H}_k + \frac{\Delta P \sigma^2}{P_T} \mathbf{I}_{\Delta P}\right\}^{-1} \mathbf{H}_k^H$$

$$= \sqrt{\frac{8}{P_T}} \left\{\mathbf{H}_k^H \mathbf{H}_k + \frac{8\sigma^2}{P_T} \mathbf{I}_8\right\}^{-1} \mathbf{H}_k^H, \tag{12.40}$$

where $\left[\mathbf{g}_i\right]_k$ is the ith column of $\mathbf{G}_k$, $\mathbf{I}_8$ is the 8×8 identity matrix and we have considered that $\Delta = 4$ and $P = 2$.

At this stage, we define a vector, Ω as $\mathrm{diag}\left[\{\mathbf{H}_k^H \mathbf{H}_k\}^{-1}\right]$ for ZF and $\mathrm{diag}\left[\{\mathbf{H}_k^H \mathbf{H}_k + \frac{8\sigma^2}{P_T}\mathbf{I}_8\}^{-1}\right]$ for MMSE, where $\mathrm{diag}[\mathbf{X}]$ is the vector containing all the diagonal components of matrix $\mathbf{X}$.

After that, the ordering for layer-by-layer detection is done, by determining the strongest received signal branch. For this, we calculate the component of a vector,

$$\mathbf{d}(p) = \Omega(p) + \Omega(p + P) + \Omega(p + 2P) + \cdots$$
$$+ \Omega(p + (\Delta - 1)P), \tag{12.41}$$

where $\Omega(p)$ and $\mathbf{d}(p)$ are the pth element of vector Ω and $\mathbf{d}$, respectively. Clearly, $\Omega \in \mathbb{C}^{[4P \times 1]}$ and $\mathbf{d} \in \mathbb{C}^{[P \times 1]}$. For example, when $\Delta = 4$ and $P = 2$, we can write that $\mathbf{d}(1) = \Omega(1) + \Omega(3) + \Omega(5) + \Omega(7)$ and $\mathbf{d}(2) = \Omega(2) + \Omega(4) + \Omega(6) + \Omega(8)$.

Using the vector $\mathbf{d}$, we define a new vector Φ, which indicates the received strength of transmitted branches in descending order, i.e. the SM branches are arranged in descending order in this vector. We define, $\Phi(1) = \arg\min_l\{\mathbf{d}(l)\}, \forall\, l \in [1, \ldots, P]$. Similarly, $\Phi(2) = \arg\min_l\{\mathbf{d}(l)\}, \forall\, l \in [1, \ldots, P]\backslash\Phi(1)$. This is continued until all P transmit branches are ordered. The metric to define the descending order of received SM branches in Φ is found based on OSIC approach as explained in [139].

If $\Phi(1) = 1$, then we extract the first receive stream by using ZF or MMSE nulling criterion, i.e. $\widehat{\mathbf{m}}_k^{(1)} = \mathbf{G}_{1,k}\mathbf{z}_k$, where $\widehat{\mathbf{m}}_k^{(1)} = \left[\widehat{m}_{4k-3}^{(1)} \ \widehat{m}_{4k-2}^{(1)} \ \widehat{m}_{4k-1}^{(1)} \ \widehat{m}_{4k}^{(1)}\right]_k^T$ and $\mathbf{G}_{1,k} = [\mathbf{g}_1 \ \mathbf{g}_2 \ \mathbf{g}_3 \ \mathbf{g}_4]_k^T$.

Once the strongest branch is extracted, then the contribution of this branch is subtracted from the remaining signal, and the new equivalent CTF is obtained:

$$\mathbf{z}_k' = \mathbf{z}_k - [\mathbf{h}_1 \ \mathbf{h}_2 \ \mathbf{h}_3 \ \mathbf{h}_4]_k \ \widehat{\mathbf{m}}_k^{(1)} = \mathbf{H}_k'\left[\mathbf{m}_k^{(2)} \ \cdots \ \mathbf{m}_k^{(P)}\right]^T, \tag{12.42}$$

where $\mathbf{H}_k' = [\mathbf{h}_5 \ \mathbf{h}_6 \ \cdots \ \mathbf{h}_{4P}]_k$ and $[\mathbf{h}_i]_k$ is the ith column of $\mathbf{H}_k$.

The next strongest branch is found as $\Phi(2)$. We define a new nulling operator as $\mathbf{G}_{\Phi(2),k} = \left[\mathbf{g}_{4\Phi(2)-3} \ \mathbf{g}_{4\Phi(2)-2} \ \mathbf{g}_{4\Phi(2)-1} \ \mathbf{g}_{4\Phi(2)}\right]_k^T$. With this nulling operator and new CTF matrix as in (12.42), a new nulling criterion is defined and the second strongest branch is extracted from the remaining signal as shown below for ZF and MMSE solutions, respectively:

$$\widehat{\mathbf{m}}_k^{(\Phi(2))} = \mathbf{G}_{\Phi(2),k}\mathbf{z}_k' = \sqrt{\frac{\Delta P}{P_T}}\left\{\mathbf{H}_k''{}^H\mathbf{H}_k''\right\}^{-1}\mathbf{H}_k''{}^H\mathbf{z}_k'$$

$$= \sqrt{\frac{8}{P_T}}\left\{\mathbf{H}_k''{}^H\mathbf{H}_k''\right\}^{-1}\mathbf{H}_k''{}^H\mathbf{z}_k' \qquad (12.43)$$

and for MMSE

$$\widehat{\mathbf{m}}_k^{(\Phi(2))} = \sqrt{\frac{\Delta P}{P_T}}\left\{\mathbf{H}_k''{}^H\mathbf{H}_k'' + \frac{\Delta P\sigma^2}{P_T}\mathbf{I}_\Delta\right\}^{-1}\mathbf{H}_k''{}^H\mathbf{z}_k'$$

$$= \sqrt{\frac{8}{P_T}}\left\{\mathbf{H}_k''{}^H\mathbf{H}_k'' + \frac{8\sigma^2}{P_T}\mathbf{I}_4\right\}^{-1}\mathbf{H}_k''{}^H\mathbf{z}_k', \qquad (12.44)$$

where $\mathbf{H}_k'' = \left[\mathbf{h}_{4\Phi(2)-3}\ \mathbf{h}_{4\Phi(2)-2}\ \mathbf{h}_{4\Phi(2)-1}\ \mathbf{h}_{4\Phi(2)}\right]_k^T$ in both (12.43) and (12.44), $\mathbf{I}_4$ is the 4×4 identity matrix and we have considered that $\Delta = 4$ and $P = 2$. When there are more than $P = 2$ transmitted streams, the whole procedure is continued until the last branch is detected.

12.3   SNR Distribution for ZF Detection

We can show that the SNR experienced by any spatially multiplexed stream at the receiver is inversely proportional to $\mathbf{H}^H\mathbf{H}$. This is true, because the thermal noise seen at any receive antenna has a zero mean and a variance of σ^2. The post-detection SNR in the receiver when a linear ZF receiver is implemented, can be written as [139]:

$$\gamma_k = \frac{\Gamma}{\left[\mathbf{H}^H\mathbf{H}\right]_{kk}^{-1}}, \qquad k = 1,\ldots,P, \qquad (12.45)$$

where $\Gamma = \frac{E_s}{\sigma^2}$, i.e. average energy transmitted per symbol divided by noise variance per antenna. Without loosing any generality, we can safely assume that the rows of the channel matrices (as shown in (12.27) and (12.37) for SM-OSFBC and SM-QSFBC, respectively) are independently drawn from multi-variate normal distribution with zero mean [44]. This is possible when it is assured that the transmit and receive antennas are largely separated in space. Thus, $\mathbf{H}''{}^H\mathbf{H}$ has a complex Wishart distribution for both SM-OSFBC and SM-QSFBC systems with ZF detection. The SNR distribution can now be written as

$$p_{\gamma_k} = \frac{\exp\left(\frac{-\gamma_k}{\Gamma}\right)}{\Gamma\Psi(\Delta + (Q-P)\Delta - 1)}\left(\frac{\gamma_k}{\Gamma}\right)^{\Delta+(Q-P)\Delta-1}, \qquad (12.46)$$

where $\Psi(x)$ is the Gamma function. The achievable diversity order at the stage of nonlinear receiver when ZF detection is used is $\Delta + (Q - P)\Delta - 1$. In addition to this, it can also be shown that the SNR distribution is similar in all the multiplexing streams [139]. Also, we can express the *cumulative distribution function* (cdf) of γ_k as

$$\mathcal{F}(\gamma_k) = \frac{\Upsilon(\Delta + (Q-P)\Delta - 1, \frac{\gamma_k}{\Gamma})}{\Psi(\Delta + (Q-P)\Delta - 1)}, \qquad (12.47)$$

where $\Upsilon(a,b)$ represents the incomplete Gamma function defined by $\Upsilon(a,b) = \int_0^b e^{-x} x^{a-1} dx$.

Figure 12.3 gives us the *probability distribution function* (pdf) of SNRs for SM and JDM schemes. As expected, we can see that JDM schemes benefit from additional diversity options offered by SFBC (or STBC) applied on each multiplexed stream. For SM-QSFBC schemes, the diversity benefits are more compared to SM-OSFBC systems. Both JDM schemes obtain higher diversity gains compared to simple SM scheme.

12.4 Numerical Results

12.4.1 System Parameters

We have used simulation scenarios with following system parameters: System bandwidth, $B = 1.25$ MHz, Carrier frequency, $f_c = 3.5$ GHz, OFDM subcarriers, $N = 128$, CP length, $N_{CP} = 16$, Sampling rate, $f_s = 1.429$ Msps, Symbol duration, $T_s = (N + N_{CP})/f_s = 100.77\,\mu$s, number of symbols in a frame, $N_f = 5$ and Frame duration, $T_f = N_f T_s = 503.85\,\mu$s. QPSK modulation with $\frac{1}{2}$-rate convolutional coding is used for all the schemes. For all our analysis and simulations, we have two SM branches and dual receive antennas in all systems, i.e. $P = Q = 2$. We have confined ourselves to the case of $\Delta = 2$ and 4 for each pth SM branch in SM-OSFBC and SM-QSFBC systems, respectively. Thus, we have 4×2 and 8×2 systems for SM-OSFBC and SM-QSFBC schemes, respectively. The tags used to denote different transmission and reception configurations are shown in Table 12.1.

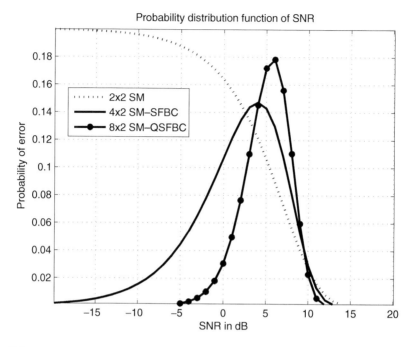

Fig. 12.3 Probability distribution functions of SNRs when different combinations of JDM schemes are used.

Table 12.1 Tags used in figures for corresponding schemes.

Scheme	Tag
2 × 1 Alamouti SFBC scheme [176]	2 × 1 *SFBC*
2 × 2 VBLAST with ZF nulling [29]	2 × 2 *ZF-BLAST*
2 × 2 VBLAST with MMSE nulling [29]	2 × 2 *MMSE-BLAST*
2 × 2 SM system optimum Maximum Likelihood (ML) reception	2 × 2 *ML*
4 × 2 SM-OSFBC transmitter and ZF based linear receiver	4 × 2 *SM-OSFBC, ZF-Lin*
4 × 2 SM-OSFBC transmitter and MMSE based linear receiver	4 × 2 *SM-OSFBC, MMSE-Lin*
4 × 2 SM-OSFBC transmitter and ZF based OSIC receiver	4 × 2 *SM-OSFBC, ZF-OSIC*
4 × 2 SM-OSFBC transmitter and MMSE based OSIC receiver	4 × 2 *SM-OSFBC, MMSE-OSIC*
8 × 2 SM-QSFBC transmitter and ZF based OSIC receiver	8 × 2 *SM-QSFBC, ZF-OSIC*
8 × 2 SM-QSFBC transmitter and MMSE based OSIC receiver	8 × 2 *SM-QSFBC, MMSE-OSIC*

We assume that perfect channel estimation values for each sub-carrier for all the spatial channels are available at the receiver. We use the exponential model to generate the corresponding CIR and CTF of the channel. In our model, the power delay profile of the channel is exponentially distributed with decay between the first and the last impulse as $-40\,$dB.

12.4.2 FER Performance

Figure 12.4 shows us the coded FER results for the schemes in WiMAX scenario. For various transmit antenna configurations, the total transmit power is kept constant, thus, the SNR at the x-axis reflects total SNR of the systems. We have used Quadrature Phase Shift Keying (QPSK) modulation for all the systems. We know that 2×2 SM [29] performs worse in terms of FER compared to 2×1 SFBC system: in fact in the SM system, we get an higher rate, but we lose in diversity [31]. Considering this, we can see that 4×2 SM-OSFBC with OSIC receiver performs better than SFBC system in terms of FER. In this case, not only the diversity gain is achieved, but spatial multiplexing gain is also realized. We can see an even better gain in FER is obtained when 8×2 SM-QSFBC system is used. This clearly shows the benefits of increased spatial dimensions at the transmitter and OSFBC or QSFBC used in the transmission system. For instance, SM-QSFBC-MMSE-OSIC and SM-OSFBC-MMSE-OSIC achieve a gain of 9 and 5 dB, respectively, compared to MMSE-Bell labs LAyered Space Time (BLAST) at an FER of 10^{-2}.

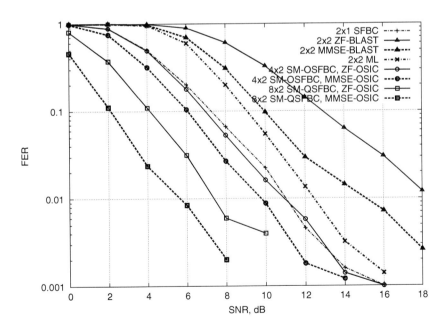

Fig. 12.4 FER performance for the schemes.

12.4.3 10% Outage Spectral Efficiency

Outage analysis is a form of reliability analysis. We define 10% outage channel capacity as the information rate that is guaranteed for 90% of the channel realizations, such that the probability that outage rate falls below the certain threshold rate is at most 10% [139]. The 10% outage spectral efficiencies for SM, SM-OSFBC, and SM-QSFBC with different receiver configurations are given in Figure 12.5. At 0 dB of SNR, 10% outage spectral efficiency of SM-QSFBC-MMSE-OSIC and SM-QSFBC-ZF-OSIC schemes are 1.75 and 1 bps/Hz, respectively. The maximum achievable spectral efficiency with the set of parameters that we have used for these simulations is 2 bps/Hz (achieved with spatial rate of 2, QPSK modulation, channel coding rate of 1/2). For SM-QSFBC-OSIC schemes, the 10% outage spectral efficiency is almost close to maximum at any SNR more than 4 dB. At that SNR, SM system obtains very low efficiency, whereas SM-OSFBC schemes obtain between 1.7 and 1.9 bps/Hz, which is quite impressive also. One conclusion that can be easily made is that added spatial dimensions at the transmitter side are exploited well

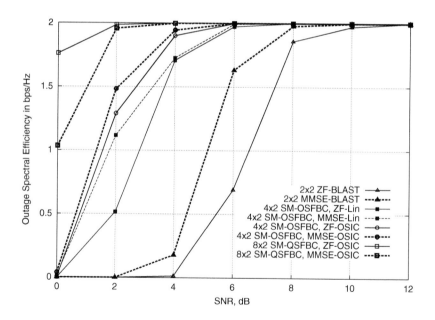

Fig. 12.5 10% outage spectral efficiency in indoor scenario.

in JDM schemes to improve system performance in terms of FER and outage spectral efficiency as seen in Figures 12.4 and 12.5, respectively.

12.4.4 Effect of Spatial Correlation

We model the spatial correlation in $\mathbf{H}$ based on the inter-element distances in the transmit and receive antenna arrays, as it is done in [211]. We can model the spatial correlation across all sub-carriers as

$$\mathbf{H}_k = \sqrt{\mathbf{R}_{k,rx}}\mathbf{H}_{w,k}\sqrt{\mathbf{R}_{k,tx}}, \qquad (12.48)$$

where $\mathbf{R}_{k,rx}$, $\mathbf{H}_{w,k}$, and $\mathbf{R}_{k,tx}$ are receive correlation matrix, uncorrelated channel matrix and transmit correlation matrix, respectively. The correlation coefficient between p_1 and p_2 transmit antennas can be written as

$$\left[\mathbf{R}_{k,tx}\right]_{p_1,p_2} = \mathcal{J}_0\left(\frac{2\pi(p_1 - p_2)d}{\lambda}\right), \qquad (12.49)$$

where $p_1, p_2 \in [1, \ldots, P\Delta]$. $\mathcal{J}_0$ denotes the 0th order Bessel function of 1st kind, d is the distance between the elements, and λ is the wavelength corresponding to the carrier frequency. Similar spatial correlation can be defined for receiver also.

In this section, we investigate the impact of spatial correlation on FER and on outage spectral efficiency. For this, the receive (transmit) is fixed to 0.3 and transmit (receive) correlation is varied as shown in Table 12.2. This 0.3 of receive spatial correlation corresponds to off-diagonal components in $\mathbf{R}_{k,rx}$, i.e. the spatial correlation between the neighboring elements is 0.3. The same applies for transmit side. The results are obtained for a particular SNR value, i.e. 12 dB.

In Figures 12.6 and 12.7, we produce the FER results of the schemes, in presence of spatial correlation due to insufficient spatial separation between antenna elements. In [213], impact of spatial correlation is studied for a system similar to SM-OSTBC scheme: however it is not clear in [213], whether the

Table 12.2 Correlation for corresponding spatial separation among antennas at 3.5 GHz of carrier frequency. R and d denote spatial correlation and separation in cm across two neighboring elements respectively.

R	0.90	0.82	0.70	0.61	0.51	0.40	0.29	0.22	0.15
d (cm)	0.56	0.78	1.00	1.17	1.33	1.50	1.67	1.78	1.89

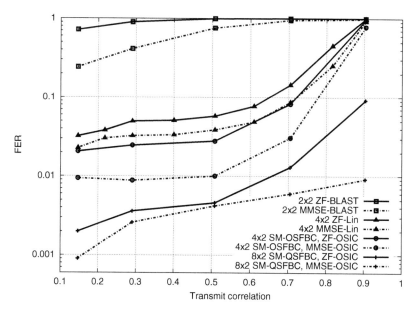

Fig. 12.6 FER with respect to increasing transmit correlation (i.e. decreasing antenna spacing) and fixed receive correlation at system SNR of 12 dB.

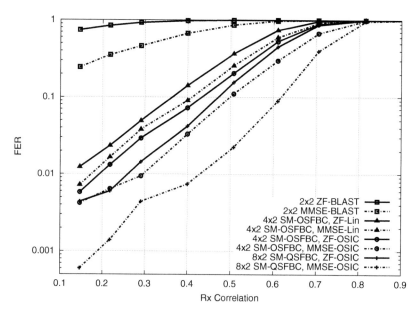

Fig. 12.7 FER with respect to increasing receive correlation (i.e. decreasing antenna spacing) and fixed transmit correlation at system SNR of 12 dB.

spatial correlation is considered at the transmitter or at the receiver or at both locations. For Figure 12.6, we varied the transmit correlation and fixed the receive correlation to 0.3, and the opposite is done for Figure 12.7. Note that we have dual receive antennas for all the schemes. One interesting conclusion that can be drawn from these figures is that receive correlation affects these systems much more than transmit correlation. This is because no extra diversity measures are taken at the receiver. From Figure 12.7, we see that the FERs increase very fast when the receive correlation increases. This is true for all the systems. From Figure 12.6, for increasing transmit correlation with a fixed receive correlation, the degradation in JDM schemes is gradual. Up to 60% of spatial correlation between the transmit antennas, the JDM schemes perform quite consistently, they worsen only when the spatial correlation is increased more than this. Thus, as long as the transmit elements are separated by a spacing of at least 1.17 cm for our system (according to Table 12.2), the effect of spatial correlation in FER performance is not quite significant for JDM systems. While SM-QSFBC-OSIC with ZF and MMSE equalization perform the best for increasing transmit correlation, ZF-BLAST and MMSE-BLAST perform quite badly, when the spatial correlation is increased. Thus, it is evident that original VBLAST systems are not so robust in spatially correlated scenario as it was previously reported in [139].

To obtain more insight into the impact of transmit correlation, we have also performed simulations to measure the 10% outage spectral efficiency for all the schemes when transmit spatial correlation is increased steadily with a fixed receive spatial correlation. In Figure 12.8, the antenna spacing at the transmitter side is shown on x-axis: the higher the antenna spacing, the lower the spatial correlation experienced in the system. As shown in Figure 12.8, as soon as the antenna spacing at the transmitter starts to increase (i.e. the correlation starts to decrease), the outage efficiency for SM-QSFBC schemes reaches the highest spectral efficiency. Other JDM schemes also have high outage spectral efficiency. The reason for good outage spectral efficiency is the diversity benefits obtained in JDM schemes in parallel with the multiplexing gains. We know that any form of diversity benefits is bound to improve the outage scenario. It is worth noticing here that SM schemes provide very low outage efficiency even at low level of transmit correlation.

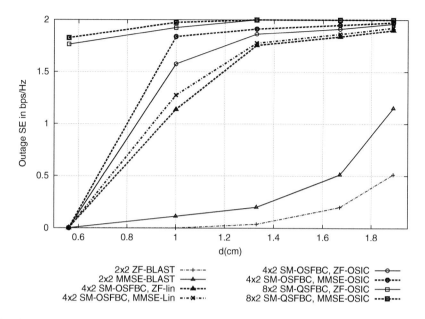

Fig. 12.8 Outage spectral efficiency with respect to decreasing transmit correlation (i.e. increasing antenna spacing) and fixed receive correlation at system SNR of 12 dB.

12.4.5   Performance in Presence of LOS

We model the wireless channel response having two distinct components [216],

1. a specular component (or LOS component) that illuminates the arrays uniformly and is thus spatially deterministic from antenna to antenna, we denote such a component as $\mathbf{H}_c$;
2. a scattered Rayleigh-distributed component that varies randomly from antenna to antenna, this component is denoted as $\mathbf{H}_w$.

The MIMO channel model for each sub-carrier with the effect of LOS component can now be written as

$$\mathbf{H}_k = \sqrt{\frac{K}{K+1}}\mathbf{H}_{c,k} + \sqrt{\frac{1}{K+1}}\mathbf{H}_{w,k}, \qquad (12.50)$$

where K is the Ricean K-factor of the system that is essentially the ratio of the power in the LOS component of the channel to the power in the fading component. $K = 0$ corresponds to pure Rayleigh fading, while $K = \infty$

corresponds to a nonfading channel. The elements of normalized scattering component $\mathbf{H}_{w,k}$ are modeled as Zero Mean Circularly Symmetric Complex Gaussian (ZMCSCG) with unit variance [139]. The specular component is given by

$$\mathbf{H}_c = \mathbf{a}(\theta_r)^T \mathbf{a}(\theta_t), \tag{12.51}$$

where $\mathbf{a}(\theta_t)$ and $\mathbf{a}(\theta_r)$ are the specular array responses at the transmitter and receiver, respectively. The array response corresponding to the transmit $P\Delta = 4P$ (or receive Q)-element linear array is given by $\left[1 \; e^{j2\pi d \cos(\theta)} \ldots e^{j2\pi d(P\Delta-1)\cos(\theta)}\right] = \left[1 \; e^{j2\pi d \cos(\theta)} \ldots e^{j2\pi(4P-1)\cos(\theta)}\right]$, where θ is the Angle of Departure (AoD) in the transmitter (or Angle of Arrival (AoA) in the receiver) of the specular component and d is the antenna spacing in wavelengths [216].

This investigation is intended for outdoor microcell scenario, therefore θ_t and θ_r are taken to be 120 and 360 degrees [189]. Spacing between arrays at both ends is taken as 0.56 cm (i.e. spatial correlation of 0.3). After obtaining the FER, we calculated the corresponding average spectral efficiency loss when LOS component becomes stronger, compared to Non Line Of Sight (NLOS) scenario (corresponding FER results for NLOS scenario is given in Figure 12.4). The average Spectral Efficiency (SE) is defined as $SE = R(1 - FER_{coded})$, where $R = 2 \, \text{bps/Hz}$ is the maximum achievable SE. For brevity, we have presented the results for MMSE based receivers only. The impact of LOS on SM-QSFBC systems is much less compared to BLAST and SM-OSFBC schemes. Figure 12.9 shows the loss because of the LOS component in average spectral efficiency. It is seen that spectral efficiency loss in *SM-QSFBC* system is 25% less compared to other schemes at $K = 50$ (i.e. $\approx 17 \, \text{dB}$).

12.5 Chapter Summary

In this chapter, we have studied different joint schemes where both diversity and multiplexing benefits are exploited at the same time. We have analyzed the performance of different receiver strategies in realistic wireless conditions. JDM schemes are in general robust to spatial correlation caused by the inadequate spatial separation between antenna elements. When spatial correlation is caused by the LOS scenario, then only SM-QSFBC type JDM schemes show

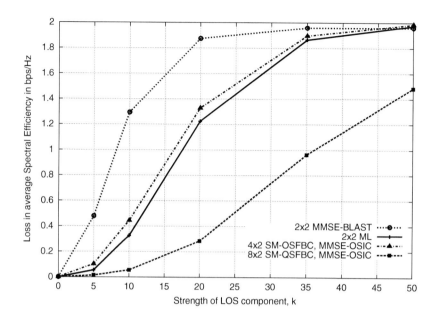

Fig. 12.9 Loss in average spectral efficiency with respect to increasingly strong LOS component.

robustness in performance. All other JDM schemes fail with little increment in strength of LOS component in the wireless channel. For both ZF and MMSE, the VBLAST-based SM schemes [29] perform poorly in realistic wireless conditions.

It is clear that JDM schemes are a suitable solution when SM is used at any cellular access point, as JDM schemes in effect increase the coverage in terms of radius where SM can be supported. Thus, it can be said that JDM schemes can be effectively used to increase the system throughput. In future, more system level analysis of the studied schemes is required. This will give us concrete understanding of these schemes' impact on multi-user throughput.

13

MIMO Design in SC-FDE/SC-FDMA Systems: Diversity and Multiplexing

As for both Downlink (DL) and Uplink (UL) the candidates for future systems are Frequency Domain Equalization (FDE) based, i.e. Orthogonal Frequency Division Multiplexing (OFDM) for DL and Single Carrier-Frequency Domain Equalization (SCFDE) for UL, and the multiple antenna techniques will be a key component of next generation wireless systems, the study of multiple antenna schemes for FDE-based systems is a timely and important issue. While quite some work has been done on the application of multiple antenna techniques to OFDM, there are much less studies for SCFDE case.

The extension of multi-antenna techniques to SCFDE is complex for some schemes such as the ones involving coding in frequency domain. In future wireless systems, the application of diversity techniques where coding is done in space and frequency domains instead of space and time, is important for two main reasons:

- Space–Frequency (SF) schemes can fulfill the requirement of low latency, as Space–Time (ST) schemes processing might not be suitable, since the receiver has to wait for more than one symbol period before processing the received stream. This extra-delay can be reduced when the coding is done across sub-carriers.
- SF schemes can work efficiently over high mobility environments, where ST schemes might fail.

Several works have recently investigated the combination of MIMO schemes and SCFDE modulation, from different points of view: equalizer [217, 218] and receiver design [219], interaction with Trellis Coded Modulation (TCM) and interleaving [220], investigation in multi-user scenario [221].

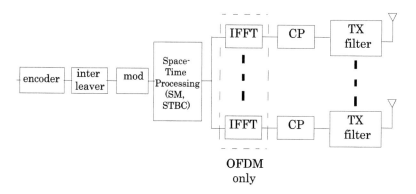

Fig. 13.1 OFDM/SCFDE MIMO transmitter.

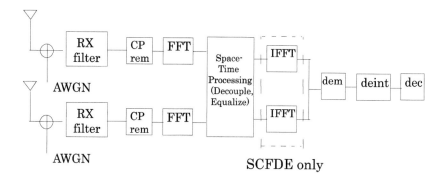

Fig. 13.2 OFDM/SCFDE MIMO receiver.

The key architectural difference between a MIMO–OFDM system and a MIMO-SCFDE system, namely the order in which the Inverse Fast Fourier Transform (IFFT) operation is executed, as outlined in Figures 13.1 and 13.2. Other differences between the SCFDE and OFDM architectures include the implementation of the Viterbi decoder and the design of transmit and receive filters [222].

13.1 Space Diversity

Space–Time Block Code-Single Carrier Frequency Domain Equalization (STBC-SCFDE) system model is shown in Figure 13.3. By following the

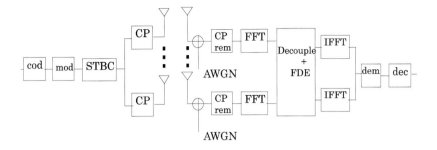

Fig. 13.3 STBC for SCFDE.

encoding rule [223]:

$$\mathbf{x}_1(t) = [x_{1,0}, x_{1,1}, \ldots, x_{1,N-1}]^T \tag{13.1}$$

$$\mathbf{x}_2(t) = [x_{2,0}, x_{2,1}, \ldots, x_{2,N-1}]^T \tag{13.2}$$

$$\mathbf{x}_1(t+T) = \left[-x_{2,0}^*, -x_{2,N-1}^*, \ldots, -x_{2,1}^* \right]^T \tag{13.3}$$

$$\mathbf{x}_2(t+T) = \left[x_{1,0}^*, x_{1,N-1}^*, \ldots, x_{1,1}^* \right]^T \tag{13.4}$$

and because of the Fast Fourier Transform (FFT) properties, we obtain in the frequency domain the same Alamouti structure than for OFDM. The equalized symbol vector has the expression:

$$\begin{bmatrix} \widehat{\mathbf{X}}_1 \\ \widehat{\mathbf{X}}_2 \end{bmatrix} = \begin{bmatrix} \mathbf{W} & \mathbf{0} \\ \mathbf{0} & \mathbf{W} \end{bmatrix} \begin{bmatrix} \widetilde{\mathbf{X}}_1 \\ \widetilde{\mathbf{X}}_2 \end{bmatrix}, \tag{13.5}$$

where $\widetilde{\mathbf{X}}_1$ and $\widetilde{\mathbf{X}}_2$ are the symbol vectors after Maximal Ratio Combining (MRC) and in the case of Minimum Mean Square Error (MMSE) equalizer we have:

$$\mathbf{W}(i, i) = \frac{\sqrt{N_T/P_T}}{|\Lambda_1(i, i)|^2 + |\Lambda_2(i, i)|^2 + \frac{N_T}{\rho}}, \tag{13.6}$$

where Λ_i, $i = 1, 2$ are $N \times N$ diagonal matrices whose (l, l)th entries are equal to the lth FFT coefficients of the Channel Impulse Responses (CIRs) $\mathbf{h}_i$, $i = 1, 2$, and $\rho = P_T/\sigma^2$, with P_T total power of the transmitted signal, and σ^2 variance of the complex Gaussian noise process at one receive antenna. We then go back to the time domain by applying the IFFT:

$$\begin{bmatrix} \widehat{\mathbf{x}}_1 \\ \widehat{\mathbf{x}}_2 \end{bmatrix} = \begin{bmatrix} \mathbf{Q}^H & \mathbf{0} \\ \mathbf{0} & \mathbf{Q}^H \end{bmatrix} \begin{bmatrix} \widehat{\mathbf{X}}_1 \\ \widehat{\mathbf{X}}_2 \end{bmatrix}, \tag{13.7}$$

where $\mathbf{Q}$ is the orthonormal $N \times N$ FFT matrix, i.e. $\mathbf{Q}(k,l) = \frac{1}{\sqrt{N}}e^{-j\frac{2\pi}{N}(k-1)(l-1)}$, $1 \leq k,l \leq N$.

13.2 Spatial Multiplexing

Spatial Multiplexing-Single Carrier Frequency Domain Equalization (SM-SCFDE) system model is shown in Figure 13.4. The SCFDE-SM equalized vectors are

$$\begin{bmatrix} \widehat{\mathbf{X}}_1 \\ \widehat{\mathbf{X}}_2 \end{bmatrix} = \mathbf{W}\begin{bmatrix} \mathbf{Y}_1 \\ \mathbf{Y}_2 \end{bmatrix}, \tag{13.8}$$

where $\mathbf{Y}_1$ and $\mathbf{Y}_2$ are the vectors received at the two receive antennas and $\mathbf{W}$ is the equalizer which for MMSE is

$$\mathbf{W} = \sqrt{\frac{N_T}{P_T}}\Lambda^H\left(\Lambda\Lambda^H + \frac{N_T}{\rho}\mathbf{I}_{2N}\right)^{-1}, \tag{13.9}$$

where

$$\Lambda = \begin{bmatrix} \Lambda_{1,1} & \Lambda_{1,2} \\ \Lambda_{2,1} & \Lambda_{2,2} \end{bmatrix} \tag{13.10}$$

is the MIMO channel matrix with $\Lambda_{i,j}$ diagonal matrix containing the frequency responses of the N sub-carriers between jth transmit and ith receive antenna and ρ is defined as before. At this point there is the IFFT block, as in Equation (13.7).

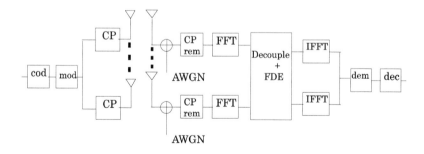

Fig. 13.4 Spatial multiplexing for SCFDE.

13.3 Space–Frequency Block Coding for SCFDE

It has been shown that SCFDE is comparable to OFDM with respect to implementation effort and performance if MIMO extensions are applied [224]. However, between SCFDE and OFDM there are some important differences, i.e.

- The sub-carriers access in SCFDE is not directly available: this creates problems for techniques that require sub-carriers manipulation, such as SFBC.
- The trade-off diversity-multiplexing depends also on the sensitivity of the transmission technique with respect to coding in time or frequency, and since SCFDE and OFDM have different behaviors from this point of view, a different trade-off diversity-multiplexing for SCFDE and OFDM is expected.

In this section, firstly the issue of extending SFBC to SCFDE systems, which is not straightforward as in case of STBC, is addressed and a low-complexity solution is proposed. Moreover, the proposed processing is used for applying a JDM SM-SFBC scheme to SCFDE system. To be able to apply SFBC and SFBC-based schemes to SCFDE systems is very important, since SFBCs show a great robustness to the time selectivity caused by users' mobility, and they work well for a wider range of propagation scenarios with respect to STBCs.

The STBC extension to SCFDE proposed in [223] cannot be directly mapped to space and frequency domains. To overcome this problem, in this section a new low-complexity SCFDE-SFBC scheme will be introduced which allows to code across antennas and sub-carriers. One has to notice that the sub-carriers domain is not directly accessible at the transmitter of a SCFDE system, but the novel scheme introduced hereafter makes possible the access to frequency domain by properly organizing the information in the time domain over the two antennas. SCFDE-SFBC is shown to outperform SCFDE-STBC technique over fast fading channels caused by high mobility speed. We compare these two schemes in terms of uncoded BER for different levels of time and frequency selectivity. Uncoded BER was here chosen as a metric to study the behavior of the proposed scheme in a simplified case, i.e. to see the potential gain due to the scheme itself, rather than to the impact of channel coding.

13.3.1 System Model

Figures 13.5 and 13.6 show the transmitter and the receiver respectively, of the proposed 2×1 transmit diversity scheme. The data stream to be transmitted is organized in 4 sequences by properly combining the even and the odd sequences of the original data stream, as it is described in this section; 2 of the

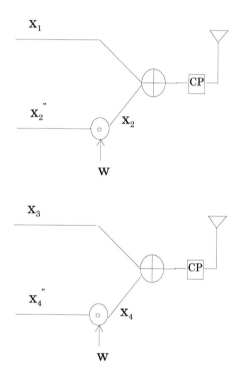

Fig. 13.5 Transmitter model of the proposed 2×1 SCFDE transmit diversity scheme.

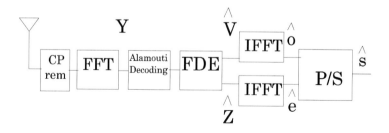

Fig. 13.6 Receiver model of the proposed 2×1 SCFDE transmit diversity scheme.

4 sequences are multiplied by a *phase vector* **w** and then contemporarily trans-mitted over two transmit antennas after the insertion of the Cyclic Prefix (CP).

At the receiver, after the CP removal, the FFT operation is performed. Then, MMSE-FDE is performed each two consecutive sub-carriers. Two $N/2$-points IFFT convert back to the time domain the even and odd sequences, which are then multiplexed to obtain the samples used for the decision process.

Transmitter: Let us consider a data stream **s** of N symbols:

$$\mathbf{s} = [s_1, s_2, \ldots, s_N]^T, \tag{13.11}$$

which is split in the odd sequence **o** and even sequence **e**:

$$\mathbf{o} = [s_1, s_3, \ldots, s_{N-1}]^T \tag{13.12}$$

$$\mathbf{e} = [s_2, s_4, \ldots, s_N]^T. \tag{13.13}$$

Let us form the following four N-long sequences: $\mathbf{x}_1 = [\mathbf{o}^T, \mathbf{o}^T]^T$, $\mathbf{x}_2' = [-\mathbf{e}^{*T}, -\mathbf{e}^{*T}]^T$, $\mathbf{x}_3 = [\mathbf{e}^T, \mathbf{e}^T]^T$, $\mathbf{x}_4' = [\mathbf{o}^{*T}, \mathbf{o}^{*T}]^T$, and let us rewrite them in terms of the original sequence **s**:

$$\mathbf{x}_1 = [s_1, s_3, \ldots, s_{N-1}, s_1, s_3, \ldots, s_{N-1}]^T \tag{13.14}$$

$$\mathbf{x}_2' = [-s_2^*, -s_4^*, \ldots, -s_N^*, -s_2^*, -s_4^*, \ldots, -s_N^*]^T \tag{13.15}$$

$$\mathbf{x}_3 = [s_2, s_4, \ldots, s_N, s_2, s_4, \ldots, s_N]^T \tag{13.16}$$

$$\mathbf{x}_4' = [s_1^*, s_3^*, \ldots, s_{N-1}^*, s_1^*, s_3^*, \ldots, s_{N-1}^*]^T. \tag{13.17}$$

It is evident that $\mathbf{x}_2' = -\mathbf{x}_3^*$ and $\mathbf{x}_4' = \mathbf{x}_1^*$. Starting from $\mathbf{x}_2'$ and $\mathbf{x}_4'$, the vectors $\mathbf{x}_2''$ and $\mathbf{x}_4''$ are built as

$$\mathbf{x}_2'' = \left[-s_2^*, -s_N^*, \ldots, -s_4^*, -s_2^*, -s_N^*, \ldots, -s_4^*\right]^T \tag{13.18}$$

$$\mathbf{x}_4'' = \left[s_1^*, s_{N-1}^*, \ldots, s_3^*, s_1^*, s_{N-1}^*, \ldots, s_3^*\right]^T. \tag{13.19}$$

The sequences transmitted simultaneously on the first transmit antenna (see Equations (13.14)–(13.18)) are $\mathbf{x}_1$ and

$$\mathbf{x}_2 = \mathbf{w} \odot \mathbf{x}_2'', \tag{13.20}$$

while the sequences transmitted from the second transmit antenna (see Equations (13.16)–(13.19)) are $\mathbf{x}_3$ and

$$\mathbf{x}_4 = \mathbf{w} \odot \mathbf{x}_4'', \tag{13.21}$$

where $\odot$ means component-wise product and $\mathbf{w}$ is the $N \times 1$ vector:

$$\mathbf{w} = \left[1, e^{j\pi/\frac{N}{2}}, \dots, e^{j(\frac{N}{2}-1)\pi/\frac{N}{2}}, -1, -e^{j\pi/\frac{N}{2}}, \dots, -e^{j(\frac{N}{2}-1)\pi/\frac{N}{2}}\right]^T. \quad (13.22)$$

Receiver: The received vector is

$$\mathbf{y} = \mathbf{H}_1(\mathbf{x}_1 + \mathbf{x}_2) + \mathbf{H}_2(\mathbf{x}_3 + \mathbf{x}_4) + \mathbf{n}, \quad (13.23)$$

where $\mathbf{H}_1$ and $\mathbf{H}_2$ are $N \times N$ circulant channel matrices. Using the eigen-value decomposition:

$$\mathbf{y} = \mathbf{Q}^H \Lambda_1 \mathbf{Q}(\mathbf{x}_1 + \mathbf{x}_2) + \mathbf{Q}^H \Lambda_2 \mathbf{Q}(\mathbf{x}_3 + \mathbf{x}_4) + \mathbf{n}, \quad (13.24)$$

where $\mathbf{Q}$ is the orthonormal $N \times N$ FFT matrix, i.e. $\mathbf{Q}(i,j) = \frac{1}{\sqrt{N}}e^{-j\frac{2\pi}{N}(i-1)(j-1)}$, $1 \leq i, j \leq N$. After the FFT block, it can be written (frequency domain):

$$\mathbf{Y} = \mathbf{Qy} \quad (13.25)$$
$$= \Lambda_1 \mathbf{Q}(\mathbf{x}_1 + \mathbf{x}_2) + \Lambda_2 \mathbf{Q}(\mathbf{x}_3 + \mathbf{x}_4) + \mathbf{Qn} \quad (13.26)$$
$$= \Lambda_1(\mathbf{X}_1 + \mathbf{X}_2) + \Lambda_2(\mathbf{X}_3 + \mathbf{X}_4) + \mathbf{N}, \quad (13.27)$$

where Λ_i, $i = 1, 2$ are $N \times N$ diagonal matrices whose (l,l)th entries are equal to the lth FFT coefficients of the CIRs $\mathbf{h}_i$, $i = 1, 2$, and $\mathbf{X}_j$, $j = 1, \dots, 4$ and $\mathbf{N}$ are $N \times 1$ vectors representing the FFT transforms of transmitted vectors $\mathbf{x}_j$, $j = 1, \dots, 4$ and noise $\mathbf{n}$, respectively. In particular, for the FFT transforms of the transmitted vectors:

$$\mathbf{X}_1 = \mathbf{Qx}_1 = \left[V_1, 0, \dots, V_{N/2}, 0\right]^T \quad (13.28)$$
$$\mathbf{X}_2 = \mathbf{Qx}_2 = \left[0, -Z_1^*, \dots, 0, -Z_{N/2}^*\right]^T \quad (13.29)$$
$$\mathbf{X}_3 = \mathbf{Qx}_3 = \left[Z_1, 0, \dots, Z_{N/2}, 0\right]^T \quad (13.30)$$
$$\mathbf{X}_4 = \mathbf{Qx}_4 = \left[0, V_1^*, \dots, 0, V_{N/2}^*\right]^T \quad (13.31)$$

from which it is possible to recognize the Alamouti-like structure in the space and frequency domains [30, 176]. For the kth couple of sub-carriers, with $k = 1, \dots, \frac{N}{2}$:

$$\begin{bmatrix} Y_{2k-1} \\ Y_{2k}^* \end{bmatrix} = \begin{bmatrix} \Lambda_{1,2k-1} & \Lambda_{2,2k-1} \\ \Lambda_{2,2k}^* & -\Lambda_{1,2k}^* \end{bmatrix} \begin{bmatrix} V_k \\ Z_k \end{bmatrix} + \begin{bmatrix} N_{2k-1} \\ N_{2k}^* \end{bmatrix}. \quad (13.32)$$

Let us assume the two channels constant for two adjacent sub-carriers:

$$\Lambda_{j,2k-1} = \Lambda_{j,2k} = \Lambda_{j,k}, \quad j = 1,2, \quad k = 1,\dots,\frac{N}{2} \tag{13.33}$$

then (13.32) can be rewritten as

$$\begin{bmatrix} Y_{2k-1} \\ Y_{2k}^* \end{bmatrix} = \begin{bmatrix} \Lambda_{1,k} & \Lambda_{2,k} \\ \Lambda_{2,k}^* & -\Lambda_{1,k}^* \end{bmatrix} \begin{bmatrix} V_k \\ Z_k \end{bmatrix} + \begin{bmatrix} N_{2k-1} \\ N_{2k}^* \end{bmatrix}. \tag{13.34}$$

By doing a MRC processing:

$$\begin{aligned} \begin{bmatrix} \tilde{Y}_{2k-1} \\ \tilde{Y}_{2k} \end{bmatrix} &= \begin{bmatrix} \Lambda_{1,k}^* & \Lambda_{2,k} \\ \Lambda_{2,k}^* & -\Lambda_{1,k} \end{bmatrix} \begin{bmatrix} Y_{2k-1} \\ Y_{2k}^* \end{bmatrix} \\ &= \begin{bmatrix} \tilde{\Lambda} & 0 \\ 0 & \tilde{\Lambda} \end{bmatrix} \begin{bmatrix} V_k \\ Z_k \end{bmatrix} + \begin{bmatrix} \tilde{N}_1 \\ \tilde{N}_2 \end{bmatrix}, \end{aligned} \tag{13.35}$$

where $\tilde{\Lambda} = |\Lambda_{1,k}|^2 + |\Lambda_{2,k}|^2$, $\tilde{N}_1 = \Lambda_{1,k}^* N_{2k-1} + \Lambda_{2,k} N_{2k}^*$, and $\tilde{N}_2 = \Lambda_{2,k}^* N_{2k-1} - \Lambda_{1,k} N_{2k}^*$. After the MMSE-FDE equalization [222]:

$$\begin{bmatrix} \widehat{V}_k \\ \widehat{Z}_k \end{bmatrix} = \sqrt{\frac{N_T}{P_T}} \begin{bmatrix} 1/\left(\tilde{\Lambda} + \frac{N_T}{\rho}\right) & 0 \\ 0 & 1/\left(\tilde{\Lambda} + \frac{N_T}{\rho}\right) \end{bmatrix} \begin{bmatrix} \tilde{Y}_{2k-1} \\ \tilde{Y}_{2k} \end{bmatrix},$$

where $\rho = P_T/\sigma^2$, and P_T is the total power of the transmitted signal, while σ^2 is the total variance of the complex Gaussian noise process. Therefore, we have recovered:

$$\widehat{\mathbf{V}} = [\widehat{V}_1, \widehat{V}_2, \dots, \widehat{V}_{N/2}]^T \tag{13.36}$$

$$\widehat{\mathbf{Z}} = [\widehat{Z}_1, \widehat{Z}_2, \dots, \widehat{Z}_{N/2}]^T \tag{13.37}$$

and we go back into the time-domain through two $\frac{N}{2}$-points IFFT, with a coefficient of normalization $\frac{1}{2}$:

$$\widehat{\mathbf{o}} = [\widehat{s}_1, \widehat{s}_3, \dots, \widehat{s}_{N-1}]^T = \frac{1}{2} IFFT(\widehat{\mathbf{V}}) \tag{13.38}$$

$$\widehat{\mathbf{e}} = [\widehat{s}_2, \widehat{s}_4, \dots, \widehat{s}_N]^T = \frac{1}{2} IFFT(\widehat{\mathbf{Z}}). \tag{13.39}$$

After a Parallel to Serial (P/S) converter, and a proper multiplexing of the two sequences, the estimation of the transmitted sequence **s** is achieved:

$$\widehat{\mathbf{s}} = [\widehat{s}_1, \widehat{s}_2, \dots, \widehat{s}_N]^T. \tag{13.40}$$

13.3.2 Simulations and Discussions

We present here the comparison of SCFDE-STBC and the proposed SCFDE space–frequency transmit scheme in terms of uncoded BER and complexity. Simulation parameters are shown in Table 13.1. Ideal time and frequency synchronization is assumed. We also assume perfect channel estimation at the receiver. An L-taps frequency selective channel is assumed, with exponential decay. In particular, we consider: a low frequency selective channel characterized by $L = 8$ taps, corresponding to a maximum delay spread of $\tau_{max} = 4.90\,\mu s$; a high frequency selective channel with $L = 30$, i.e. a maximum delay spread of $\tau_{max} = 20.29\,\mu s$; a quasi-static channel with normalized Doppler frequency $f_d T_s = 0.001$, corresponding to a velocity of 3 kmph; a high time selective channel with normalized Doppler frequency $f_d T_s = 0.03$, i.e. velocity of 90 kmph.

Figure 13.7 shows the uncoded BER of a SCFDE-STBC transmit diversity scheme with respect to the proposed scheme in a low frequency selective channel. As expected, when the channel is quasi-static the two schemes show almost identical performance. However, in a fast varying channel ($f_d T_s = 0.03$) the proposed scheme outperforms the SCFDE-STBC transmit diversity scheme, which shows a floor in the uncoded BER. As in a OFDM-SFBC scheme, the proposed scheme assumes the channel constant over two consecutive sub-carriers. Therefore, the performance of the proposed scheme are expected to show a degradation over high frequency selective channels. More insight into sensitivity of the two transmit diversity schemes over different propagation conditions are provided by Figures 13.8 and 13.9. In particular, Figure 13.8 shows the performance degradation of the two schemes when the normalized coherence bandwidth $\left(\text{defined as } B_{c,norm} = \frac{1}{\tau_{max}\Delta f}\right)$ decreases, i.e. when the

Table 13.1 Simulation parameters.

Parameters	Value
System bandwidth, B	1.25 MHz
Sampling frequency, f_s	1.429 MHz
Carrier frequency, f_c	3.5 GHz
Symbols per block, N	128
Sub-carrier spacing, Δf	9.77 KHz
CP length, N_{CP}	16
Data symbol mapping	BPSK
FDE Equalizer	MMSE
Channel coding scheme	1/2-rate convolutional coding

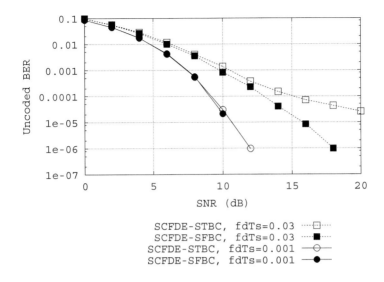

Fig. 13.7 SCFDE-STBC vs the proposed scheme in low and high time selective channels.

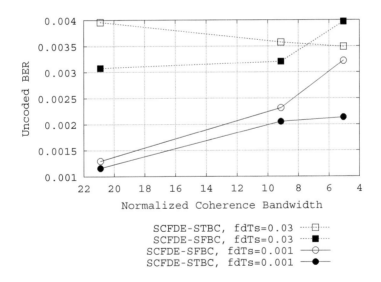

Fig. 13.8 Uncoded BER vs normalized coherence bandwidth at the SNR of 8 dB.

channel frequency selectivity increases. First of all, we observe that the STBC scheme is not very sensitive to the increase of the frequency selectivity. Moreover, when the channel is quasi-static the STBC scheme always outperforms the proposed scheme. However, the performance gap of the two schemes is

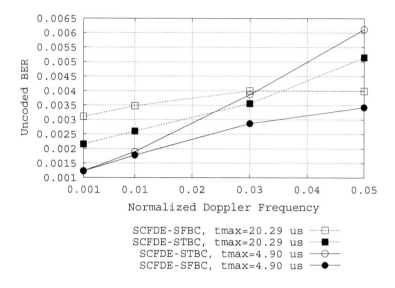

Fig. 13.9 Uncoded BER vs normalized Doppler frequency at the SNR of 8 dB.

quite low for a wide range of frequency selectivity conditions. In presence of a high time selective channel, the proposed scheme noticeably outperforms the STBC scheme for a big range of frequency selectivity.

While STBC has very low sensitivity with respect to the frequency selectivity of the channel, the time selectivity has a greater impact on the performance of the proposed scheme, especially when the frequency selectivity is low. This is shown in Figure 13.9, which plots the uncoded BER of the two schemes for increasing normalized Doppler frequency, i.e. increasing time-variability. However, the sensitivity of the STBC scheme with respect to the time-variability is much greater and the proposed scheme outperforms the STBC scheme over highly time-varying channels also when the frequency selectivity is high.

It is interesting to notice from Figures 13.8 and 13.9 that when SCFDE-STBC scheme fails because of high velocity (high normalized Doppler frequency), this scheme recovers in BER performance in presence of frequency selectivity (high maximum delay spread), since SCFDE is able to exploit the frequency diversity without channel coding, differently from OFDM. The same does not hold for SFBC since for this scheme the frequency selectivity is detrimental (hypothesis of channel constance over adjacent sub-carriers).

13.3.3 Computational Complexity

The computational complexities, in terms of the overall number of complex multiplications (considering transmitter and receiver) of SCFDE-SFBC and SCFDE-STBC are, respectively:

$$C^{SC-SF} = N_T N^2 + (N_R + 1)\frac{N}{2}\log_2 N - \frac{N}{2}$$

$$C^{SC-ST} = N_T N^2 + (N_R + 1)\frac{N}{2}\log_2 N.$$

The complexity for the 2×1 case is shown in Table 13.2. The parameters in this case are $N_T = 2$, $N_R = 1$.

Therefore, the presented transmit diversity scheme has a slightly less complexity than STBC. As SCFDE is considered as a possible alternative to OFDM transmission in the uplink of future broadband wireless systems, the possibility of implementing SCFDE-SFBC without increasing the complexity at the mobile is a positive aspect.

13.4 Combining Diversity and Multiplexing in SCFDE

The trade-off diversity-multiplexing, which has been studied for OFDM systems [209, 225, 226], to the best of our knowledge has not yet been investigated for SCFDE. Basically, techniques that try to combine diversity and multiplexing, use the available antennas to get either more diversity gain or more multiplexing (rate) gain, according to the desired trade-off. Since this trade-off depends also on the sensitivity of the transmission technique with respect to coding in time or frequency, and SCFDE and OFDM have different behaviors from this point of view, we also expect a different trade-off diversity-multiplexing for SCFDE and OFDM. The pre-processing technique which allows a low complexity extension of SFBC technique to SCFDE is used to implement a 4×2 scheme which combines SM with SFBC, where the four antennas are divided in to two groups of two antennas: the SM is applied to

Table 13.2 Complexity in terms of complex multiplications for 2×1 case.

Scheme	$N = 64$	$N = 128$
SCFDE-SF	8544	33600
SCFDE-ST	8576	33664

the two groups of antennas, while SFBC is applied to the two antennas of each group.

13.4.1 System Model

In the presented 4×2 SCFDE SF-JDM system, the four transmit antennas are divided in to two groups of two antennas each. Signals over the two groups of antennas are sent simultaneously and recovered via a SM approach, whilst over the two antennas of each group, a SFBC processing is applied. The system model is depicted in Figure 13.10, considering also the ST case. The objective is to get both multiplexing and diversity gain.

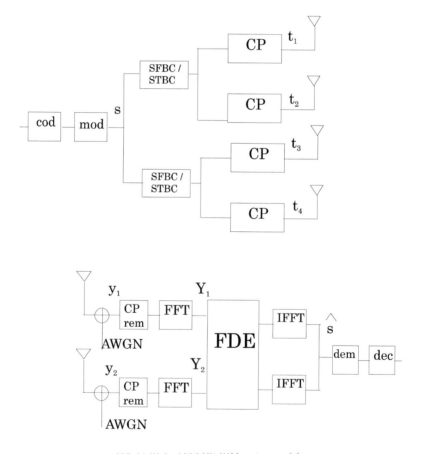

Fig. 13.10 4×2 SCFDE JDM system model.

4 × 2 *SCFDE SF-JDM*: Four sequences are transmitted through the first two antennas, implementing a SCFDE-SFBC scheme, and other four sequences are transmitted in the same way via the third and fourth transmit antennas.

The vector received at the ith receive antenna is $\mathbf{y}_i = \sum_{j=1}^{4} \mathbf{H}_{ij}\mathbf{t}_j = \mathbf{H}_{i1}(\mathbf{x}_{1,1} + \mathbf{x}_{1,2}) + \mathbf{H}_{i2}(\mathbf{x}_{1,3} + \mathbf{x}_{1,4}) + \mathbf{H}_{i3}(\mathbf{x}_{2,1} + \mathbf{x}_{2,2}) + \mathbf{H}_{i4}(\mathbf{x}_{2,3} + \mathbf{x}_{2,4}) + \mathbf{n}_i$, $i = 1, 2$ where $\mathbf{H}_{ij}$, $i = 1, 2$, $j = 1, \ldots, 4$ are $N \times N$ circulant channel matrices and $\mathbf{x}_{i,j}$, $i = 1, 2$, $j = 1, \ldots, 4$ are the SFBC transmitted vectors, defined as in (13.14), (13.16), (13.20), and (13.21). After eigen-value decomposing the circulant channel matrices, i.e. $\mathbf{H}_{ij} = \mathbf{Q}^H \Lambda_{ij} \mathbf{Q}$ and the FFT block at the receiver, we have $\mathbf{Y}_i = \mathbf{Q}\mathbf{y}_i$. For the kth couple of sub-carriers, with $k = 1, \ldots, \frac{N}{2}$, after some simple linear algebra manipulation, it is possible to obtain:

$$\mathbf{Y}'_k = \Lambda_k \mathbf{X}_k + \mathbf{N}_k, \tag{13.41}$$

where $\mathbf{Y}'_k = \left[Y_{1,2k-1}, Y_{2,2k-1}, Y^*_{1,2k}, Y^*_{2,2k} \right]^T$, $\mathbf{X}_k = \left[V_{1,2k-1}, Z_{1,2k-1}, V_{2,2k-1}, Z_{2,2k-1} \right]^T$, $\mathbf{N}_k = \left[N_{1,2k-1}, N_{2,2k-1}, N^*_{1,2k}, N^*_{2,2k} \right]^T$, and

$$\Lambda_k = \begin{bmatrix} \Lambda_{11,k} & \Lambda_{12,k} & \Lambda_{13,k} & \Lambda_{14,k} \\ \Lambda_{21,k} & \Lambda_{22,k} & \Lambda_{23,k} & \Lambda_{24,k} \\ \Lambda^*_{12,k} & -\Lambda^*_{11,k} & \Lambda^*_{14,k} & -\Lambda^*_{13,k} \\ \Lambda^*_{22,k} & -\Lambda^*_{21,k} & \Lambda^*_{24,k} & -\Lambda^*_{23,k} \end{bmatrix}, \tag{13.42}$$

where the CTF is assumed identical over two adjacent sub-carriers. At this point, $\mathbf{Y}'_k$ is MMSE equalized, i.e.

$$\widehat{\mathbf{X}}_k = \mathbf{G}_k^{\text{MMSE}} \mathbf{Y}'_k, \tag{13.43}$$

where

$$\mathbf{G}_k^{\text{MMSE}} = \sqrt{\frac{N_T}{P_T}} \left(\Lambda_k^H \Lambda_k + \frac{N_T}{\rho} \mathbf{I}_{N_T} \right)^{-1} \Lambda_k^H, \tag{13.44}$$

where $\rho = P_T/\sigma^2$, with P_T total power of the transmitted signal and σ^2 noise variance at one receive antenna. After this, the IFFT operation is performed.

4 × 2 *SCFDE ST-JDM*: Let us consider a 4 × 2 system, with 2 SM branches and on each SM branch an STBC block, such that both multiplexing and diversity gains can be realized, through SM and STBC, respectively. At time t, the vectors $\mathbf{x}_i(t) = [x_{i,1}, x_{i,2}, \ldots, x_{i,N}]^T$, $i = 1, \ldots, 4$ are sent from the four transmit antennas. At time $t + T$, the special encoding rule introduced in [223] is applied, i.e. $\mathbf{x}_i(t + T) = [-x_{j,1}^*, -x_{j,N}^*, \ldots, -x_{j,2}^*]^T$, $i = 1, 3$, $j = 2, 4$ are transmitted from antennas 1 and 3, and $\mathbf{x}_i(t + T) = [x_{j,1}^*, x_{j,N}^*, \ldots, x_{j,2}^*]^T$, $i = 2, 4$, $j = 1, 3$ from antennas 2 and 4. Analogously to SCFDE SF-JDM, and following the same procedure as in [209], it can be written, for kth sub-carrier, $k = 1, \ldots, N$:

$$\mathbf{Y}'_k = \Lambda_k \mathbf{X}_k + \mathbf{N}_k, \tag{13.45}$$

where $\quad \mathbf{Y}'_k = [Y_{1,k}(n), Y_{2,k}(n), Y_{1,k}^*(n + 1), Y_{2,k}^*(n + 1)]^T$, $\quad \mathbf{X}_k = [V_{1,k}(n),$ $Z_{1,k}(n), V_{2,k}(n), Z_{2,k}(n)]^T$, $\mathbf{N}_k = [N_{1,k}(n), N_{2,k}(n), N_{1,k}^*(n+1), N_{2,k}^*(n+1)]^T$, and

$$\Lambda_k(n) = \begin{bmatrix} \Lambda_{11,k}(n) & \Lambda_{12,k}(n) & \Lambda_{13,k}(n) & \Lambda_{14,k}(n) \\ \Lambda_{21,k}(n) & \Lambda_{22,k}(n) & \Lambda_{23,k}(n) & \Lambda_{24,k}(n) \\ \Lambda_{12,k}^*(n + 1) & -\Lambda_{11,k}^*(n + 1) & \Lambda_{14,k}^*(n + 1) & -\Lambda_{13,k}^*(n + 1) \\ \Lambda_{22,k}^*(n + 1) & -\Lambda_{21,k}^*(n + 1) & \Lambda_{24,k}^*(n + 1) & -\Lambda_{23,k}^*(n + 1) \end{bmatrix} \tag{13.46}$$

with n identifying the nth time instant. Assuming $\Lambda_{i,j}(n) = \Lambda_{i,j}(n + 1)$, i.e., assuming a quasi-static channel, Equations (13.42)–(13.44) still hold, but instead of $\frac{N}{2}$ couples of sub-carriers, this time N sub-carriers have to be considered.

4×2 SCFDE QOD schemes: In the next section, the SF and ST-JDM schemes previously presented are compared with 4×2 antennas SF and ST-Quasi-Orthogonal Design (QOD) schemes. The QOD transmitter is shown in Figure 13.11, with the receiver part that is the same as in Figure 13.10. The QOD schemes use the 4-antennas code proposed in [208]. The SF-QOD code can be achieved in SCFDE in a similar way to the 2-antennas code SFBC presented in Section 13.3. It is worth noting that QOD codes provide a full rate of 1, but present a loss in orthogonality. The rate of 1 is provided by sending four symbols over the transmit antennas in four sub-carriers, for SF-QOD, and in four block durations, for ST-QOD.

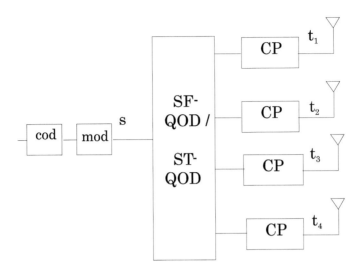

Fig. 13.11 4×2 SCFDE QOD system model: Transmitter.

13.4.2 Simulations and Discussions

Simulation parameters have been shown in Table 13.1. Ideal time and frequency synchronization and perfect channel estimation are assumed at the receiver. The channel is a L-taps frequency selective channel with exponentially distributed power decay. This section presents results achieved in terms of average SE, without and with channel coding ($\frac{1}{2}$-rate convolutional coding). The average spectral efficiency is defined in terms of rate R and FER, as SE $= R(1 - \text{FER})$, where R considers modulation order, spatial rate, and channel coding rate. The average SE is a good metric of the achievable trade-off diversity-multiplexing, since it considers both rate and reliability.

Space–frequency schemes: Figure 13.12 shows the uncoded average SE of SF-JDM schemes with respect to SF-QOD diversity schemes, with the same 4×2 antennas configuration and same rate $R = 2$ (i.e. BPSK modulation for SF-JDM schemes that have spatial rate of 2, and QPSK for SF-QOD that have spatial rate of 1), in highly time-variant channels ($f_d T_s = 0.03$ corresponding to $v = 90\,\text{kmph}$) with low frequency selectivity. Both SCFDE and OFDM systems are considered in the comparison. Figure 13.12 shows that for low frequency selective channels, it is more efficient to use the redundancy introduced by the 4 transmit antennas in order to maximize the diversity gain. This

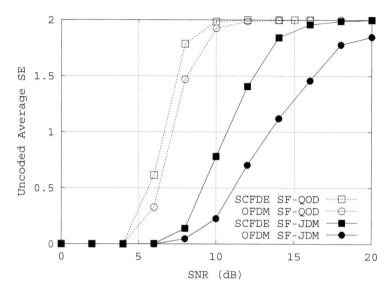

Fig. 13.12 Uncoded average SE for low frequency selectivity ($\tau_{max} = 2.80\,\mu s$) and high time selectivity ($f_d T_s = 0.03$).

holds both for SCFDE and OFDM systems. It is expected that the frequency selectivity of the channel severely degrades the performance of SF schemes, which are based on the assumption that the channel is constant over two or more consecutive sub-carriers. In particular, for the 4×2 SF-QOD scheme, the channel is supposed to be constant over 4 sub-carriers.

However, as it is shown in Figures 13.13 and 13.14, in SCFDE systems, the combination of SM and SFBC is more robust to the frequency selectivity with respect to QOD. Already for moderate frequency selective channels, the SF-JDM scheme begins to outperform the SF-QOD scheme in terms of SE. Moreover, JDM keeps on working for high frequency selective channels where the QOD performance are severely degraded and the SE is almost half of the maximum that is achievable for that configuration. Also for OFDM systems, the SF-JDM scheme is more robust to the frequency selectivity of the channel, since it requires the channel to be constant over less consecutive sub-carriers. However, the performance of both OFDM schemes are severely degraded by the frequency selectivity of the channel. This is expected since the uncoded OFDM is not able to exploit the frequency selectivity of the channel as it is a SCFDE transmission scheme. Therefore, for uncoded OFDM the

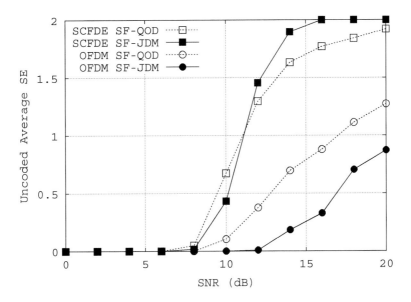

Fig. 13.13 Uncoded average SE for moderate frequency selectivity ($\tau_{max} = 14.70\,\mu s$) and high time selectivity ($f_d T_s = 0.03$).

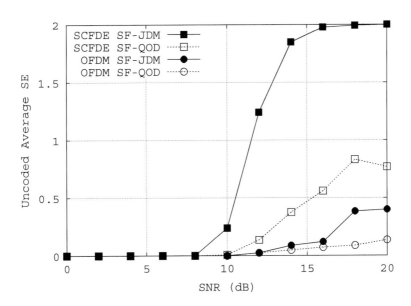

Fig. 13.14 Uncoded average SE for high frequency selectivity ($\tau_{max} = 18.20\,\mu s$) and high time selectivity ($f_d T_s = 0.03$).

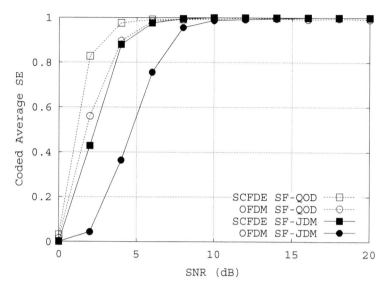

Fig. 13.15 Coded average SE for high frequency selectivity ($\tau_{max} = 18.20\,\mu s$) and high time selectivity ($f_d T_s = 0.03$).

frequency selectivity has only a negative impact, even if it is less negative for the JDM scheme.

Figure 13.15 shows the coded average SE comparison. The maximum achievable rate is 1, since also the channel coding rate of $\frac{1}{2}$ is taken into account. The negative effect of frequency selectivity on SF schemes is very much mitigated by channel coding, particularly for OFDM schemes and SCFDE SF-QOD. Therefore, in the coded case, it is more convenient to use the available antennas to provide only diversity, using higher order constellations, instead of providing both diversity and multiplexing, using lower order symbol mappings.

Finally, we also consider the sensitivity of the JDM scheme with respect to spatial correlation. The spatial correlation is modeled according to the inter-element distances in the transmit and receive antenna arrays, as it is done in [211]. Correlation at both the transmit and receive arrays is considered, with values between 0, for no correlation, and 1, for full correlation. The impact of spatial correlation on the performance of SCFDE-SF-JDM and SCFDE-SF-QOD for moderate frequency selectivity in terms of uncoded average SE is shown in Figure 13.16. As expected, the JDM scheme is quite sensitive to

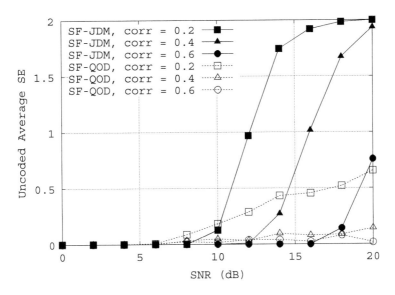

Fig. 13.16 Uncoded average SE in presence of spatial correlation (at both transmit and receive arrays) and moderate frequency selectivity ($\tau_{max} = 14.70\,\mu$s) and high time selectivity ($f_d T_s = 0.03$).

the spatial correlation. Nevertheless, QOD scheme is also affected by spatial correlation, though its drop in performance is less impressive than for JDM. If channel coding is used (see Figure 13.17), QOD scheme is much less impacted by the spatial correlation than JDM scheme.

Space–time schemes: The dual propagation scenarios are considered when comparing ST schemes, i.e. a high frequency selectivity ($\tau_{max} = 18.20\,\mu$s) and increasing time selectivity. In Figure 13.18, the case for low time selectivity is shown ($f_d T_s = 0.001$ corresponding to $v = 3\,$kmph): the QOD schemes outperform the JDM schemes, and the ones with SCFDE modulation are better than the ones with OFDM.

As shown in Figures 13.19 and 13.20, corresponding to moderate ($f_d T_s = 0.015$, $v = 45\,$kmph) and severe ($f_d T_s = 0.03$, $v = 90\,$kmph) time selectivity, the SCFDE ST-JDM scheme performs the best, in uncoded case, as SCFDE SF-JDM was doing among the SF schemes in the harsh scenario of high time and frequency variability. In Figure 13.20, also the SCFDE SF-JDM and QOD schemes have been included, to compare them with the correspondent ST schemes. The SF and ST-JDM schemes perform practically the same, whilst SF-QOD is performing much worse than ST-QOD, suggesting that the impact

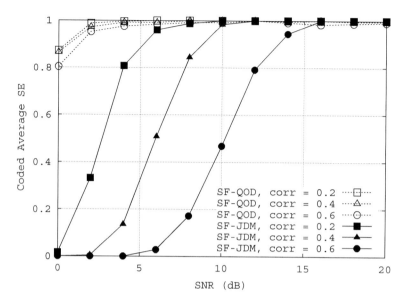

Fig. 13.17 Coded average SE in presence of spatial correlation (at both transmit and receive arrays) and moderate frequency selectivity ($\tau_{max} = 14.70\,\mu s$) and high time selectivity ($f_d T_s = 0.03$).

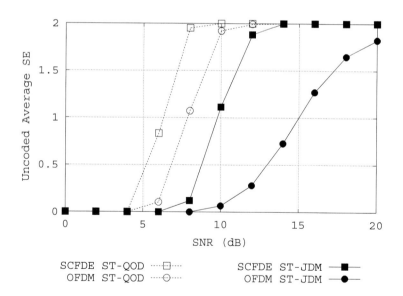

Fig. 13.18 Uncoded average SE for low time selectivity ($f_d T_s = 0.001$) and high frequency selectivity ($\tau_{max} = 18.20\,\mu s$).

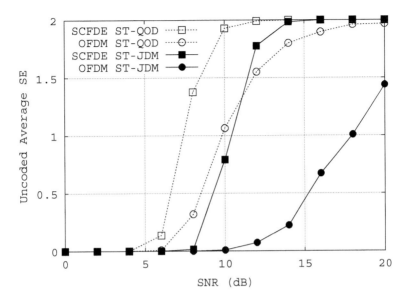

Fig. 13.19 Uncoded average SE for moderate time selectivity ($f_d T_s = 0.015$) and high frequency selectivity ($\tau_{max} = 18.20\,\mu s$).

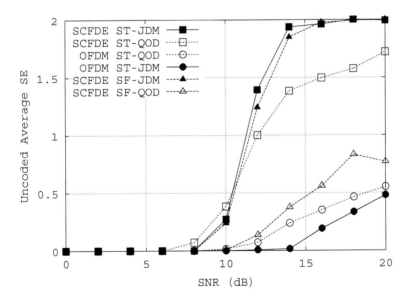

Fig. 13.20 Uncoded average SE for high time selectivity ($f_d T_s = 0.03$) and high frequency selectivity ($\tau_{max} = 18.20\,\mu s$).

of frequency selectivity is more important than the impact of time variability. The OFDM schemes perform the worst, since as already mentioned, OFDM does need channel coding, differently from SCFDE that can get frequency diversity gain without coding. Once again, the introduction of channel coding suggests a very different conclusion with respect to the uncoded case (see Figure 13.21): the QOD schemes perform better than JDM schemes. Comparing SF and ST-QOD, and SF and ST-JDM, ST schemes perform slightly better. Still SCFDE modulation shows better behavior than OFDM.

An important aspect to mention is that by using ST schemes an extra processing delay is introduced with respect to SF schemes. In fact, ST-QOD needs to wait four symbol periods at the receiver before completing the detection process, while ST-JDM needs to wait for two periods. In real-time applications this could be an obstacle to the use of ST schemes, even if they are the ones shown to have the best performance. Therefore the SCFDE SF-QOD and SF-JDM, which process the symbols period by period, without added delay, appear to be appealing solutions especially for real-time applications, preferring the SF-QOD for coded case, and preferring SF-JDM for uncoded case.

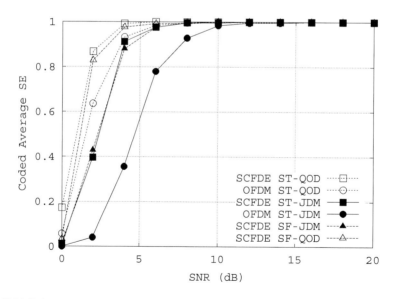

Fig. 13.21 Coded average SE for high time selectivity ($f_d T_s = 0.03$) and high frequency selectivity ($\tau_{\max} = 18.20\,\mu s$).

13.4.3 Conclusions

A new low-complexity SCFDE-SFBC scheme has been presented which allows to code across antennas and sub-carriers, even if the sub-carriers domain is not directly accessible at the transmitter of a SCFDE system. The proposed scheme has been shown to be more robust than SCFDE-STBC in highly time-varying channels.

The above mentioned SCFDE-SFBC scheme has then been applied to a JDM scheme (combined SM and SFBC). It has been shown that for SCFDE it is very efficient to use the available antennas to get multiplexing gain, since the resulting JDM scheme is robust to both the frequency and time selectivity of the channel. In case channel coding is present, it is more convenient to use the available antennas to provide only diversity instead of both diversity and multiplexing.

13.5 Linear Dispersion Codes for SCFDE

Linear Dispersion Codes (LDC) are considered as one of attractive advanced multiple antenna techniques for future wireless systems [8]. Proposed in [10], they represent a space–time transmission scheme that has many of the coding and diversity advantages of previously designed codes, but has also the decoding simplicity of VBLAST at high data rates. Moreover, with respect to VBLAST, they work with arbitrary number of transmit and receive antennas and both VBLAST and pure transmit diversity schemes can be considered as a special case of LDC.

In the present chapter, the application of space–frequency LDC to SCFDE systems is addressed. As for SFBC schemes (see Sections 13.3 and 13.4), which are a special case of LDC, the application of space–frequency LDC schemes to SCFDE is not straightforward. Moreover, in case of LDC, the dispersion in frequency instead of in time might be even more beneficial than in other multi-antenna techniques since the hypothesis of channel constant over a certain number of symbol intervals might be more stringent. The main contribution of this work is the definition of the dispersion matrices to be applied at the transmitter of a SCFDE system to achieve the desired SF dispersion. To the best of our knowledge, so far no work has investigated SF-LDC in SCFDE systems.

13.5.1 Linear Dispersion Codes

Multiple-antenna systems that operate at high rates require simple yet effective space–time transmission schemes to handle the large traffic volume in real time. At rates of tens of bits per second per hertz, VBLAST, where every antenna transmits its own independent substream of data, has been shown to have good performance and simple encoding and decoding. Yet, VBLAST suffers from its inability to work with fewer receive antennas than transmit antennas — this deficiency is especially important for modern cellular systems, where a base station typically has more antennas than the mobile handsets. Furthermore, because VBLAST transmits independent data streams, on its antennas there is no built-in spatial coding to guard against deep fades from any given transmit antenna.

On the other hand, there are many previously proposed space–time codes that have good fading resistance and simple decoding, but these codes generally have poor performance at high data rates or with many antennas.

In [10], one proposes a high-rate coding scheme that can handle any config-uration of transmit and receive antennas and that subsumes both VBLAST and many proposed space–time block codes as special cases. The scheme breaks the data stream into substreams that are dispersed in linear combinations over space and time. These codes are designed to optimize the mutual informa-tion between the transmitted and the received signals. Because of their linear structure, the codes retain the decoding simplicity of VBLAST, and because of their information-theoretic optimality, they possess many coding advantages. The Hassibi and Hochwald [10] refer to these codes as LDC. The LDC have the following properties:

1. They subsume, as special cases, both VBLAST [227] and the space block codes of [228].
2. They generally outperform both VBLAST and the space block codes.
3. They can be used for any number of transmit and receive antennas.
4. They are very simple to encode.
5. They can be decoded in a variety of ways.
6. They are designed with the numbers of both the transmit and the receive antennas in mind.

7. They satisfy the following information-theoretic optimality criterion: *the codes are designed to maximize the mutual information between the transmit and the receive signals.*

Let us assume N_T number of transmit antennas, N_R number of receive antennas, and an interval of T symbols where the propagation channel is constant and known to the receiver. The transmitted signal can then be written as a $T \times N_T$ matrix $\mathbf{S}$ that governs the transmission over the N_T antennas during the interval. Let us assume that the data sequence has been broken into Q substreams and that $s_1, \ldots, s_Q$ are the complex symbols chosen from an arbitrary, say r-PSK or r-QAM, constellation. A rate $R = \frac{Q}{T} \log_2 r$ linear dispersion code is a code for which $\mathbf{S}$ obeys

$$\mathbf{S} = \sum_{q=1}^{Q} \left(\alpha_q \mathbf{A}_q + j\beta_q \mathbf{B}_q \right), \tag{13.47}$$

where the scalars $\left(\alpha_q, \beta_q \right)$ are determined by

$$s_q = \alpha_q + j\beta_q. \tag{13.48}$$

The design of LDC depends crucially on the choices of the parameters T, Q and of the dispersion matrices $\left(\mathbf{A}_q, \mathbf{B}_q \right)$. To choose $\left(\mathbf{A}_q, \mathbf{B}_q \right)$, Hassibi and Hochwald [10] propose to optimize the mutual information between the transmitted signals $\left(\alpha_q, \beta_q \right)$ and the received signal.

13.5.2 Space–Frequency LDC for SCFDE

The application of LDC is problematic for highly variable channels. In fact, as it has been shown in [10], LDC are more efficient than pure transmit diversity schemes for a wide range of SNRs, when the required rate is high. To get high data rate, when the dispersion is done over time and space, the channel must be supposed constant over a number T of symbol intervals that is usually higher than 2 (as it is in a simple Alamouti scheme), hypothesis that might not hold in realistic propagation environment. In fact, when applied to OFDM systems T symbol intervals mean T OFDM symbols, that can be quite a long time.

This motivated the study of SF-LDC in OFDM systems: originally designed for flat fading channels, LDC have been extended to frequency selective channels in combination with OFDM. In [229], the transmitted symbols are dispersed over uncorrelated subchannels instead of adjacent subchannels, to take advantage of multi-path diversity, provided that the sub-carriers inside a subchannel behave the same. Actually, when extended to frequency domain, LDC could also be used to disperse the symbols over time and frequency instead of time and space, in order to exploit the multi-path diversity of the channel [230].

Some effort has recently been spent, where LDC are applied in the time and frequency domains for SCFDE systems [231]. However, as stated in [231], application of LDC to frequency-domain for SCFDE systems is not straightforward since there is no direct access to sub-carriers at the transmitter. The same problem has been observed in Section 13.3 for the easier case of SFBC and a solution has been provided.

This chapter proposes a general solution for applying SF-LDC to SCFDE systems, which includes as special case the processing proposed in Section 13.3. As it is shown in this chapter, SF-LDC are more efficient solutions in high-data rate and high-mobility environment with respect to ST-LDC or pure transmit diversity schemes. The sensitivity of SF-LDC with respect to the frequency selectivity and time selectivity of the channel is also investigated, to better identify the range of propagation conditions in which SF-LDC are a suitable solution.

The computational complexity of the proposed solution (SCFDE SF-LDC) is analyzed and compared with SCFDE ST-LDC and OFDM SF-LDC.

System model: First, in Section 13.5.2 the system model for the original LDC proposed in [10] is recalled, as a starting point for developing a general system model for SF-LDC in SCFDE, presented afterwards. Then, some specific examples on the application of the proposed procedure are presented, such as 2-antennas Alamouti [30, 176] and Hassibi codes [10], and 4-antennas Jafarkhani [208] and Hassibi [10] codes.

Original LDC: The design method proposed in [10] consists of the following two steps:

1. Choose $Q \leq N_R T$ (typically $Q = \min(N_T, N_R) \cdot T$).
2. Choose $(\mathbf{A}_q, \mathbf{B}_q)$ that solve the optimization problem

$$C_{\mathrm{LD}}(\rho, T, N_T, N_R)$$

$$= \max_{\mathbf{A}_q, \mathbf{B}_q, q=1,\dots,Q} \frac{1}{2T} E \log \det \left(\mathbf{I}_{2N_R T} + \frac{\rho}{N_T} \mathcal{H} \mathcal{H}^T \right) \quad (13.49)$$

for an SNR ρ of interest, subject to one of the following constraints:

(a) $\sum_{q=1}^{Q}\left(\text{tr}\mathbf{A}_q^*\mathbf{A}_q + \text{tr}\mathbf{B}_q^*\mathbf{B}_q\right) = 2T N_T$

(b) $\text{tr}\mathbf{A}_q^*\mathbf{A}_q = \text{tr}\mathbf{B}_q^*\mathbf{B}_q = \frac{T N_T}{Q}, q = 1, \ldots, Q$

(c) $\mathbf{A}_q^*\mathbf{A}_q = \mathbf{B}_q^*\mathbf{B}_q = \frac{T}{Q}\mathbf{I}_{N_T}, q = 1, \ldots, Q$

where $\mathcal{H}$ is given by (13.50) with the $\mathbf{h}_n$ having independent $\mathcal{N}\left(0, \frac{1}{2}\right)$ entries

$$\mathcal{H} = \begin{bmatrix} \mathcal{A}_1\mathbf{h}_1 & \mathcal{B}_1\mathbf{h}_1 & \cdots & \mathcal{A}_Q\mathbf{h}_1 & \mathcal{B}_Q\mathbf{h}_1 \\ \vdots & \vdots & \ddots & \vdots & \vdots \\ \mathcal{A}_1\mathbf{h}_N & \mathcal{B}_1\mathbf{h}_N & \cdots & \mathcal{A}_Q\mathbf{h}_N & \mathcal{B}_Q\mathbf{h}_N \end{bmatrix}, \tag{13.50}$$

where

$$\mathcal{A}_q = \begin{bmatrix} \mathbf{A}_{R,q} & -\mathbf{A}_{I,q} \\ \mathbf{A}_{I,q} & \mathbf{A}_{R,q} \end{bmatrix} \tag{13.51}$$

$$\mathcal{B}_q = \begin{bmatrix} -\mathbf{B}_{I,q} & -\mathbf{B}_{R,q} \\ \mathbf{B}_{R,q} & -\mathbf{B}_{I,q} \end{bmatrix} \tag{13.52}$$

with

$$\mathbf{A}_q = \mathbf{A}_{R,q} + j\mathbf{A}_{I,q} \tag{13.53}$$

$$\mathbf{B}_q = \mathbf{B}_{R,q} + j\mathbf{B}_{I,q} \tag{13.54}$$

and

$$\mathbf{h}_n = \begin{bmatrix} \mathbf{h}_{R,n} \\ \mathbf{h}_{I,n} \end{bmatrix}, \quad n = 1, \ldots, N_R \tag{13.55}$$

with $\mathbf{h}_{R,n}$ and $\mathbf{h}_{I,n}$ nth columns of $\mathbf{H}_R$ and $\mathbf{H}_I$, and $\mathbf{H} = \mathbf{H}_R + j\mathbf{H}_I$ ($N_T \times N_R$ matrix).

For $T = 2$, $N_T = 2$, $N_R = 1$, one solution to (13.49), for any of the constraints (a)–(c), is the orthogonal design (Alamouti structure [30]). This holds because the mutual information of this particular orthogonal design achieves the actual channel capacity $C(\rho, N_T = 2, N_R = 1)$, thus Alamouti for the $N_T \times N_R = 2 \times 1$ cannot be defeated by LDC. Let us move to more than one receive antenna ($N_R > 1$).

When $N_R \geq N_T$ and $Q = N_T T$, Hassibi and Hochwald [10] provides an explicit solution to (13.49) subject to the constraint (c). For $T = N_T$, one such set of matrices is given by

$$\mathbf{A}'_{N_T(k-1)+l} = \mathbf{B}'_{N_T(k-1)+l}$$
$$= \frac{1}{\sqrt{N_T}}\mathbf{D}^{k-1}\Pi^{l-1}, \quad k = 1,\ldots,N_T, \; l = 1,\ldots,N_T, \quad (13.56)$$

where

$$\mathbf{D} = \begin{bmatrix} 1 & 0 & \cdots & & 0 \\ 0 & e^{j\frac{2\pi}{N_T}} & 0 & \cdots & \\ \vdots & & \ddots & & \\ 0 & & & e^{j\frac{2\pi(N_T-1)}{N_T}} & \end{bmatrix}_{N_T \times N_T} \quad (13.57)$$

and

$$\Pi = \begin{bmatrix} 0 & \cdots & & 0 & 1 \\ 1 & 0 & \cdots & & 0 \\ 0 & 1 & 0 & \cdots & 0 \\ \vdots & & \ddots & & \vdots \\ 0 & \cdots & 0 & 1 & 0 \end{bmatrix}_{N_T \times N_T}. \quad (13.58)$$

General space–frequency LDC for SCFDE: Let us consider a SCFDE system with an N-points FFT at the equalizer, N_T transmit and N_R receive antennas. LDC in the space and frequency domains that cannot be directly applied to such a SCFDE system since its transmit sequence is processed in the time domain. Direct access to the sub-carriers at the transmitter would be possible by performing FFT/IFFT at the transmitter side. However, this would imply a price to pay in complexity. In this section, we propose a transmitter that allows to induce a given dispersion in frequency and space without performing any FFT pre-coding at the transmitter. The proposed transmitter and receiver are depicted in Figure 13.22. The transmitter consists of: a pre-coder block that performs simple linear operations over the data symbols; a LDC block that disperses N_{tot} symbols across N time intervals and N_T antennas according to specific $N \times N_T$ ST dispersion matrices, the rate of the LDC is $(N_{tot}/N)\log_2 r$, where r is the modulation order; a CP insertion block to cope with the inter-block interference. The CP is removed at the receiver before the FFT processing. We assume that the CP duration is longer than the maximum

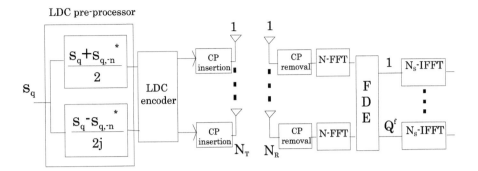

Fig. 13.22 Space–frequency LDC transceiver.

delay spread of the channel and hence, the equivalent channel matrix is circulant and diagonalizable by the FFT matrix at the receiver.

The objective of this section is to design the transmitter of a SCFDE system that allows to disperse the transmitted symbols over N_T transmit antennas and over a given number B of consecutive sub-carriers, as opposed to a conventional ST-LDC which disperse symbols over N_T transmit antennas and over a given number of consecutive time intervals. The designed SF-LDC will work as expected, thus getting the desired diversity and/or multiplexing gain, if the channel can be supposed to be constant over the B consecutive sub-carriers. Therefore, let us suppose we want to disperse Q^f symbols over B sub-carriers according to some target $B \times N_T$ SF dispersion matrices. Let us re-arrange the N_{tot} symbols to be transmitted $\mathbf{s} = [s_1, s_2, \ldots, s_{N_{\text{tot}}}]^T$ as the concatenation of Q^f vectors of length $N_S = N_{\text{tot}}/Q^f$, i.e. $\mathbf{s} = [\mathbf{s}_1, \ldots, \mathbf{s}_{Q^f}]^T$, where $\mathbf{s}_q = [s_{q,1}, \ldots, s_{q,N_S}]^T$, $q = 1, \ldots, Q^f$. Let us denote with $\mathbf{s}_q^f = FFT(\mathbf{s}_q) = \alpha_q^f + j\beta_q^f$. In the following derivations we also need to define the vector α_q:

$$\alpha_q = [\alpha_{q,1}, \ldots, \alpha_{q,N_S}]^T = \text{IFFT}(\alpha_q^f) = \text{IFFT}([\alpha_{q,1}^f, \ldots, \alpha_{q,N_S}^f]^T), \quad (13.59)$$

which is related to the sequence $\mathbf{s}_q$ as $\alpha_q = \frac{1}{2}(\mathbf{s}_q + \mathbf{s}_{q,-n}^*)$, where $\mathbf{s}_{q,-n}^*$ is the conjugate of the vector that is symmetric to $\mathbf{s}_q$ w.r.t. the origin. Similarly, it can be shown that $\beta_q = \frac{1}{2j}(\mathbf{s}_q - \mathbf{s}_{q,-n}^*)$. We have to process these Q^f blocks of N_S symbols $\mathbf{s}_q$ in a way that the real and the imaginary parts of the corresponding symbols in the frequency domain, which have been denoted by α_q^f and β_q^f, are dispersed in space and frequency according to desired $B \times N_T$ dispersion matrices $\mathbf{A}_q^f$ and $\mathbf{B}_q^f$. For the sake of simplicity, we consider only the dispersion of the real part.

Let us write the SF $B \times N_T$ desired dispersion matrices for the real part as

$$
\mathbf{A}_q^f = \begin{bmatrix} a_{11}^q & a_{12}^q & \cdots & a_{1N_T}^q \\ a_{21}^q & a_{22}^q & \cdots & a_{2N_T}^q \\ \vdots & \vdots & \vdots & \vdots \\ a_{B1}^q & a_{B2}^q & \cdots & a_{BN_T}^q \end{bmatrix}, \quad q = 1, \ldots, Q^f. \tag{13.60}
$$

The dispersion matrix (13.60) is supposed to be the same over all the N_S groups of symbols.

To disperse the real part of the N_S frequency domain symbols, α_q^f, according to the dispersion matrix $\mathbf{A}_q^f$, the $N \times 1$ vector that should be sent over the nth antenna is

$$
\sum_{q=1}^{Q^f} \sum_{i=1}^{B} a_{i,n}^q \mathbf{x}_{q,i}^f, \tag{13.61}
$$

where

$$
\mathbf{x}_{q,1}^f = \left[\alpha_{q,1}^f, \underbrace{0, \cdots}_{(B-1) \ 0s}, \cdots, \alpha_{q,N_S}^f, \underbrace{0, \cdots}_{(B-1) \ 0s} \right]^T. \tag{13.62}
$$

Equation (13.61) is related to Equation (16) in [1]. The difference is related to the fact that in this SCFDE system, N symbols are simultaneously processed at the receiver (N-points FFT). Therefore, the signal coming out from the LD encoder must be N-long and contains N_S groups of Q_f symbols and each group is dispersed over B consecutive sub-carriers.

In SCFDE systems, it is not possible to access directly to the symbols α_q^f without performing a FFT at the transmitter. However, we can get the same dispersion in frequency and space by sending over the nth antenna the signal:

$$
\sum_{q=1}^{Q^f} \sum_{i=1}^{B} a_{i,n}^q \mathrm{IFFT}(\mathbf{x}_{q,i}^f), \tag{13.63}
$$

which is the IFFT of Equation (13.61). It is straightforward to show that the IFFT of a signal such as $\mathbf{x}_{q,i}^f$, which contains a number of N_S symbols different from zero, equally spaced by consecutive 0's, can be written as follows:

$$\text{IFFT}(\mathbf{x}_{q,i}^f) = \sum_{k=1}^{N_S} \alpha_{q,k} \mathbf{p}_{k,i}. \tag{13.64}$$

The $N \times 1$ phase vectors $\mathbf{p}_{k,i}$ in (13.64) are given by

$$\mathbf{p}_{1,i} = \left[\underbrace{1, 0, \cdots,}_{1^{\text{st}}} \cdots, \underbrace{e^{j\frac{(N-N_S)(i-1)\pi}{N/2}}, 0, \cdots}_{B^{\text{th}}} \right]^T \tag{13.65}$$

$$\mathbf{p}_{2,i} = \left[\underbrace{0, e^{j\frac{(i-1)\pi}{N/2}}, \cdots,}_{1^{\text{st}}} \cdots, \underbrace{0, e^{j\frac{(N-N_S+1)(i-1)\pi}{N/2}}, \cdots}_{B^{\text{th}}} \right]^T \tag{13.66}$$

$$\vdots$$

$$\mathbf{p}_{N_S,i} = \left[\underbrace{\cdots, 0, e^{j\frac{(N_S-1)(i-1)\pi}{N/2}},}_{1^{\text{st}}} \cdots, \underbrace{\cdots, 0, e^{j\frac{(N-1)(i-1)\pi}{N/2}}}_{B^{\text{th}}} \right]^T \tag{13.67}$$

and the complex vectors $\alpha_{q,k}$, $q = 1, \ldots, Q^f$, $k = 1, \ldots, N_S$ are defined as in Equation (13.59).

Therefore, the signal to be sent over the nth antenna (in the time domain) in Equation (13.63) can be written as

$$\sum_{q=1}^{Q^f} \sum_{i=1}^{B} a_{i,n}^q \text{IFFT}(\mathbf{x}_{q,i}^f) = \sum_{q=1}^{Q^f} \sum_{i=1}^{B} a_{i,n}^q \sum_{k=1}^{N_S} \alpha_{q,k} \mathbf{p}_{k,i} \tag{13.68}$$

and by inverting the order of sums (index i and k):

$$\sum_{q=1}^{Q^f} \sum_{i=1}^{B} a_{i,n}^q \text{IFFT}(\mathbf{x}_{q,i}^f) = \sum_{q=1}^{Q^f} \sum_{k=1}^{N_S} \alpha_{q,k} \sum_{i=1}^{B} a_{i,n}^q \mathbf{p}_{k,i}. \tag{13.69}$$

Equation (13.69) means that it is possible to define the $N \times N_T$ space–time dispersion matrix $\mathbf{A}_{q,k}$, to be applied to the $Q^f N_S$ symbols at the transmitter, as the matrix where the nth column, with $n = 1, \ldots, N_T$, is given by $\mathbf{A}_{q,k}(:, n) = \sum_{i=1}^{B} a_{i,n}^q \mathbf{p}_{k,i}$. Therefore, given the desired dispersion matrices in the space and frequency domains, it is possible to define the dispersion matrices to be applied at the transmitter in the space and time domains. Note that the derived dispersion matrices are not applied directly to the real and imaginary parts of s_q but to α_q and β_q, which are easily generated by s_q by the pre-coding block shown in Figure 13.22. This slightly increases the complexity of the transmitter w.r.t. ST-LDC. However, the straightforward application of the dispersion in frequency in SCFDE systems would involve FFT and IFFT operations at the transmitter which are avoided by the proposed procedure. Therefore, this slight complexity increase at the transmitter is well compensated by the reduction of the complexity achieved by the proposed implementation of SF-LDC w.r.t. the straightforward implementation with FFT/IFFT processing. It is also worth noting that pure diversity or pure multiplexing space–frequency techniques can be seen as special cases of LDC and hence this paper presents a general low complexity implementation of MIMO space–frequency techniques in SCFDE systems.

Some Examples: The general procedure just introduced will be applied to some known LDC. The features of the considered schemes are listed up in Table 13.3.

2×1 *Space frequency block coding as an example*:   In this section, the proposed transmitter for LDC is used for implementing an Alamouti-like SFBC scheme over a SCFDE system. We show that the proposed implementation is equivalent to the low complexity transmission schemes that has been proposed in [232, 233] to overcome the problem of the access of sub-carriers at the transmitter. However, the presented procedure is much more general and can be applied to any desired dispersion in frequency and space domains.

The parameters for the SFBC scheme are $N_T = 2$, $B = 2$, and $Q^f = 2$. In other terms, the SFBC scheme disperses $Q^f = 2$ symbols over $B = 2$

Table 13.3 Schemes characteristics.

Scheme	2×2 LDC	2×2 OD	4×4 LDC	4×4 QOD
No. transmit antennas, N_T	2	2	4	4
No. transmit antennas, N_R	2	2	4	4
No. symbols dispersed per subchannel (period), Q^f (Q^t)	4	2	16	4
No. "sub-carriers" per subchannel, B (symbols per period, T)	2	2	4	4
Rate, $R = \frac{Q^f}{B}\log_2 r$ $\left(R = \frac{Q^t}{T}\log_2 r\right)$, r = mod. order	$2\log_2 r$	$\log_2 r$	$4\log_2 r$	$\log_2 r$
Reference equation	(31) [10]	Table 1 [30]	(31) [10]	(5) [208]

consecutive sub-carriers and $N_T = 2$ antennas. The space–frequency dispersion matrices of this Alamouti-like scheme are

$$\mathbf{A}_1^f = \begin{bmatrix} 1 & 0 \\ 0 & 1 \end{bmatrix}, \; \mathbf{B}_1^f = \begin{bmatrix} 1 & 0 \\ 0 & -1 \end{bmatrix}, \; \mathbf{A}_2^f = \begin{bmatrix} 0 & 1 \\ -1 & 0 \end{bmatrix}, \; \mathbf{B}_2^f = \begin{bmatrix} 0 & 1 \\ 1 & 0 \end{bmatrix}. \quad (13.70)$$

Note that in this case $N_{\text{tot}} = Q^f N_S = B N_S = N$ (there is no multiplexing gain). Therefore, the N symbols block $\mathbf{s}$ to be transmitted is divided in $N_S = N/2$ groups and passed through the pre-coding block in Figure 13.22. According to the proposed procedure, the first symbol of the first group in output from that pre-coding block, which is denoted by $\alpha_{1,1}$, must be dispersed in the time and space domain by the following matrix:

$$\mathbf{A}_{1,1} = \begin{bmatrix} a_{1,1}^1 \mathbf{p}_{1,1} + a_{2,1}^1 \mathbf{p}_{1,2} & a_{1,2}^1 \mathbf{p}_{1,1} + a_{2,2}^1 \mathbf{p}_{1,2} \end{bmatrix}$$

$$= \begin{bmatrix} \mathbf{p}_{1,1} & \mathbf{p}_{1,2} \end{bmatrix} \quad (13.71)$$

$$= \begin{bmatrix} 1 & 0 & \cdots & 1 & 0 & \cdots \\ 1 & 0 & \cdots & e^{j\frac{(N-N/2)\pi}{N/2}} & 0 & \cdots \end{bmatrix}^T. \quad (13.72)$$

Therefore, the signal to be transmitted over the antenna $n = 1$, to get the desired frequency dispersion of the real part is

$$\sum_{q=1}^{2} \sum_{k=1}^{N/2} \left(\frac{s_{q,k} + s_{q,(-(k-1)N/2+1)}^*}{2} \right) \sum_{i=1}^{2} a_{i,1}^q \mathbf{p}_{k,i}, \quad (13.73)$$

while, the signal to be sent over the same antenna to get the desired frequency dispersion of the imaginary part is

$$\sum_{q=1}^{2} \sum_{k=1}^{N/2} \left(\frac{s_{q,k} - s_{q,(-(k-1)N/2+1)}^*}{2j} \right) \sum_{i=1}^{2} b_{i,1}^q \mathbf{p}_{k,i}, \quad (13.74)$$

where $(\cdot)_{N/2}$ indicates the modulo-$N/2$ operation. Therefore, the vector to be transmitted from the antenna $n = 1$ is $(\mathbf{x}_1 + \mathbf{x}_2)$, where

$$\mathbf{x}_1 = \frac{1}{2} \begin{bmatrix} s_{1,1}, s_{1,2}, \ldots, s_{1,N/2}, s_{1,1}, s_{1,2}, \ldots, s_{1,N/2} \end{bmatrix}^T$$

$$\mathbf{x}_2 = \frac{1}{2} \begin{bmatrix} -s_{2,1}^*, -s_{2,N/2}^* e^{j\frac{\pi}{N/2}}, \ldots, -s_{2,2}^* e^{j\frac{(N/2-1)\pi}{N/2}}, \end{bmatrix}$$

$$\begin{bmatrix} s_{2,1}^*, s_{2,N/2}^* e^{j\frac{\pi}{N/2}}, \ldots, s_{2,2}^* e^{j\frac{(N/2-1)\pi}{N/2}} \end{bmatrix}^T$$

and from the second antenna the sequence $(\mathbf{x}_3 + \mathbf{x}_4)$ is transmitted, where

$$\mathbf{x}_3 = \frac{1}{2j}\left[s_{2,1}, s_{2,2}^t, \ldots, s_{2,N/2}, s_{2,1}^t, s_{2,2}, \ldots, s_{2,N/2}\right]^T$$

$$\mathbf{x}_4 = \frac{1}{2j}\left[s_{1,1}^*, s_{1,N/2}^* e^{j\frac{\pi}{N/2}}, \ldots, s_{1,2}^{t*}e^{j\frac{(N/2-1)\pi}{N/2}},\right.$$

$$\left. -s_{1,1}^*, -s_{1,N/2}^* e^{j\frac{\pi}{N/2}}, \ldots, -s_{1,2}^* e^{j\frac{(N/2-1)\pi}{N/2}}\right]^T,$$

which correspond to the SCFDE-SFBC scheme introduced in Section 13.3.

2×2 *SCFDE SF-LDC*:   For the 2-transmit antennas Hassibi code (Equation (31) [10]) it is $N_S = \frac{N}{2}$, $B = 2$, and $Q^f = 4$. The space–frequency dispersion matrices are

$$\mathbf{A}_1^f = \mathbf{B}_1^f = \begin{bmatrix} 1 & 0 \\ 0 & 1 \end{bmatrix}, \quad \mathbf{A}_2^f = \mathbf{B}_2^f = \begin{bmatrix} 0 & 1 \\ 1 & 0 \end{bmatrix},$$

$$\mathbf{A}_3^f = \mathbf{B}_3^f = \begin{bmatrix} 1 & 0 \\ 0 & -1 \end{bmatrix}, \quad \mathbf{A}_4^f = \mathbf{B}_4^f = \begin{bmatrix} 0 & 1 \\ -1 & 0 \end{bmatrix} \tag{13.75}$$

and the phase vectors of Equations (13.65)–(13.67) are, for $i = 1$:

$$\mathbf{p}_{1,1} = \left[\underbrace{1,0,\cdots}_{1^{\text{st}}}, \underbrace{1,0,\cdots}_{2^{\text{nd}}}\right]^T \tag{13.76}$$

$$\mathbf{p}_{2,1} = \left[\underbrace{0,1,\cdots}_{1^{\text{st}}}, \underbrace{0,1,\cdots}_{2^{\text{nd}}}\right]^T \tag{13.77}$$

$$\vdots$$

$$\mathbf{p}_{N/2,1} = \left[\underbrace{\cdots,0,1}_{1^{\text{st}}}, \underbrace{\cdots,0,1}_{2^{\text{nd}}}\right]^T \tag{13.78}$$

and for $i = B = 2$:

$$\mathbf{p}_{1,2} = \left[\underbrace{1,0,\cdots}_{1^{\text{st}}}, \underbrace{e^{j\frac{(N-N/2)\pi}{2}},0,\cdots}_{2^{\text{nd}}}\right]^T \tag{13.79}$$

$$\mathbf{p}_{2,2} = \left[\underbrace{0, e^{j\frac{\pi}{2}}, \cdots,}_{1^{st}} \underbrace{0, e^{j\frac{(N-N/2+1)\pi}{2}}, \cdots}_{2^{nd}} \right]^{T} \tag{13.80}$$

$$\vdots$$

$$\mathbf{p}_{N/2,2} = \left[\underbrace{\cdots, 0, e^{j\frac{(N/2-1)\pi}{2}},}_{1^{st}} \underbrace{\cdots, 0, e^{j\frac{(N-1)\pi}{2}}}_{2^{nd}} \right]^{T}. \tag{13.81}$$

2×2 *SCFDE SF-OD*: For the 2-transmit antennas Orthogonal Design (OD) it is $N_S = \frac{N}{2}$, $B = 2$, and $Q^f = 2$. The space–frequency dispersion matrices are

$$\mathbf{A}_1^f = \begin{bmatrix} 1 & 0 \\ 0 & 1 \end{bmatrix}, \ \mathbf{B}_1^f = \begin{bmatrix} 1 & 0 \\ 0 & -1 \end{bmatrix}, \ \mathbf{A}_2^f = \begin{bmatrix} 0 & 1 \\ -1 & 0 \end{bmatrix}, \ \mathbf{B}_2^f = \begin{bmatrix} 0 & 1 \\ 1 & 0 \end{bmatrix} \tag{13.82}$$

and the phase vectors are the same as for 2×2 SCFDE SF-LDC.

4×4 *SCFDE SF-QOD and SF-LDC*: The 4-transmit antennas SF schemes considered (Equation (31) [10] and Equation (5) [208]) are straightforward extensions of the $N_T = 2$ case. The QOD code of [208] was chosen because it provides a sufficient high spatial rate of 1, with no need to go for too high (unrealistic) modulation orders, with respect to OD (e.g. the OD (38) of [228] has spatial rate of 1/2 and needs 256-QAM to achieve $R = 4$, while LDC only BPSK, and needs 65536-QAM (!!!) to achieve $R = 8$, while LDC only QPSK). As an example the space–frequency dispersion matrices for Jafarkhani QOD code (in this case it is $N_S = \frac{N}{4}$, $B = 4$, and $Q^f = 4$) are

$$\mathbf{A}_1^f = \begin{bmatrix} 1 & 0 & 0 & 0 \\ 0 & 1 & 0 & 0 \\ 0 & 0 & 1 & 0 \\ 0 & 0 & 0 & 1 \end{bmatrix}, \ \mathbf{B}_1^f = \begin{bmatrix} 1 & 0 & 0 & 0 \\ 0 & -1 & 0 & 0 \\ 0 & 0 & -1 & 0 \\ 0 & 0 & 0 & 1 \end{bmatrix} \tag{13.83}$$

$$\mathbf{A}_2^f = \begin{bmatrix} 0 & 1 & 0 & 0 \\ -1 & 0 & 0 & 0 \\ 0 & 0 & 0 & 1 \\ 0 & 0 & -1 & 0 \end{bmatrix}, \ \mathbf{B}_2^f = \begin{bmatrix} 0 & 1 & 0 & 0 \\ 1 & 0 & 0 & 0 \\ 0 & 0 & 0 & -1 \\ 0 & 0 & -1 & 0 \end{bmatrix} \tag{13.84}$$

$$\mathbf{A}_3^f = \begin{bmatrix} 0 & 0 & 1 & 0 \\ 0 & 0 & 0 & 1 \\ -1 & 0 & 0 & 0 \\ 0 & -1 & 0 & 0 \end{bmatrix}, \quad \mathbf{B}_3^f = \begin{bmatrix} 0 & 0 & 1 & 0 \\ 0 & 0 & 0 & -1 \\ 1 & 0 & 0 & 0 \\ 0 & -1 & 0 & 0 \end{bmatrix} \quad (13.85)$$

$$\mathbf{A}_4^f = \begin{bmatrix} 0 & 0 & 0 & 1 \\ 0 & 0 & -1 & 0 \\ 0 & -1 & 0 & 0 \\ 1 & 0 & 0 & 0 \end{bmatrix}, \quad \mathbf{B}_4^f = \begin{bmatrix} 0 & 0 & 0 & 1 \\ 0 & 0 & 1 & 0 \\ 0 & 1 & 0 & 0 \\ 1 & 0 & 0 & 0 \end{bmatrix} \quad (13.86)$$

and the phase vectors of Equations (13.65)–(13.67) are, for $i = 1$:

$$\mathbf{p}_{1,1} = \left[\underbrace{1, 0, \cdots,}_{1^{\text{st}}} \cdots, \underbrace{1, 0, \cdots}_{4^{\text{th}}} \right]^T \quad (13.87)$$

$$\mathbf{p}_{2,1} = \left[\underbrace{0, 1, \cdots,}_{1^{\text{st}}} \cdots, \underbrace{0, 1, \cdots}_{4^{\text{th}}} \right]^T \quad (13.88)$$

$$\vdots$$

$$\mathbf{p}_{N/4,1} = \left[\underbrace{\cdots, 0, 1,}_{1^{\text{st}}} \cdots, \underbrace{\cdots, 0, 1}_{4^{\text{th}}} \right]^T \quad (13.89)$$

up to $i = B = 4$:

$$\mathbf{p}_{1,4} = \left[\underbrace{1, 0, \cdots,}_{1^{\text{st}}} \cdots, \underbrace{e^{j\frac{(N-N/4)3\pi}{4}}, 0, \cdots}_{4^{\text{th}}} \right]^T \quad (13.90)$$

$$\mathbf{p}_{2,4} = \left[\underbrace{0, e^{j\frac{3\pi}{4}}, \cdots,}_{1^{\text{st}}} \cdots, \underbrace{0, e^{j\frac{(N-N/4+1)3\pi}{4}}, \cdots}_{4^{\text{th}}} \right]^T \quad (13.91)$$

$$\vdots$$

$$\mathbf{p}_{N/4,4} = \left[\underbrace{\cdots, 0, e^{j\frac{(N/4-1)3\pi}{4}},}_{1^{\text{st}}} \cdots, \underbrace{\cdots, 0, e^{j\frac{(N-1)3\pi}{4}}}_{4^{\text{th}}} \right]^T \quad (13.92)$$

Equalization: Let us denote with $\mathbf{Y}_{j,k} = \mathbf{Y}_{R,j,k} + j\mathbf{Y}_{I,j,k}$ the kth group $(k = 1, 2, \ldots, N_S)$ of B symbols (also referred as sub-carriers since after the

FFT we are in the frequency domain), which is received by the jth antenna after the CP removal and the FFT block.

$$
\begin{bmatrix}
\mathbf{Y}_{R,1,k} \\
\mathbf{Y}_{I,1,k} \\
\vdots \\
\mathbf{Y}_{R,N_R,k} \\
\mathbf{Y}_{I,N_R,k}
\end{bmatrix}
= \sqrt{\frac{P_T}{N_T}}
\begin{bmatrix}
\mathcal{A}_1^f \mathbf{h}_{1,k} & \mathcal{B}_1^f \mathbf{h}_{1,k} & \cdots & \mathcal{A}_{Q^f}^f \mathbf{h}_{1,k} & \mathcal{B}_{Q^f}^f \mathbf{h}_{1,k} \\
\vdots & \vdots & \ddots & \vdots & \vdots \\
\mathcal{A}_1^f \mathbf{h}_{N_R,k} & \mathcal{B}_1^f \mathbf{h}_{N_R,k} & \cdots & \mathcal{A}_{Q^f}^f \mathbf{h}_{N_R,k} & \mathcal{B}_{Q^f}^f \mathbf{h}_{N_R,k}
\end{bmatrix}
$$

$$
\times
\begin{bmatrix}
\alpha_{1,k}^f \\
\beta_{1,k}^f \\
\vdots \\
\alpha_{Q^f,k}^f \\
\beta_{Q^f,k}^f
\end{bmatrix}
+
\begin{bmatrix}
\mathbf{N}_{R,1,k} \\
\mathbf{N}_{I,1,k} \\
\vdots \\
\mathbf{N}_{R,N_R,k} \\
\mathbf{N}_{I,N_R,k}
\end{bmatrix}
\tag{13.93}
$$

The channel is supposed to be constant over the kth group of sub-carriers. Let us denote with $\mathbf{H}_{m,n}(i)$ the Channel Transfer Function (CTF) between mth transmit antenna and the nth receive antenna for the ith sub-carrier. Assuming that the CP is longer than the maximum delay spread of the channel and following the notation in [10], we can write Equation (13.93), where matrices $\mathcal{A}_q^f$ and $\mathcal{B}_q^f$, $q = 1, \ldots, Q^f$ are defined as in [10] as a function of the $B \times N_T$ matrices $\mathbf{A}_q^f$ and $\mathbf{B}_q^f$, $q = 1, \ldots, Q^f$; the $2N_T \times 1$ vector $\mathbf{h}_{j,k} = [\mathbf{h}_{R,j,k}^T, \mathbf{h}_{I,j,k}^T]^T$ is such that $\mathbf{h}_{R,j,k}$ and $\mathbf{h}_{I,j,k}$ are the jth columns of $\mathbf{H}_{R,k}$ and $\mathbf{H}_{I,k}$, respectively, which are given by Equation (13.94).

Equation (13.93) can be written in a more compact form as $\mathbf{Y}_k = \sqrt{\frac{P_T}{N_T}} \mathcal{H}_k^f \mathbf{t}_k^f + \mathbf{N}_k$, $k = 1, \ldots, N_S$. The dimensions of $\mathcal{H}_k^f$ are $2BN_R \times 2Q^f$.

$$
\mathbf{H}_k = \mathbf{H}_{R,k} + j\mathbf{H}_{I,k}
$$

$$
=
\begin{bmatrix}
\mathbf{H}_{1,1}((k-1)B+1) & \cdots & \mathbf{H}_{1,N_R}((k-1)B+1) \\
\vdots & \ddots & \vdots \\
\mathbf{H}_{N_T,1}((k-1)B+1) & \cdots & \mathbf{H}_{N_T,N_R}((k-1)B+1)
\end{bmatrix}_{N_T \times N_R} .
\tag{13.94}
$$

We assume a linear Minimum Mean Square Error (MMSE) receiver and the estimated vector is given by $\widehat{\mathbf{t}}_k^{f'} = \mathbf{G}_k \mathbf{Y}_k$, where $\mathbf{G}_k = \sqrt{\frac{N_T}{P_T}} \left(\mathcal{H}_k^H \mathcal{H}_k + \frac{N_T}{\rho} \mathbf{I}_{N_T} \right)^{-1} \mathcal{H}_k^H$ and $\rho = P_T/\sigma^2$ is the total SNR per receive antenna.

Table 13.4 Simulation parameters.

Parameters	Value
System bandwidth, B	1.25 MHz
Sampling frequency, f_s	1.429 MHz
Carrier frequency, f_c	3.5 GHz
Symbols per block, N	128
Subcarrier spacing, Δf	9.77 KHz
CP length, N_{CP}	16
Max delay spread, τ_{max}	2.51 μs
Velocity	30 Kmph
FDE equalizer	MMSE
Channel coding scheme	1/2-rate convolutional coding

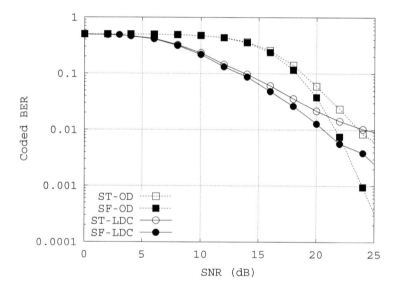

Fig. 13.23 Coded BER performance for 2 × 2 ST and SF-LDC vs ST and SF-OD, rate = 8, B = T = 2, ITU vehicular A channel.

Simulations and discussions: The simulations parameters are shown in Table 13.4, and consider a WiMAX scenario [171] with values for maximum delay spread and velocity taken from the ITU Vehicular A channel model [234]. In Figure 13.23, we show the uncoded BER for 2 × 2 schemes, LDC and OD, both ST and SF. The rate is $R = 8$, which means 16-QAM for LDC and 256-QAM for OD. As we can see in ITU Vehicular A channel model, the ST schemes show worse performance than the SF schemes, since the assumption they make of channel identical over two time periods [30] does not hold anymore, already at a moderate velocity of 30 kmph. The SF schemes instead, show a much more robust behavior to the time-variability of the channel. We can note that even though the SF-LDC scheme outperforms SF-OD up to SNR = 22 dB, OD behaves better for higher SNR: this diversity loss of the LDC can be explained by the fact that the code used satisfies more the mutual information criterion than the diversity criterion [10].

In Figure 13.24, we compare 4×4 LDC and QOD pure-diversity systems, both ST and SF. It is interesting to notice that the impact of time selectivity is much higher for 4×4 ST-LDC than for any other scheme considered. This is due to the fact that the assumption of channel constant over four time periods is not realistic already for moderate velocities. This is also confirmed by Figure 13.25, which shows the impact of increasing time-selectivity on the studied systems. The velocity considered in the ITU Vehicular A model is correspondent to a normalized Doppler frequency $f_d T_s = 0.01$.

Finally, the impact of frequency selectivity is shown in Figure 13.26. The normalized coherence bandwidth is defined as $B_{c,\mathrm{norm}} = \frac{1}{\tau_{\max} \Delta f}$. The dual behavior of Figure 13.25 can be seen, and the big degradation in performance of SF-LDC is due to the nonrealistic assumption of channel constant over four

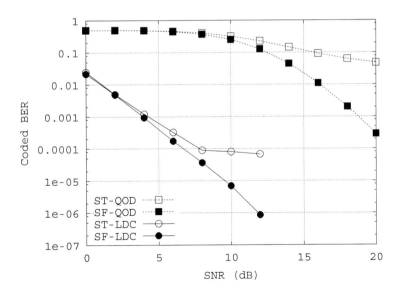

Fig. 13.24 Coded BER performance for 4×4 ST and SF-LDC vs ST and SF-QOD, rate = 8, B = T = 4, ITU vehicular A channel.

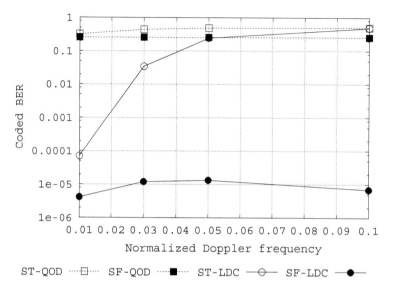

Fig. 13.25 Impact of time-selectivity for 4 × 4 ST and SF-LDC vs ST and SF-QOD, rate = 8, B = T = 4, SNR = 10 dB.

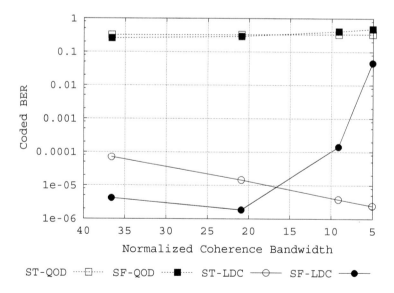

Fig. 13.26 Impact of frequency-selectivity for 4 × 4 ST and SF-LDC vs ST and SF-QOD, rate = 8, B = T = 4, SNR = 10 dB.

sub-carriers, in case of high frequency-selectivity. The maximum delay spread considered in the ITU Vehicular A model is correspondent to a normalized coherence bandwidth $B_{c,\text{norm}} = 36.58$. From Figure 13.26, we can notice that both ST and SF-LDC for SCFDE systems show a higher frequency diversity gain as the coherence bandwidth decreases. However, in the case of SF-LDC, the diversity gain is compensated by the performance degradation due to the hypothesis of channel constance over B consecutive sub-carriers. This explains the minimum in the curve of SF-LDC.

The influence of correlation, due either to insufficient antennas spacing or to the presence of LOS component, was not investigated for LDC and is left for future investigations.

Computational complexity: The complexities for SCFDE-SF-LDC, SCFDE-ST-LDC, and OFDM-SF-LDC are, respectively:

$$C^{\text{SC-SF}} = 2\frac{Q^f N_S N N_T}{B} + N_R \frac{N}{2} \log_2 N + Q^f \frac{N_S}{2} \log_2 N_S \quad (13.95)$$

$$C^{\text{SC-ST}} = 2\frac{Q^t \frac{N_P}{T} N N_T}{T} + N_R \frac{N}{2} \log_2 N + \frac{Q^t}{T} \frac{N_P}{2} \log_2 N_P \quad (13.96)$$

$$C^{\text{MC-SF}} = 2\frac{N_S Q^f N_S B N_T}{B} + N_T \frac{N}{2} \log_2 N + N_R \frac{N}{2} \log_2 N, \quad (13.97)$$

where the number of subchannels for SF codes is $N_S = N/B$ and the number of symbol periods for ST codes is $N_P = N$.

The complexity for the 2×2 case is shown in Table 13.5. The parameters in this case are: $N_T = N_R = 2$, $B = T = 2$. Moreover, it is $Q^f = Q^t = B = T = 2$ for Alamouti, and $Q^f = Q^t = N_T B = N_T T = 4$ for Hassibi.

The complexity for the 4×4 case is shown in Table 13.6. The parameters in this case are: $N_T = N_R = 4$, $B = T = 4$. Moreover, it is $Q^f = Q^t = B = T = 4$ for Jafarkhani, and $Q^f = Q^t = N_T B = N_T T = 16$ for Hassibi.

Therefore, the proposed SCFDE-SF schemes are slightly less complex than the correspondent SCFDE-ST and OFDM-SF schemes.

Table 13.5 Complexity in terms of complex multiplications for 2×2 case.

Scheme	Hassibi $N = 64$	Hassibi $N = 128$	Alamouti $N = 64$	Alamouti $N = 128$
SCFDE-SF	17088	67200	8736	34048
SCFDE-ST	17152	67328	8768	34112
OFDM-SF	17152	67328	8960	34560

Table 13.6 Complexity in terms of complex multiplications for 4×4 case.

Scheme	Hassibi $N = 64$	Hassibi $N = 128$	Jafarkhani $N = 64$	Jafarkhani $N = 128$
SCFDE-SF	34048	134144	9088	34880
SCFDE-ST	34304	134656	9152	35008
OFDM-SF	34304	134656	9728	36352

13.5.3 SD, SM, and JDM as Special Cases of LDC

In this section it is explicitly shown how the schemes considered in the former chapters, i.e. 2-transmit antennas Alamouti SD, 2-transmit antennas SM, and 4-transmit antennas JDM, can all be incorporated in the LDC theoretical framework.

In the case of Alamouti SD with $N_T = T = 2$ (or $N_T = B = 2$), the matrix $\mathbf{S}$ of Equation (13.47) can be written as

$$
\mathbf{S}_{\text{SD}} = \begin{bmatrix} s_1 & s_2 \\ -s_2^* & s_1^* \end{bmatrix} \tag{13.98}
$$

$$
= \begin{bmatrix} \alpha_1 + j\beta_1 & \alpha_2 + j\beta_2 \\ -\alpha_2 + j\beta_2 & \alpha_1 - j\beta_1 \end{bmatrix} \tag{13.99}
$$

$$
= \alpha_1 \begin{bmatrix} 1 & 0 \\ 0 & 1 \end{bmatrix} + j\beta_1 \begin{bmatrix} 1 & 0 \\ 0 & -1 \end{bmatrix} + \alpha_2 \begin{bmatrix} 0 & 1 \\ -1 & 0 \end{bmatrix} + j\beta_2 \begin{bmatrix} 0 & 1 \\ 1 & 0 \end{bmatrix} \tag{13.100}
$$

$$
= \alpha_1 \mathbf{A}_1 + j\beta_1 \mathbf{B}_1 + \alpha_2 \mathbf{A}_2 + j\beta_2 \mathbf{B}_2. \tag{13.101}
$$

For SM with $N_T = 2$ and $T = 1$ (or $B = 1$), the matrix $\mathbf{S}$ of Equation (13.47) can be written as

$$
\mathbf{S}_{\text{SM}} = \begin{bmatrix} s_1 & s_2 \end{bmatrix} \tag{13.102}
$$

$$
= \begin{bmatrix} \alpha_1 + j\beta_1 & \alpha_2 + j\beta_2 \end{bmatrix} \tag{13.103}
$$

$$
= \alpha_1 \begin{bmatrix} 1 & 0 \end{bmatrix} + j\beta_1 \begin{bmatrix} 1 & 0 \end{bmatrix} + \alpha_2 \begin{bmatrix} 0 & 1 \end{bmatrix} + j\beta_2 \begin{bmatrix} 0 & 1 \end{bmatrix} \tag{13.104}
$$

$$
= \alpha_1 \mathbf{A}_1 + j\beta_1 \mathbf{B}_1 + \alpha_2 \mathbf{A}_2 + j\beta_2 \mathbf{B}_2. \tag{13.105}
$$

In the case of JDM with $N_T = 4$ and $T = 2$ (or $B = 2$), the matrix $\mathbf{S}$ of Equation (13.47) can be written as

$$
\mathbf{S}_{\text{JDM}} = \begin{bmatrix} s_1 & s_2 & s_3 & s_4 \\ -s_2^* & s_1^* & -s_4^* & s_3^* \end{bmatrix} \tag{13.106}
$$

$$= \begin{bmatrix} \alpha_1 + j\beta_1 & \alpha_2 + j\beta_2 & \alpha_3 + j\beta_3 & \alpha_4 + j\beta_4 \\ -\alpha_2 + j\beta_2 & \alpha_1 - j\beta_1 & -\alpha_4 + j\beta_4 & \alpha_3 - j\beta_3 \end{bmatrix} \quad (13.107)$$

$$= \alpha_1 \begin{bmatrix} 1 & 0 & 0 & 0 \\ 0 & 1 & 0 & 0 \end{bmatrix} + j\beta_1 \begin{bmatrix} 1 & 0 & 0 & 0 \\ 0 & -1 & 0 & 0 \end{bmatrix} \quad (13.108)$$

$$+ \alpha_2 \begin{bmatrix} 0 & 1 & 0 & 0 \\ -1 & 0 & 0 & 0 \end{bmatrix} + j\beta_2 \begin{bmatrix} 0 & 1 & 0 & 0 \\ 1 & 0 & 0 & 0 \end{bmatrix}$$

$$+ \alpha_3 \begin{bmatrix} 0 & 0 & 1 & 0 \\ 0 & 0 & 0 & 1 \end{bmatrix} + j\beta_3 \begin{bmatrix} 0 & 0 & 1 & 0 \\ 0 & 0 & 0 & -1 \end{bmatrix}$$

$$+ \alpha_4 \begin{bmatrix} 0 & 0 & 0 & 1 \\ 0 & 0 & -1 & 0 \end{bmatrix} + j\beta_4 \begin{bmatrix} 0 & 0 & 0 & 1 \\ 0 & 0 & 1 & 0 \end{bmatrix}$$

$$= \alpha_1 \mathbf{A}_1 + j\beta_1 \mathbf{B}_1 + \alpha_2 \mathbf{A}_2 + j\beta_2 \mathbf{B}_2 + \alpha_3 \mathbf{A}_3$$
$$+ j\beta_3 \mathbf{B}_3 + \alpha_4 \mathbf{A}_4 + j\beta_4 \mathbf{B}_4 \quad (13.109)$$

13.5.4 Conclusions

A general procedure has been introduced to derive the dispersion matrices to be applied at the transmitter of a SCFDE system, to achieve the desired space-frequency dispersion. To the best of Authors' knowledge, no other work before has addressed the issue of applying LDC in the frequency domain to SCFDE systems. It has been shown that the dispersion in frequency is more suitable than the dispersion in time, already at moderate terminal speeds and for a wide range of frequency selectivity of the channel.

13.6 Multi-antenna in SC-FDMA

MIMO is an essential tool to achieve the high data rate and increased system capacity required by The Third Generation Partnership Project (3GPP) (E-UTRALTE), for both uplink and downlink. While OFDMA is the chosen modulation for dowlink, Single Carrier-Frequency Division Multiple Access (SC-FDMA) seems to be the candidate for uplink. The Evolved–UMTS Terrestrial Radio Access (E-UTRA) should support an instantaneous uplink peak data rate of 50 Mbps within a 20 MHz uplink spectrum allocation (2.5 bps/Hz) [235]. Although theoretically a UE with a single transmitter can almost achieve 50 Mbps using 16-QAM digital modulation, it has been concluded that at least 2×2 MIMO is necessary to realistically achieve the required uplink throughput [236, 237]. It has also been shown that to achieve the highest throughput in uplink transmission the use of precoding is a necessity [238, 239]. An additional advantage to having two transmitters on the UE is the possibility to use beamforming to enhance Multi-User MIMO [240] and also better transmit diversity schemes such as ST-FD [241].

In [241], a pre-coding scheme for UL SU-MIMO for SC-FDMA modulation is proposed; the pre-coding is based on transmit beamforming using, e.g. eigen-beamforming based on SVD. For pre-coding using eigen-beamforming the channel matrix is decomposed using a SVD or equivalent operation as

$$\mathbf{H} = \mathbf{U}\mathbf{D}\mathbf{V}^H, \qquad (13.110)$$

where $\mathbf{H}$ is the channel matrix. The 2D transform for spatial multiplexing, beamforming, etc. can be expressed as $\mathbf{x} = \mathbf{T}\mathbf{s}$, where $\mathbf{s}$ is the data vector and $\mathbf{T}$ is a generalized transform matrix. In the case when transmit eigen-beamforming is used, the transform matrix $\mathbf{T}$ is chosen to be a beamforming matrix $\mathbf{V}$ which is obtained from Equation (13.110). This is similar to eigen-beamforming for OFDMA, but it is modified in order to be applied to SC-FDMA.

Single Carrier-Frequency Division Multiple Access (SC-FDMA) with pre-coding will increase the PAPR because each transmit signal becomes a composite signal due to spatial processing. Expressed in terms of a transmit signal in the time domain, applying pre-coding in the frequency domain is equivalent to a convolution and summation of the data symbols in the time domain. Thus pre-coding will increase the PAPR of the composite transmitted signal [241].

In the literature some conclusions on uplink SU-MIMO for SC-FDMA have been drawn [241]:

- Pre-coding at the UE can be based on SVD or a comparable algorithm performed at the eNode-B.
- Feedback of the pre-coding matrix index can be performed efficiently using combined differential and nondifferential feedback.
- PAPR due to transmit beamforming can be effectively mitigated using amplitude clipping.
- Simulations showed that SU-MIMO can almost double (186%) the uplink data rate compared with SIMO.

13.7 Chapter Summary

In this chapter, several space–frequency coding schemes that required to achieve any desired amount of the possible trade-off diversity-multiplexing in SCFDE systems are introduced and numerically evaluated. The SCFDE versions of Alamouti and Jafarkhani codes, JDM schemes and LDC are introduced. The impact of channel time and frequency selectivity on the proposed schemes is also tested. The computational complexity of the proposed schemes is discussed. Some considerations about multiple antenna techniques for SC-FDMA conclude the chapter.

14

Conclusions and Perspectives

To speed up the introduction of emerging broadband and multimedia services, as well as to support the corresponding growth in traffic this would generate, it is important that wireless broadband access to information networks is available evenly and for everyone. This increasing demand for high-performance future wireless systems imposes requirements that new technologies must effectively deal with: heterogeneity, dynamic behavior, higher capacity, better performance, low power consumption, reconfigurability and reuse. Thus, the main objectives of all ongoing research efforts are to develop a radio system concept based on a high-speed transmission scheme (OFDM) enhanced to fulfill a set of evolved future-generation requirements, in particular for wide-area and short-range communication scenarios, with the multiple-input, multiple-output (MIMO) concept. The future systems should provide significantly improved capabilities in terms of supported data rates, spectral efficiency, delays, QoS, support of mobility, cost and power consumption. These systems should exploit the actual propagation and interference conditions in an intelligent way to comply with the need for more efficient use of the bandwidth, scalability and flexibility. The idea should be developed through a systematic approach based on technology assessment, simulation and optimization, including advanced methodology, such as cross-layer optimization.

The overall future target is cellular systems in which there will also be a support for the short-range communication among the wireless terminals. The rationale for introducing short-range communication can be seen through two arguments:

1. The need to support peer-to-peer (P2P) high-speed wireless links among the terminals.

2. To enhance the communication between a terminal and the BS by fostering cooperative communication protocols among spatially proximate wireless terminals.

The communication enhancement primarily refers to the increase of the cell edge capacity, but it can also be used to increase the link reliability, spectral efficiency, reduced power consumption, and Quality of Service (QoS) provisioning.

14.1 Cross-layer Design and Optimization

In all recent system design efforts in industry and academia, it is now clear that system design across several layers of protocol stack is a must.

It is well-known that layering is desirable since it helps in the creation of modular software components. The layering paradigm has greatly simplified network design and has led to robust and scalable protocols for the Internet. However, protocol stack implementations based on layering do not function efficiently in wireless environments due to inefficiency, inflexibility, and suboptimality. To meet these requirements a cross layer protocol design supporting adaptability and optimizing the use of resources over multiple layers of the protocol stack is required [242]. The concept of cross layer design introduces the idea of jointly optimizing two or more of the layers of communication. Although this concept can be employed in all communication networks, it is especially important in wireless networks because of the unique challenge of the wireless environment described above. The main idea of cross layer design is to exchange information across the layers that would not be possible with the traditional layer interfaces. This additional information is used by the layers to better adapt to varying transmission conditions. The information is passed in both directions, from the lower layers to the upper layers as well as vice versa. Inter layered processing was first introduced in [243]. Two different kinds of cross layer feedback can be distinguished

Upper to lower layers: This feedback may be application QoS requirements to lower layers, user feedback, and Transmission Control Protocol (TCP) timer information for lower layers.

Lower to upper layers: This feedback may be Channel State Information (CSI), or other link characteristics and network connectivity information to upper layers.

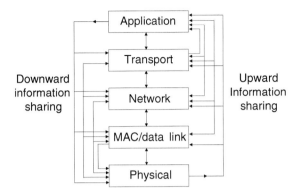

Fig. 14.1 The cross-layer design model: performance improvement through information sharing.

The adaptation model of cross layer design in Figure 14.1 depicts these ideas [244]. Basically the interactions are no longer limited to the layer directly above or below a particular layer. But information exchanges between the different layers can be defined, for example by additional fields in packets, extra control messages, etc.

Cross-layer design is becoming a popular design methodology for next-generation wireless systems. Two streams can be distinguished, the *implicit* and *explicit* approach. In case of implicit cross-layer design, there is no exchange of information between the layers during run-time, but in the design criteria cross-layer interactions are taken into account. Layers are designed to complement each other and unnecessary duplicate implementations of func-tionalities are eliminated, preventing for example MAC level collisions in case of network flooding, and applying only one retransmission policy and not one at transport layer and one at link layer. In the explicit case a cross-layer signal-ing scheme exists by which the different layers can communicate and optimize their parameters [245, 246].

In the implicit approach, cross layer optimization constraints such as flow classification and optimization policy definitions can be based on the infor-mation present in the packet headers [247]. Packet headers can be checked for protocol analysis. For example, if the header is belonging to a Real Time Transport Protocol (RTP) [248] the policy should take real time sensitivity of the application into account. As another example, the packet headers of the video encoder can be checked to differentiate between I, B, and P frames. I frames are intra-coded frames and contain more information than inter coded frames such as B and P frames. The loss of an I frame can cause distortion of multiple frames in its group. The detailed information on I, B, and P frames can be found in [249]. Therefore, if transmission of an I frame is detected by checking the encoder header, then the cross-layer optimization policy can increase the number of retransmissions assigned to I frames in case of failure and it can decrease the number of retransmissions of B or P frames [247] or the optimization policy can assign more priority to I frames in the scheduling policy [250]. The packet headers can also be used to understand whether the network infrastructure is QoS enabled. Then optimization policy should not implement duplicate functionalities which are already provided by *Diffserv* protocol in the network, i.e. traffic categorization.

The studies in [70, 251, 252, 253] are using explicit cross-layer signaling scheme for throughput optimization and capacity increase for OFDMA systems. The allocation of sub-carriers is done dynamically with respect to channel characteristics and the allocation is signaled to users using the in-band signaling scheme. The disadvantage of this scheme is the need for signaling the allocated sub-carriers to users. Therefore, the design of blind scheduling algorithms has attracted wide interest from researchers.

Recently also papers indicating negative consequences of cross-layer optimization have been published, for example [254]. A trade-off is seen between short term gain in terms of for example throughput or delay optimization when using cross-layer optimization versus a long term negative effect for the architectural design of letting go of a systematic/layer approach to the design. Especially in an early phase of the system design a sound architectural design is of the utmost importance.

From the above argumentations it becomes clear that there is a gain that can be made by using information across different layers, but that care should be taken when doing so. In the cross-layer design concept understanding and exploiting the interactions between the different layers and their impact on the overall system are crucial.

Advantages of cross layer design are

- Exploits interactions between different layers.
- Enables adaptability at all layers based on mutual information exchange.
- In wireless networks a tight interdependence between the protocol layers exists.

and disadvantages are

- Care should be taken to maintain a sound overall architectural design.
- Difficult to characterize the interactions between protocols at different layers.
- Interdependency between functionalities at different layers.
- Joint optimization across layers may lead to complex algorithms.

The use of cross-layer optimization does not mean complete abolishment of the ISO/OSI model, but rather the definition and implementation of specific, well defined interfaces between different layers. For this purpose for example in [255] and inter-layer signaling framework is proposed.

The research community needs to identify the parameters involved in cross-layer design and attempt to develop some general guideline for the use of cross-layer design. The main focus should be the identification of different parameters at different layers that could be suitable for cross-layer optimization. The identification part will provide a starting point for discussion of estimation and embedding of such parameters. It is worthy to mention here that cross-layer design can optimize the performance, but should be handled with care as it has major architectural consequences and could result in a "spaghetti" type design.

14.1.1 Cross-layer Opportunities

Several main areas for cross-layer optimization can be defined:

1. Application layer to MAC.
2. NET to MAC.
3. Link Layer/MAC to PHY.

These interactions are illustrated in Figure 14.2. As examples, here we present three possible applications of cross-layer optimization illustrating these cross-layer interactions:

1. The first example is an interaction between application layer and the OFDM scheduling, where cross-layer interaction can be used to optimize the communication of video data over wireless system.

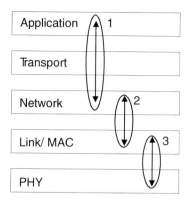

Fig. 14.2 General cross-layer opportunities.

2. A second example can be the interaction between MAC and PHY, using multiple base stations to communicate to wireless terminal using a set of sub-carriers from different base stations selected based on available CSI for the communication.
3. Another example can be given of an inter-layer signaling structure that optimizes the scheduling and Radio Link Protocols (RLP) to the user requirements and the channel state.

14.1.2 Cross-layer Design Related to System Integration

Based on cross-layer considerations, it becomes obvious that implementing systems that can utilize CL information calls for innovative methods. CL information will impact functionality at different system levels, in principle meaning that system parts can have their functionality altered based on CL information. Important implementation related issues become flexibility, re-configurability, and reuse. One implementation candidate is Software Define Radio (SDR), which is widely used in today's base stations. SDR's has previously been considered too performance and energy inefficient for handheld terminals, but as embedded technologies achieves higher performance and becomes energy aware, SDR is believed to become practical in near future. In today's mobile terminals the hardware software partitioning also indicate, that there is movement toward software implementations, where only limited functionality is performed in hardware.

Another very important issue for system integration is the power consumption. In today's embedded systems Power Management is a huge issue, and will become even more important if SDR's based terminals are introduced. The primary function of the PM, is to reduce the overall system power consumption, and has traditional be done by turning unused hardware parts off. In the last decades research on Dynamic Power Management (DPM) methods has been of great interest. The philosophy is to obtain a more optimal power policy based on dynamic information. The challenge is to figure out what information could be usable, and make decision policies that can utilize this information. An example assuming that channel information is obtainable, utilization of the most optimal baseband modulation scheme is selected, either for maximum throughput or for minimum power consumption. The effort is to place parameters in the system (software, hardware), so the CL/DPM mechanism obtains the wanted performance.

14.2 Technological Aspects: Future Perspective

Current 3G systems based on the Wideband Code Division Multiple Access (WCDMA) and cdma2000 radio-access technologies are gaining momentum in the mobile-communication market. Over several years to come new enhancements and improvements will be gradually introduced for the 3G systems, as part of the 3G evolution. In parallel to this work on the 3G evolution there is also an increasing activity on defining a new radio-access technology, often referred to as 4G, intended to be part of a future radio-communication network with expected introduction in the 2013–2015 time span. In January 2004, the European WINNER project was launched with a clear focus on defining a future ubiquitous radio-access technology [256]. Extensive work on future radio-access technologies is also carried out in other parts of the world [257, 258].

A key requirement on a new radio-access technology is the provision of significantly enhanced services by means of higher data rates, lower latency, etc., at a substantially reduced cost, compared to current mobile-communication systems and their direct evolution. As an example, within the WINNER project, data rates in the order of 100 Mbps for wide-area coverage and up to 1 Gbps for local-area coverage are set as targets. At the same time, the delay/latency of the radio-access network should be limited to a few milliseconds.

The services that are expected to dominate the future radio-access systems are likely to evolve from the services we see in the fixed networks already today, i.e. best effort and rate-adaptive Transmission Control Protocol (TCP)-based services. In current 3G systems, services are typically divided into different QoS classes. However, in order to support TCP data rates in the range 100–1000 Mbps, the requirements on delay and packet-loss rates must be kept so stringent that for future radio-access systems it is meaningless to set different requirements for real-time, quasi real-time, and nonreal-time services.

Currently, it is not clear what spectrum a future radio-access technology is to be operated in. It is anticipated that new spectrum will be allocated for a future radio-access technology, e.g. in the 4–5 GHz band. However, it is highly desirable that a new radio-access technology should also be deployable in other frequency bands, including existing 2G and 3G spectrum. Thus, a new radio-access technology should support a high degree of spectrum flexibility including, among other things, flexibility in terms of frequency-band-of-operation, size of available spectrum, and duplex arrangement.

Based on the requirements discussed above it is possible to give some suggestions about a future radio-access system:

- The DL transmission is OFDM-based with support for fast link adaptation and time- and frequency-domain scheduling.
- The UL modulation is SC-based and with different prefiltering operations it is possible to operate in both a pure OFDM mode and a pure SC mode.
- Advanced antennas are integrated in the system design by means of LDC [10].

14.2.1 Single-Carrier vs Multi-Carrier and Frequency-Domain vs Time-Domain Equalization

IEEE 802.16 has considered the following multiple access schemes: Multi-Carrier (MC)-based (OFDMA) and SC-based (TDMA). Some recent studies have clearly shown that the basic issue is not OFDM (OFDMA) vs SC (TDMA) but rather FDE vs TDE. FDE has several advantages over TDE in high mobility propagation environments (usually with long tail channel impulse responses).

We can also have a SC transmission with FDE so more generally, the chosen approach should be FDE-based, keeping an eye on other possibilities.

For channels with severe delay spread, FDE is computationally simpler than corresponding TDE for the same reason OFDM [11] is simpler: because equalization is performed on a block of data at a time, and the operations on this block involve an efficient FFT operation and a simple channel inversion operation. Sari *et al.* [12, 13] pointed out that when combined with FFT processing and the use of a cyclic prefix, an SC system with FDE (SCFDE) has essentially the same performance and low complexity as an OFDM system. It should also be noticed that a frequency domain receiver processing SC modulated data shares a number of common signal processing functions with an OFDM receiver.

14.2.2 Multi-antenna Issues

In next generation radio access, a very important role will be played by advanced multiple antenna systems. The transmitter will have one or more antennas and will use an algorithm consisting of four parts:

- Multi-user channel-dependent scheduling over several cells, which schedules the User Terminals (UTs) connected to an Access-Point (AP) in a coordinated way, taking their average/instantaneous channel properties and quality into account.
- Stream multiplexing with per-stream rate control (similar to Selective Per-Antenna Rate Control (S-PARC), which is used when the transmitter has more than one antenna and when multiplexing is more desirable than diversity.
- Transmit Diversity (TD) using Linear Dispersion Codes (LDC) [10]. Transmit diversity requires that the transmitter has more than one antenna and is used when more diversity is needed, at the expense of possibly sacrificing some multiplexing gain. Each data stream can then be spread in space and time to obtain the desired amount of diversity and multiplexing.
- Higher order of sectorization (implemented using, e.g. fixed multibeam), which is used when the transmitter has more than one antenna and when directivity is more desirable than diversity or multiplexing.

The selection and/or combination of different MIMO techniques, such as Space Division Multiplexing (SDM), Space–Time Transmit Diversity (STTD), Cyclic Delay Diversity (CDD), etc. need to be considered. New and innovative algorithms are required to cope with the terminal heterogeneity (in terms of number of antennas and processing capabilities) and the procedures for adaptive selection of the multiplexing/diversity schemes based on the requirements set by the upper layers. Therefore, a cross-layer optimization should be performed considering the interactions with the MAC layer, with the final goal of a QoS-aware scheduler with reduced complexity with respect to the current systems. The goals will be to maximize the system capacity while providing QoS for the individual users. The proposed multiuser MIMO schemes will be engineered in a way to be compatible with the proposed multiple access scheme. A special care should be put on the practical implementability of the proposed algorithms.

14.2.3 Frequency Overlay

The air-interface need to be designed in a future-safe manner, such that the system can operate when new chunks of spectrum will be available for use in the future. This requires different approach to the design of the access scheme, since the operating bandwidth for which the system is designed will be complemented with additional (non)adjacent bandwidth in the future. In such case, the new terminals that will use the extended bandwidth should not prevent the legacy terminals of conducting their communication within the legacy bandwidth in an unchanged manner. If the bandwidth extension is not accounted for at this phase, then the overlaying might not be feasible or it might require a huge processing complexity in the terminals that will use the extended bandwidth (for example, interference cancelation).

Therefore, the future systems will be both backward and forward compatible. As an example, future IMT-Advanced system will be based on a multi-carrier transmission and it is assumed that the bandwidth will be allocated in chunks of 20 MHz. The legacy terminal should be able to communicate with the new entities that are using larger bandwidth (the access points and the terminals). The transmission schemes need to be well-designed with respect to the communication performance (BER, outage probability, capacity) when both legacy and new terminals are coexisting in the system. Another important aspect is the required (and/or expected) complexity for both the legacy and the new terminals.

14.2.4 Radio Resource Management and Packet Scheduling

One of the main advantages of the multi-carrier schemes is that they possess enough granularity to take advantage of the multi-user diversity effect in independently faded frequency-selective channels. Hence, one of the main objectives in the radio resource management (RRM) will be to design multi-user resource allocation algorithms that exploit the multi-user diversity. However, accounting only for the multi-user diversity can maximize the system capacity, while causing degradation of the individually perceived QoS. Therefore, we need to design a joint cross-layer procedure for resource allocation and packet scheduling. Furthermore, the procedure should be designed by accounting for the resource allocation over different neighboring cells, but also to fit with the concepts of short-range wireless links support, frequency overlay, the proposed multiple access scheme as well as the MIMO-related algorithms.

Resource allocation algorithms, defined as bit/sub-carrier/power allocation algorithms in an OFDM-based system, have been extensively studied throughout the literature. However, their design framework is far from being realistic as only basic assumptions have been used such as: perfect CSI, infinite incoming data flow, minimum rate requirement as QoS, etc. Besides, suboptimal algorithms as well as optimal algorithms in the literature have fairly high complexity and are vulnerable to imperfect CSI. Moreover, a packet scheduler operating on top of the resource allocation algorithm has to be considered. The task of this packet scheduler is to assign priorities to the incoming packets according to available PHY layer resource (which depends on the time-varying user channels), user QoS requirements and traffic queue status. The sorted packets are then mapped to the OFDM data frames by the resource allocation algorithm. Link adaptation is performed by adapting the modulation and coding and consequently the packet length to the varying channel states. Along with the current state-of-the-art, more works need to be performed on joint optimization of resource allocation algorithm and packet scheduler for OFDMA. The best performance can be expected for a cross-layer optimized scheme between the PHY resource allocation and the scheduler, where even the framing of the packets would be influenced by the resource allocation algorithm.

14.2.5   Cooperative Communication

Recently, much research effort has been put to understand and utilize the benefit of cooperative behavior in wireless networks. The cooperation is expected to achieve high spectral efficiency, large coverage, high link reliability, and less power consumption of terminals through the use of exclusive cooperative stations (e.g., relay station deployed by operators) or short-range communications among different mobile terminals. In order to achieve these goals, research need to focus on the design of different cooperative mechanisms in lower layers of protocol stacks, i.e., PHY and MAC layers.

- *PHY layer*: The research in PHY-layer cooperation has already promised a great potential. Cooperative diversity protocols exploiting the feature of wireless broadcast medium have shown potential to achieve similar effect to conventional MIMO transmission. However, theoretical analysis of capacity increase brought by cooperative behavior has been the main topic in the literature, and more efforts are needed on practical protocol designs to achieve the high diversity gain. More focus should be put on the design of coding, modulation, receiver algorithms and forwarding mechanisms such as amplify and forward (AF) and decode and forward (DF) in order to achieve high diversity gain in practical cooperative scenarios.
- *MAC layer*: An important issue that needs to be addressed is the formation of the cooperating group for different targets (coverage enhancement, less power consumption, etc.), taking account of the inter-terminal channel conditions and the spatial distribution of the terminals. The design of protocols such as radio resource management and handoff also need to be addressed to coordinate transmissions between short-range links (links within cooperative groups) and long-range links (links between BS and terminals) so that the most appropriate links can be used by terminals with the least interference conditions.

We need to understand the practical gains obtained through the addition of relay/cooperation functionalities to cellular networks. The gains to be expected are: link reliability improvement, coverage enhancement, spectral efficiency improvement, and reduction of power consumption of mobile terminals. Possible technologies that can be applied to achieve these goals are: cooperative diversity (combined with coding, modulation, forwarding), packetization, retransmission protocols, radio resource management, packet scheduling, handoff, etc. Detailed works are required on mapping of these technologies into different gains and detailed protocol design in each technology to achieve the goals. The costs such as complexity and signaling overhead should also be taken into account for the evaluation of applicability to actual systems. The protocols to be developed should comply with overall system settings and target in the project, such as detailed air interface design, manners to use frequency spectrum (frequency overlay), etc. Different criteria can be used for evaluating different targets. Some representative measures are as follows:

- Reliability: BER, PER, required SINR to achieve certain BER/PER.
- Coverage: spatial/temporal outage probability.
- Power consumption: Energy per bit transmission, life time of mobile terminal.
- Spectral efficiency: cell throughput, user throughput.
- Complexity and signaling overhead will be also evaluated.

14.3 Future of Wireless Systems

14.3.1 Software Defined Radio

A Software Defined Radio (SDR) system is a radio communication system which can tune to any frequency band and receive any modulation across a large frequency spectrum by means of programmable hardware which is controlled by software.

The goal of SDR is to produce a radio that can receive and transmit when there is a change in the radio protocol, simply by running new software. In the future advanced SDR the antenna will remain the only band specific item.

Software Defined Radio (SDR) can currently be used to implement simple radio modem technologies, but in the long run, software-defined radio is expected by its proponents to become the dominant technology in radio communications: SDR will likely be the enabler of Cognitive Radio (CR).

14.3.2 Cognitive Radio

Around the year 2000 a new idea was launched concerning improvements on exploiting the frequency spectrum: CR technology. *A CR is a radio that can change its transmitter parameters based on interaction with the environment in which it operates* [259]. A CR is looking for virtual unlicensed spectrum bands, i.e. bands to be shared between primary licensed users and secondary unlicensed users, on a noninterfering basis, i.e. without any activity to or from primary users at present [260].

Cognitive Radio (CR) is a novel approach for improving the utilization of precious natural resource of radio electromagnetic spectrum. The CR, built on top of a SDR [261], is defined as an intelligent wireless communication system that is aware of its environment, uses the methodology of understanding-by-building to learn from the environment and employs this knowledge to adapt to statistical variations in the input stimuli, with two primary objectives in mind [262]:

- Highly reliable communication whenever and wherever needed.
- Efficient utilization of radio spectrum.

Cognitive Radio (CR) techniques provide the capability to use or share spectrum in an opportunistic manner. Dynamic spectrum access techniques allow the CR to operate in the best available channel [263].

14.3.3 Spectrum Sharing

IMT-A is a concept from International Telecommunication Union (ITU) for mobile communication systems with capabilities which go further those of International Mobile Telecommunications (IMT)-2000. International Mobile Telecommunications-Advanced (IMT-A) was previously known as "systems beyond IMT-2000." It is foreseen that the development of IMT-2000 will reach a limit of around 30 Mbps.

In the vision of the ITU, there may be a need for a new wireless access technology to be developed around the year 2010 capable of supporting even higher data rates than 30 Mbps with high mobility, which could be widely deployed around the year 2015 in some countries. The new capabilities of these IMT-A systems are envisaged to handle a wide range of supported data rates according to economic and service demands in multi-user environments, with target peak data rates of up to approximately 100 Mbps for high mobility such as mobile access, and up to approximately 1 Gbps for low mobility such as nomadic/local wireless access [264].

To support this wide variety of services, it may be necessary for IMT-A to have different radio interfaces and frequency bands for mobile access and for new nomadic/local area wireless access. The band which is envisaged to be needed to provide the afore-mentioned very high rates, is in the order of 100 MHz, and likely such a high band will not be available for each operator: this means that some kind of flexible and smart spectrum usage among different operators and networks will be needed. A key component enabling this "peaceful" coexistence between systems and networks in the same bandwidth is the so called Spectrum Sharing (SpS), which can be broadly classified as follows:

- Inter-system SpS, allowing the coexistence of different Radio Access Technologys (RATs).
- Inter-network SpS, allowing the coexistence of different operators/networks operating with the same RAT (this kind of sharing is also called Flexible Spectrum Usage (FSU)) [265].
- Intra-network SpS, allowing the coexistence of different cells owned by the same network, and operating with the same RAT (this kind of sharing is commonly called Multi-cell Radio Resource Management (RRM)).

14.3.4 Self-organizing Networks

Without a human administrator, an *ad-hoc* network must assemble itself from any devices that happen to be nearby, and adapt as devices move in and out of wireless range. The building blocks of *ad-hoc* mobile networks are low-power devices that must do their own wireless routing, forwarding signals from other devices that would otherwise be out of radio range [266].

A typical network could contain tens or even hundreds of these embedded systems, ranging from handheld computers down to motes: tiny units each equipped with a sensor, a microcontroller and a radio that can be scattered around an area to be monitored. Other devices could be mounted at fixed points, carried by robots, or worn as smart clothing or body area networks [266].

Wireless standards are not the issue: most mobile devices use common protocols, such as GSM, Wi-Fi, Bluetooth, and ZigBee. The real challenge, is to build self-managing networks that work reliably on a large scale, with a variety of operating systems and low-power consumption [266].

14.3.5 Low Emission — Green Systems

Wireless systems are replacing wired data transmission to an increasing extent, especially in sensor networks and control applications, where the cost of installation or modifications are a significant consideration, but this calls for ultra low power design techniques to provide extended battery life [267].

Low emission systems are also very important for medical applications: significant academic and corporate resources are being directed towards the development of novel low power circuits and systems [268]. In this respect, wireless systems hold a number of advantages over wired alternatives, including: ease of use, reduced risk of infection, reduced risk of failure, reduced patient discomfort, enhanced mobility, and lower cost of care delivery [269].

Abbreviations

2G	2nd Generation
3G	3rd Generation
3GPP	The Third Generation Partnership Project
3GPP2	The Third Generation Partnership Project 2
4G	4th Generation
4GG	4th Generation Generalized
ACI	Adjacent Channel Interference
AMC	Adaptive Modulation and Coding
AoA	Angle of Arrival
AoD	Angle of Departure
AP	Access-Point
AS	Angle Spread
AWGN	Additive White Gaussian Noise
BER	Bit Error Rate
BF	Beamforming
BICM	Bit-Interleaved Coded Modulation
BLAST	Bell labs LAyered Space Time
BLER	Block Error Rate
BO	Back Off
BPSK	Binary Phase Shift Keying
BS	Base Station
CCI	Co-Channel Interference
CDA-SM-OFDM	Cyclic Delay Assisted Spatially Multiplexed Orthogonal Frequency Division Multiplexing
CDD	Cyclic Delay Diversity
CDF	Cumulative Distribution Function
cdf	cumulative distribution function

CDMA	Code Division Multiple Access
CDS	Channel Dependent Scheduling
CFO	Carrier Frequency Offset
CIR	Channel Impulse Response
COFDM	Coded Orthogonal Frequency Division Multiplexing
CP	Cyclic Prefix
CPRI	Common Public Radio Interface
CQI	Channel Quality Index
CR	Cognitive Radio
CSA	Contiguous Subcarrier Assignment
CSI	Channel State Information
CTF	Channel Transfer Function
DAB	Digital Audio Broadcasting
DAPSK	Differential Amplitude and Phase Shift Keying
DD	Delay Diversity
DFT	Discrete Fourier Transform
DL	Downlink
DoA	Direction of Arrival
DoD	Direction of Departure
DS	Direct Sequence
DS-CDMA	Direct Sequence Code Division Multiple Access
DVB	Digital Video Broadcasting
ECC	Error Correction Coding
EGC	Equal Gain Combining
E-UTRA	Evolved-UMTS Terrestrial Radio Access
FD	Frequency Decoding
FDMA	Frequency Division Multiple Access
FDD	Frequency Division Duplex
FDE	Frequency Domain Equalization
FEC	Forward Error Correction
FER	Frame Error Rate
FFT	Fast Fourier Transform
FSU	Flexible Spectrum Usage
FWA	Fixed Wireless Access
HPA	High Power Amplifier
HSPA	High Speed Packet Access

HSDPA	High Speed Downlink Packet Access
HSUPA	High Speed Uplink Packet Access
IBI	Inter-Block Interference
ICI	Inter-Carrier Interference
IDFT	Inverse Discrete Fourier Transform
IEEE	Institute of Electrical and Electronics Engineers
IFDMA	Interleaved SC-FDMA
IFFT	Inverse Fast Fourier Transform
IMT	International Mobile Telecommunications
IMT-A	International Mobile Telecommunications-Advanced
IO	Interfering Objects
ISI	Inter-Symbol Interference
ITU	International Telecommunication Union
JDM	Joint Diversity and Multiplexing
LA	Link Adaptation
LDC	Linear Dispersion Codes
LDPC	Low Density Parity Check
LFDMA	Localized SC-FDMA
LMS	Least Mean Square
LOS	Line of Sight
LS	Least Squares
LTE	Long Term Evolution
MAI	Multiple Access Interference
MARC	Maximal Average Ratio Combining
MBS	Mobile Broadband System
MC	Multi-Carrier
MC-CDMA	Multi-Carrier Code Division Multiple Access
MIMO	Multiple Input Multiple Output
MISO	Multiple Input Single Output
ML	Maximum Likelihood
MMAC	Multimedia Mobile Access Communication
MMSE	Minimum Mean Square Error
MRC	Maximal Ratio Combining
MS	Mobile Station
MU	Multi User
NLOS	Non Line Of Sight

OD	Orthogonal Design
OFDM	Orthogonal Frequency Division Multiplexing
OFDM–CDMA	Orthogonal Frequency Division Multiplexing–Code Division Multiple Access
OFDM–TDMA	Orthogonal Frequency Division Multiplexing–Time Division Multiple Access
OFDMA	Orthogonal Frequency Division Multiple Access
OFDMA-FSCH	Orthogonal Frequency Division Multiple Access–Fast Sub-Carrier Hopping
OSFBC	Orthogonal Space-Frequency Block Code
OSIC	Ordered Successive Interference Cancellation
PAM	Pulse Amplitude Modulation
PAPR	Peak to Average Power Ratio
pdf	probability density function
PER	Packet Error Rate
PLL	Phase Locked Loop
P/S	Parallel to Serial
PSK	Phase Shift Keying
QAM	Quadrature Amplitude Modulation
QOD	Quasi-Orthogonal Design
QoS	Quality of Service
QPSK	Quadrature Phase Shift Keying
QSFBC	Quasi-orthogonal Space-Frequency Block Code
RAT	Radio Access Technology
RF	Radio Frequency
RLS	Recursive Least Squares
RMS	Root Mean Square
RRM	Radio Resource Management
SC	Single Carrier
SCo	Selection Combining
SCFDE	Single Carrier-Frequency Domain Equalization
SC-FDMA	Single Carrier-Frequency Division Multiple Access
SD	Space Diversity
SDMA	Space Division Multiple Access
SDNR	Signal to Distortion plus Noise Ratio
SDR	Software Defined Radio

SE	Spectral Efficiency
SF	Space–Frequency
SFBC	Space–Frequency Block Code
SFBC-OFDM	Space–Frequency Block Coded Orthogonal Frequency Division Multiplexing
SFC	Space–Frequency Coding
SIC	Successive Interference Cancellation
SIMO	Single Input Multiple Output
SISO	Single Input Single Output
SM	Spatial Multiplexing
SM-OFDM	Spatially Multiplexed Orthogonal Frequency Division Multiplexing
SM-OSFBC	Spatially-Multiplexed Orthogonal Space–Frequency Block Coding
SM-OSTBC	Spatially-Multiplexed Orthogonal Space–Time Block Coding
SM-QSFBC	Spatially-Multiplexed Quasi-orthogonal Space–Frequency Block Coding
SM-QSTBC	Spatially-Multiplexed Quasi-orthogonal Space–Time Block Coding
SNR	Signal to Noise Ratio
S-PARC	Selective Per-Antenna Rate Control
SpS	Spectrum Sharing
ST	Space–Time
STBC	Space–Time Block Code
STBC-OFDM	Space–Time Block Coded OFDM
STC	Space–Time Coding
STFBC	Space–Time–Frequency Block Code
STTC	Space–Time Trellis Code
SSPA	Solid State Power Amplifier
SU	Single User
SVD	Singular Value Decomposition
TCM	Trellis Coded Modulation
TCP	Transmission Control Protocol
TD	Transmit Diversity
TDe	Total Degradation

TDD	Time Division Duplex
TDE	Time Domain Equalization
TDMA	Time Division Multiple Access
TWTA	Traveling Wave Tube Amplifier
Tx	Transmitter
TxBF	Transmit Beamforming
TxDiv	Transmit Diversity
UE	User Equipment
UL	Uplink
UMTS-LTE	Universal Mobile Telecommunications Systems-Long Term Evolution
USLA	Uniform Spaced Linear Arrays
UT	User Terminal
VBLAST	Vertical-Bell Labs LAyered Space–Time Architecture
VCO	Voltage Controlled Oscillator
VSF-OFCDMA	Variable Spreading Factor-Orthogonal Frequency and Code Division Multiple Access
WCDMA	Wideband Code Division Multiple Access
WiMAX	Worldwide Interoperability for Microwave Access
WLAN	Wireless Local Area Network
XPD	Cross Polarization Discrimination
ZF	Zero Forcing
ZMCSCG	Zero Mean Circularly Symmetric Complex Gaussian

References

"Seek knowledge from the cradle to the grave."

— Probhet Muhammad (Peace be upon Him)

[1] Joshua S. Gans, Stephen P. King, and J. Wright, "Wireless communications," *Handbook of Telecommunications Economics*, vol. 2.

[2] Erik Dahlman, Stefan Parkval, Johan Skld, and Per Beming, *3G Evolution: HSPA and LTE for Mobile Broadband*, 3rd ed., Academic Press, August 2008.

[3] R. Prasad, W. Konhauser, and W. Mohr (eds), *Third Generation Mobile Communications Systems*, first ed., Artech House Publishers, June 2000.

[4] T. Ojanpera and R. Prasad, *Wideband CDMA for Third Generation Mobile Communications*, first ed., Artech House Publishers, October 2001.

[5] H. Holma and A. Toskala (eds), *HSDPA/HSUPA for UMTS: High Speed Radio Access for Mobile Communications*, first ed., Wiley Publishers, June 2006.

[6] R. Prasad and L. Munoz, *WLANs and WPANs Towards 4G Wireless*. Artech House Publishers, January 2003.

[7] Y. K. Kim and R. Prasad, *4G Roadmap and Emerging Communications Technologies*, first ed., Artech House Publishers, January 2006.

[8] D. Astèly, E. Dahlman *et al.*, "A future radio-access framework," *IEEE Journal on Selected Areas on Communications*, vol. 24, no. 3, pp. 693–706, March 2006.

[9] K. R. Santhi and G. S. Kumaran, "Migration to 4G: Mobile IP based solutions," *proceedings of IEEE AICT-ICIW'06*, February 2006.

[10] B. Hassibi and B. M. Hochwald, "High-rate codes that are linear in space and time," *IEEE Transformation on Information Theory*, vol. 48, no. 7, pp. 1804–1824, July 2002.

[11] R. Prasad, *OFDM for Wireless Communications Systems*. Artech House Publishers, September 2004.

[12] H. Sari *et al.*, "Frequency domain equalization of mobile radio and terrestrial broadcast channels," in *IEEE International Conference on Global Communications*, pp. 1–5, 1994.

[13] H. Sari, G. Karam, and I. Jeanclaude, "Transmission techniques for digital terrestrial TV broadcasting," *IEEE Communications on Magazine*, vol. 33, pp. 100–109, February 1995.

[14] H. Sari *et al.*, "An analysis of orthogonal frequency-division multiple access," in *GLOBECOM*, vol. 3, pp. 1635–1639, November 1997.

[15] H. Rohling and R. Grunheid, "Performance of an OFDM-TDMA mobile communication system," in 46th *VTC*, vol. 3, pp. 1589–1593, April–May 1996.

[16] N. Yee, J. P. Linnartz *et al.*, "Multicarrier CDMA in indoor wireless networks," in *IEEE PIMRC*, Yokohama, Japan, pp. 109–113, September 1993.

[17] H. Atarashi, S. Abeta, and M. Sawahashi, "Variable spreading factor-orthogonal frequency and code division multiplexing (VSF-OFCDM) for broadband packet wireless access," *IEICE Transformations on Communictions*, vol. E86-B, no. 1, pp. 291–299, January 2003.

[18] M. I. Rahman, S. S. Christensen, R. V. Reynisson, S. S. Das, B. Can, A. B. Olsen, J. M. Kristensen, N. Marchetti, D. V. P. Figueiredo, H. C. Nguyen, P. Popovski, and F. Fitzek, "Comparison of various modulation and access schemes under ideal channel conditions," Aalborg University, Denmark, JADE project Deliverable, D3.1[1], July 2004.

[19] W. Rhee and J. M. Cioffi, "Increase in capacity of multiuser OFDM system using dynamic subchannel allocation," in *IEEE VTC*, Tokyo, Japan, May 2000.

[20] A. T. Toyserkani, S. Naik, J. Ayan, Y. Made, and O. Al-Askary, "Sub-carrier based Adaptive Modulation in HIPERLAN/2 System," *IEEE ICC'04*, vol. 6, pp. 3460–3464, June 2004.

[21] Z. Song, K. Zhang, and Y. L. Guan, "Statistical adaptive modulation for QAM-OFDM systems," *proceedings of IEEE GLOBECOM'02*, vol. 1, pp. 706–710, November 2002.

[22] M. Siebert and O. Stauffer, "Enhanced link adaptation performance applying adaptive sub-carrier modulation in OFDM systems," *IEEE VTC Spring'03*, vol. 57, no. 2, pp. 920–924, April 2003.

[23] M. Lei, P. Zhang, H. Harada, and H. Wakana, "An adaptive power distribution algorithm for improving spectral efficiency in OFDM," *IEEE Transactions on Broadcasting*, vol. 50, no. 3, pp. 347–351, September 2004.

[24] L. Zhen *et al.*, "Link adaptation of wideband OFDM systems in multi-path fading channel," *IEEE CCECE*, vol. 3, pp. 1295–1299, May 2002.

[25] S. T. Chung and A. J. Goldsmith, "Degrees of freedom in adaptive modulation: A unified view," *IEEE Transactions on Communications*, vol. 49, no. 9, pp. 1561–1571, September 2001.

[26] J. H. Winters, "The diversity gain of transmit diversity in wireless systems with Rayleigh fading," *IEEE Transactions on Vehicular Technology*, vol. 47, no. 1, pp. 119–123, February 1998.

[27] G. L. Stuber *et al.*, "Broadband MIMO-OFDM wireless communications," in *Proceedings of the IEEE*, vol. 92, no. 2, pp. 271–294, February 2004.

[28] K. Witrisal *et al.*, "Antenna diversity for OFDM using cyclic delays," in *Proceedings of IEEE 8th Symposium on Communications and Vehicular Technology in the Benelux, SCVT-2001*, October 2001.

[29] P. W. Wolniansky *et al.*, "V-BLAST: An architecture for realizing very high data rates over the rich-scattering wireless channel," in *Proceedings of IEEE-URSI International Symposium on Signals, Systems and Electronics*, Pisa, Italy, May 1998.

[30] S. M. Alamouti, "A simple transmit diversity technique for wireless communications," *IEEE JSAC*, vol. 16, no. 8, October 1998.

[31] H. Bolcskei, D. Gesbert *et al.*, *Space-Time Wireless Systems: From Array Processing to MIMO Communications*. IEEE Press, Spring 2006.

[32] R. Heath and A. Paulraj, "Switching between diversity and multiplexing in MIMO systems," *IEEE Transactions on Communications*, vol. 53, no. 6, June 2005.

[33] R. S. Blum, "MIMO capacity with interference," *IEEE JSAC (Special Issue on MIMO Systems)*, vol. 21, no. 5, pp. 793–801, June 2003.

[34] M. Webb, M. Beach, and A. Nix, "Capacity limits of MIMO channels with co-channel interference," in *Proceedings of IEEE VTC Spring*, Milan, Italy, May 2004.

[35] S. Ye and R. S. Blum, "Optimized signaling for MIMO interference systems with feedback," *IEEE JSAC (special issue on MIMO Systems)*, vol. 51, no. 11, pp. 2839–2848, November 2003.

[36] A. Naguib *et al.*, "Applications of space-time block codes and interference suppression for high-capacity and high data rate wireless systems," in *Proceedings of Asilomar Conference on Signals, Systems and Computers*, vol. 2, Pacific Grove, CA, USA, pp. 1803–1810, 1998.

[37] J. Li *et al.*, "Co-channel interference cancellation for space-time coded OFDM systems," *IEEE Transactions on Wirless Communications*, vol. 2, no. 1, pp. 41–49, January 2003.

[38] J. H. Winters *et al.*, "The impact of antenna diversity on the capacity of wireless communication systems," *IEEE Transactions on Communications*, vol. 42, no. 2/3/4, pp. 119–123, February/March/April 1994.

[39] D. T. Emerson, "The work of Jagadis Chandra Bose: 100 years of MM-wave research," in *International Microwave Symposium Digest, 1997, IEEE MTT-S*, vol. 2. IEEE, pp. 553–556, June 1997.

[40] T. S. Rappaport, *Wireless Communications Principles and Practice*. Prentice-Hall, January 1996.

[41] V. Erceg *et al.*, "An empirically based path loss model for wireless channels in suburban environments," vol. 17, no. 7, pp. 1205–1211, July 1999.

[42] V. Erceg, "Channel model for fixed wireless application," IEEE 802.16 working group, Technical Report, 2001.

[43] Technical specification group radio access network; physical layer aspects for evolved universal terrestrial radio access (utra): Release 7, 3rd Generation Partnership Project, Draft Standard 3GPP TR 25.814 V7.1.0, 2006–09.

[44] J. G. Proakis, *Digital Communications*. Prentice-Hill, 1995.

[45] A. F. Molisch, *Wireless Communications*, first ed., John Wiley & Sons, 2005.

[46] M. C. Lawton, R. L. Davies, and J. P. McGeehan, "An analytical model for indoor multipath propagation in the picocellular environment," in *Sixth International Conference on Mobile Radio and Personal Communications*, pp. 1–8, December 1991.

[47] V. Erceg *et al.*, "A model for the multipath delay profile of fixed wireless channels," vol. 17, no. 3, pp. 399–410, March 1999.

[48] Klaus Witrisal, "OFDM Air Interface Design for Multimedia Communications," Ph.D. dissertation, Delft University of Technology, The Netherlands, April 2002.

[49] M. Patzold, *Mobile Fading Channels: Modelling, Analysis, and Simulation*, first ed., John Wiley and Sons, 2002.

[50] L. J. Greenstein *et. al.*, "A new path-gain/delay-spread propagation model for digital cellular channels," vol. 46, no. 2, pp. 477–485, May 1997.

[51] L. R. D. Cox, "Correlation bandwidth and delay spread multipath propagation statistics for 910-MHz urban mobile radio channels," vol. 23, no. 11, pp. 1271–1280, 1975.

[52] S. S. Das and R. Prasad, "Time correlation function for RMS delay spread of a channel model," in *IST Mobile Summit*, June 2006.

[53] X. Fuqin, *Digital Modulation Techniques*. Artech House Publishers, 2000.

[54] W. Y. Zou, and Y. Wu, "COFDM: An overview," *IEEE Transactions on Broadcasting*, vol. 41, no. 1, March 1995.

[55] E. P. Lawrey, "Adaptive techniques for multiuser OFDM," PhD dissertation, James Cook University, Australia, December 2001.

[56] C. R. Nassar *et al.*, *Multi-carrier Technologies for Wireless Communication*. Kluwer Academic Publishers, 2002.

[57] W. C. Y. Lee, *Mobile Cellular Telecommunications Systems*. New York, USA: McGraw Hill Publications, December 1989.

[58] S. Haykin, *Adaptive Filter Theory*, third ed., Prentice-Hall, December 1996.

[59] H. Rohling *et al.*, "Broad-band OFDM radio transmission for multimedia applications," in *Proceedings of the IEEE*, vol. 87, no. 10, pp. 1778–1789, October 1999.

[60] J. H. Scott, "The how and why of COFDM: BBC research and development," EBU Technical Review, Winter 1999.

[61] J. Geier, *Wireless LANs, Implementing High Performance IEEE 802.11 Networks*, 2nd ed., Sams Publishing, July 2001.

[62] D. Matic, "OFDM synchronization and wideband power measurements at 60 GHz for future wireless broadband multimedia communications," PhD dissertation, Aalborg University, Denmark, September 2001.

[63] U. S. Jha, "Low complexity resource efficient OFDM based transceiver design," PhD dissertation, Aalborg University, Denmark, September 2002.

[64] O. Edfors *et al.*, "An introduction to orthogonal frequency-division multiplexing," Division of Signal Processing, Lulea University of Technology, Sweden, Research Report 1996:16, September 1996.

[65] T. Pollet *et al.*, "BER sensitivity of OFDM systems to carrier frequency offset and wiener phase noise," *IEEE Transactions on Communications*, vol. 43, no. 2-3-4, pp. 191–193, February–April 1995.

[66] H. Rohling and T. May, "Comparison of PSK and DPSK modulation in a coded OFDM system," in *Proceedings of IEEE VTC*, Phoenix, Arizona, USA, pp. 5–7, 1997.

[67] T. Bruns, "Performance comparison of multiuser OFDM techniques," EE360 Class Project, Stanford University, Spring 2001.

[68] H. Rohling and R. Grunheid, "Performance comparison of different multiple access schemes for the downlink of an OFDM communication system," in 47th *VTC*, vol. 3, pp. 1365–1369, May 1997.

[69] C. Y. Wong, R. S. Cheng, K. B. Letaief, and R. D. Murch, "Multiuser OFDM with adaptive subcarrier, bit, and power allocation," *IEEE Journal on Selected Areas Communications*, October 1999.

[70] H. Sari, Y. Levy, and G. Karam, "Orthogonal frequency-division multiple access for the return channel on CATV networks," in *IEEE ICT*, Istanbul, April 1996.

[71] C. Y. Wong, C. Y. Tsui, R. S. Cheng, and K. B. Letaief, "A real-time subcarrier allocation scheme for multiple access downlink OFDM transmission," in *IEEE VTC*, Houston, USA, May 1999.

[72] D. Kivanc and H. Liu, "Subcarrier allocation and power control for OFDMA," in *IEEE Signals, Systems and Computers*, Pacific Grove, USA, October 2000.

[73] T. Muller, H. Rohling, and R. Grunheid, "Comparison of different detection algorithms for OFDM-CDMA in broadband Rayleigh fading," in 45th *VTC*, vol. 2, pp. 835–838, July 1995.

[74] IEEE, "The IEEE 802.16 working group on broadband wireless access standards," http://ieee802.org/16/, 2003.

[75] D. Galda and H. Rohling, "A low complexity transmitter structure for OFDM-FDMA uplink systems," in *IEEE VTC*, Birmingham, USA, May 2002.

[76] S. Barbarossa, M. Pompili, and G. B. Giannakis, "Time and frequency synchronization of orthogonal frequency division multiple access systems," in *IEEE ICC*, Helsinki, Finland, June 2001.

[77] Z. Cao, U. T'ureli, and Y. D. Yao, "User separation and frequency-time synchronization for the uplink of interleaved OFDMA," in *IEEE Signals, Systems and Computers*, Pacific Grove, USA, November 2002.

[78] H. Yoo, C. Kang, and D. Hong, "Edge sidelobe suppressor schemes for uplink of orthogonal frequency division multiple access systems," in *IEEE GLOBECOM*, Teipei, November 2002.

[79] H. Yoo and D. Hong, "Edge sidelobe suppressor scheme for OFDMA uplink systems," *IEEE Communications Letters*, November 2003.

[80] Y. Segal, Z. Hahad, and I. Kitroser, "Initial OFDMA proposal for the 802.16.4 PHY layer," IEEE 802.16 Broadband Wireless Access Working Group, RunCom Technologies Ltd. 14, Moshe Levi St., Rishon Lezion, Israel, Technical Report, November 2001.

[81] S. Hara and R. Prasad, *Multicarrier Techniques For 4G Mobile Communications*. Artech House Publishers, June 2003.

[82] A. Persson, T. Ottosson, and E. G. Strom, "Analysis of the downlink BER performance for coded OFDMA with fading co-channel interference," Chalmers University of Technology, Sweden, SE-412 96 Goteborg Sweden, Technical Report R001/2004 ISSN 1403-266X, January 2004.

[83] P. Carson, "Flash-OFDM technology," Flarion Whitepaper, November 2001.

[84] S. Zhou, G. B. Ginnakis, and A. Scglione, "Long codes for generalized FH-OFDMA through unknown multipath channels," *IEEE Transactions on Communications*, vol. 49, no. 4, pp. 721–733, April 2001.

[85] S. Kapoor *et al.*, "Initial contribution on a sytem meeting MBWA characteristics," *IEEE 802.20 Working Group on Mobile Broadband Wireless Access*, March 2003.

[86] F. Daffara and O. Adami, "A new frequency detector for orthogonal multicarrier transmission techniques," *IEEE VTC*, pp. 804–809, July 1995.

[87] S. De Fina, "Comparison of FH-MA communications using OFDM and DS-MA systems for wideband radio access," in *Universal Personal Communications (ICUPC)*, vol. 1, pp. 143–147, October 1998.

[88] T. Kurt and H. Deliç, "Collision avoidance in space-frequency coded FH-OFDMA," in *IEEE ICC*, Paris, France, June 2004.

[89] A. Al-Dweik and F. Xiong, "Frequency-hopped multiple-access communication with noncoherent OFDM-MASK in AWGN channels," in *Military Communications Conference (MILCOM)*, vol. 2, pp. 1355–1359, October 2001.

[90] S. U. H. Qureshi, "Adaptive equalization," in *Proceedings of IEEE*, 1985.

[91] F. Falconer, S. L. Ariyavisitakul, A. Benyamin-Seeyar, and B. Eidson, "Frequency domain equalization for single-carrier broadband wireless systems," *IEEE Communications Magazine*, April 2002.

[92] M. V. Clark, "Adaptive frequency-domain equalization and diversity combining for broadband wireless communications," *IEEE JSAC*, pp. 1385–1395, October 1998.

[93] H. Sari *et al.*, "Transmission and multiple access techniques for broadband wireless access networks," November 2005.

[94] T. Walzman and M. Schwartz, "Automatic equalization using the discrete Fourier domain," *IEEE Transaction on Information Theory*, vol. IT-19, pp. 59–68, January 1973.

[95] A. Benyamin-Seeyar *et al.*, "Harris Corporation Inc." http://www.ieee802.org/16/tg3/contrib/802163c – 01_32.pdf, 2001.

[96] H. G. Myung, "Single carrier multiple access technique for broadband wireless communications," PhD dissertation, Polytechnic University, Brooklyn, NY, USA, January 2007.

[97] H. G. Myung, J. Lim, and D. J. Goodman, "Single carrier FDMA for uplink wireless transmission," *IEEE Vehicular Technology Magazine*, vol. 1, no. 3, September 2006.

[98] H. Ekstroem *et al.*, "Technical solutions for the 3G long-term evolution," *IEEE Communications Magazine*, pp. 38–45, March 2006.

[99] E. Dahlman *et al.*, "The 3G long-term evolution — Radio interface concepts and performance evaluation," in *IEEE VTC Spring 2006*, Melbourne, Australia, May 2006.

[100] R. van Nee and R. Prasad, *OFDM for Wireless Multimedia Communications*. Artech House, 2000.

[101] U. Sorger, I. De Broeck, and M. Schnell, "Interleaved FDMA — A new spread-spectrum multiple-access scheme," in *IEEE ICC*, Atlanta, USA, June 1998.

[102] H. G. Myung, J. Lim, and D. J. Goodman, "Peak-to-average power ratio of single carrier FDMA signals with pulse shaping," in *IEEE PIMRC*, 2006.

[103] F. Adachi *et al.*, "Broadband CDMA techniques," *IEEE Wireless Comm.*, vol. 12, no. 2, April 2005.

[104] C. Chang and K. Chen, "Frequency-domain approach to multiuser detection in DS-CDMA communications," *IEEE Comm. Letters*, vol. 4, no. 11, November 2000.

[105] C. Chang and K. Chen, "Frequency-domain approach to multiuser detection over frequency-selective slowly fading channels."

[106] T. Pollet and M. Moeneclaey, "Synchronizability of OFDM signals," in *Proceedings of Globecom '95*, vol. 3, Singapore, pp. 2054–2058, November 1995.

[107] J. Kivinen and P. Vainikainen, "Phase noise in a direct sequence based channel sounder," in *Proceedings of IEEE PIMRC '97*, Helsinki, Finland, pp. 1115–1119, 1-4 Septmeber 1997.

[108] P. Robertson and S. Kaiser, "Analysis of the effects of phase-noise in orthogonal frequency division multiplex systems," in *Proceedings of IEEE VTC '95*, pp. 1652–1657, 1995.

[109] L. Tomba, "On the effects of Wiener phase noise in OFDM systems," *IEEE Transactions on Communications*, vol. 46, no. 5, pp. 580–583, May 1998.

[110] T. N. Zogakis and J. M. Cioffi, "The effect of timing jitter on the performance of a discrete multitone system," *IEEE Transactions on Communications*, vol. 44, no. 7, pp. 799–808, July 1996.

[111] T. Pollet, P. Spruyt and M. Moeneclaey, "The BER performance of OFDM systems using non-synchronized sampling," in *Proceedings of Globecom '94*, Helsinki, Finland, pp. 253–257, 1-4 Septmeber 1997.

[112] J. J. Van de Beek, M. Sandell, M. Isaksson, and P. O. Börjesson, "Low-complex frame synchronization in OFDM systems," in *Proceedings of International Conference on Universal Personal Communications ICUPC '95*, November 1995.

[113] M. Sandell, J. J. van de Beek, and P. O. Börjesson, "Timing and frequency synchronization in OFDM systems using the cyclic prefix," in *Proceedings of International Symposium On Synchronization*, Saalbau, Essen, Germany, pp. 16–19, 14-15 December 1995.

[114] R. Böhnke and T. Dölle, "Preamble structures for HiperLAN type 2 system," ETSI BRAN, Technical submission HL13SON1A, 7 April 1999.

[115] T. M. Schmidl and D. C. Cox, "Robust frequency and timing synchronization for OFDM," *IEEE Transactions on Communications*, vol. 45, no. 12, pp. 1613–1621, December 1997.

[116] P. H. Moose, "A technique for orthogonal frequency division multiplexing frequency offset correction," *IEEE Transactions on Communications*, vol. 42, no. 10, pp. 2908–2914, October 1994.

[117] W. D. Warner and C. Leung, "OFDM/FM frame synchronization for mobile radio data communization," *IEEE Transactions on Vehicular Technology*, vol. 42, no. 3, pp. 302–313, August 1993.

[118] U. Lambrette, M. Speth, and H. Meyr, "OFDM burst frequency synchronization by single carrier training data," *IEEE Communications Letters*, vol. 1, no. 2, pp. 46–48, March 1997.

[119] M. Sandell and O. Edfors, "A comparative study of pilot-based channel estimators for wireless OFDM," Research Report TULEA 1996:19, Division of Signal Processing, Luleå University of Technology, Technical Report, September 1996.

[120] O. Edfors, M. Sandell, J. J. van de Beek, S. K. Wilson, and P. O. Börjesson, "OFDM channel estimation by singular value decomposition," in *IEEE VTC*, pp. 923–927, April 1996.

[121] M. Sandell, S. K. Wilson, and P. O. Börjesson, "Performance analysis of coded OFDM on fading channels with non-ideal interleaving and channel knowledge," Research Report TULEA 1996:20, Division of Signal Processing, Luleå University of Technology, Technical Report, September 2006.

[122] J. J. van de Beek, O. Edfors, M. Sandell, S. K. Wilson, and P. O. Börjesson, "On channel estimation in OFDM systems," in *IEEE VTC*, vol. 1, Rosemont, IL, USA, pp. 715–719, July 1995.

[123] L. Scharf, *Statistical Signal Processing: Detection, Estimation and Time Series Analysis*. Addison-Wesley, 1991.

[124] P. Höher, "TCM on frequency-selective land-mobile fading channels," in *Proceedings of the Tirrenia International Workshop on Digital Communications*, September 1991.

[125] H. Takanashi and R. van Nee, "Merged physical layer specification for the 5 GHz band," Technical Report.

[126] P. Frenger and A. Svensson, "Decision directed coherent detection in multicarrier systems on Rayleigh fading channels," *IEEE Transactions on Vehicular Technology*, 1999.

[127] V. Engels and H. Rohling, "Multilevel differential modulation techniques (64-DAPSK) for multicarrier transmission techniques," *European Transactions on Telecommunication Related Technologies*, November–December 1995.

[128] V. Engels and H. Rohling, "Differential modulation techniques for a 34 Mbit/s radio channel using OFDM," *Wireless Personal Communications*, 1995.

[129] T. May, H. Rohling, and V. Engels, "Performance analysis of viterbi decoding for 64-DAPSK and 64-QAM modulated signals," *IEEE Transactions on Communications*, February 1998.

[130] P. Banelli, G. Baruffa, and S. Cacopardi, "Effect of HPA non linearity on frequency multiplexed OFDM signals," in *IEEE Transaction on Broadcasting*, vol. 47, no. 2, pp. 123–136, June 2001.

[131] S. S. Das, M. I. Rahman *et al.*, "Influence of PAPR on link adaptation algorithms in OFDM systems," in *IEEE VTC Spring'07*, Dublin, Ireland, 22–25 April 2007.

[132] G. Santella and F. Mazzenga, "A hybrid analytical-simulation procedure for performance evaluationin M-QAM-OFDM schemes in presence of nonlinear distortions," in *IEEE Transactions on Vehicular Technology*, vol. 47, no. 1, pp. 142–151, February 1998.

[133] Adel A. M. Saleh, "Frequency-independent and frequency-dependent nonlinear models of TWT amplifiers," in *IEEE Transaction on Communication*, vol. COM-29, no. 11, pp. 1715–1720, November 1981.

[134] C. Rapp, "Effects of HPA-nonlinearity on a 4-DPSK/OFDM-signal for a digital sound broadcasting signal," in *ESA, Second European Conference on Satellite Communications (ECSC-2) p 179-184 (SEE N92-15210 06-32)*, pp. 179–184, October 1991.

[135] J. Thomas, "Overview of amplifier technology choices for ground based ka-band amplifiers," in *5th Ka Band Utilization Conference*.

[136] S. H. Han and J. H. Lee, "An overview of peak-to-average power ratio reduction techniques for multicarrier transmission," *IEEE Wireless Communications*, vol. 12, no. 2, pp. 56– 65, April 2005.

[137] R. Van Nee and R. Prasad, *OFDM Wireless Multimedia Communication*. Artech House Publishers, 2000.

[138] A. J. Paulraj, R. Nabar, and D. Gore, *Introduction to Space-Time Wireless Communications*, first ed., Cambridge University Press, September 2003.

[139] J. Heiskala and J. Terry, *OFDM Wireless LANs: A Theoretical and Practical Guide*, second ed., Sams Publishing, July 2001.

[140] W. C. Jakes Jr., *Microwave Mobile Communications*. New York: John Wiley & Sons, May 1994.

[141] G. J. Foschini and M. J. Gans, "On limits of wireless communications in a fading environment when using multiple antennas," *Wireless Personnal Communications*, vol. 6, pp. 311–335, March 1998.

[142] H. Bolcskei, D. Gesbert, and A. J. Paulraj, "On the capacity of OFDM-based spatial multiplexing systems," *IEEE Transactions on Communications*, vol. 50, no. 2, pp. 225–234, February 2002.

[143] V. Tarokh, N. Seshadri, and A. R. Calderbank, "Space-time codes for high data rate wireless communications: Performance criterion and code construction," *IEEE Transactions on Information Theory*, vol. 44, pp. 744–765, March 1998.

[144] A. Paulraj and T. Kailath, "Increasing capacity in wireless broadcast systems using distributed transmission/directional reception," Patent US Patent, 5/345/599, 1994.

[145] G. J. Foschini, "Layered space-time architecture for wireless communication in a fading environment when using multiple antennas," *Bell Labs Technical Journal*, pp. 41–59, Autumn 1996.

[146] I. Telatar, "Capacity of multi-antenna Gaussian channels," *European Transactions on Telecommunications*, November 1999.

[147] R. Heath, Jr and A. Paulraj, "Switching between multiplexing and diversity based on constellation distance," in *Proceedings on Allerton Conference on Communication, Control and Computing*, October 2000.

[148] L. Zheng and D. N. C. Tse, "Diversity and multiplexing: A fundamental tradeoff in multiple antenna channels," *IEEE Transactions on Information Theory*, vol. 49, no. 4, pp. 1073–1096, May 2003.

[149] H. Bolcskei *et al.*, "Fixed broadband wireless access: State of the art, challenges, and future directions," *IEEE Commun. Magazine*, vol. 39, no. 1, pp. 100–108, January 2002.

[150] R. Sokal and F. Rohlf, *The Principles and Practice od Statistics in Biological Research.* Third ed., New York, NY: Freeman Co., 1995.

[151] R. Nabar *et al.*, "Outage performance of space-time block codes for generalized MIMO channels," *IEEE Transactions on Information Theory*, March 2002.

[152] C. Shannon, "A mathematical theory of communication," *Bell Labs Technical Journal*, vol. 27, pp. 379–423, 623–656, October 1948.

[153] D. Gesbert, M. Shafi *et al.*, "From theory to practice: An overview of MIMO space-time coded wireless systems," *IEEE JSAC*, vol. 21, no. 3, pp. 281–302, April 2003.

[154] E. Telatar, "Capacity of multiantenna Gaussian channels," AT&T Bell Labs, Tech. Memo., June 1995.

[155] G. Raleigh and J. M. Cioffi, "Spatial-temporal coding for wireless communications," *IEEE Transactions on Communications*, vol. 46, pp. 357–366, 1998.

[156] D. Shiu, "Wireless communication using dual antenna arrays," *Kluwer International Series in Engineering and Computer Science*, 1999.

[157] R. Farrokhi *et al.*, "Link-optimal space-time processing with multiple transmit and receive antennas," *IEEE Communications Letters*, vol. 5, pp. 85–87, 2001.

[158] C. N. Chuah *et al.*, "Capacity scaling in MIMO wireless systems under correlated fading," *IEEE Trans. Info. Theory*, vol. 48, pp. 637–650, March 2002.

[159] P. J. Smith and M. Shafi, "On a Gaussian approximation to the capacity of wireless MIMO systems," in *IEEE ICC*, vol. 4, pp. 406–410, 2002.

[160] C.-N. Chuah *et al.*, "Capacity on indoor multiantenna array systems in indoor wireless environment," in *IEEE GLOBECOM*, 1998.

[161] P. F. Driessen and G. J. Foschini, "On the capacity formula for multiple-input multiple-output wireless channels: A geometric interpretation," *IEEE Transactions on Communications*, vol. 47, 1999.

[162] D. Shiu, G. J. Foschini, M. J. Gans, and J. M. Kahn, "Fading correlation and its effect on the capacity of multi-element antenna systems," *IEEE Transactions on Communications*, vol. 48, pp. 502–513, March 2000.

[163] D. Chiznik *et al.*, "Effect of antenna separation on the capacity of BLAST in correlated channels," *IEEE Communications Letters*, vol. 4, pp. 337–339, November 2000.

[164] P. B. Rapajic and D. Popescu, "Information capacity of a random signature multiple-input multiple-output channel," *IEEE Transactions on Communications*, vol. 39, pp. 1245–1248, August 2000.

[165] A. M. Sengupta and P. P. Mitra, "Capacity of multivariate channels with multiplicative noise: I. Random matrix techniques and large-n expansions for full transfer matrices," Phy. Arch., no. 0 010 081, 2000.

[166] V. L. Girko, "Theory of random deteminants," *Norwell, MA: Kluwer*, 1990.

[167] V. L. Girko, *Theory of Linear Algebraic Equations With Random Coefficients*, Second ed., New York: Allerton, 1996.

[168] IST-Winner Consortium, "Assesment of advanced beamforming and MIMO technologies," IST-EU, Technical Deliverable, D2.7, ver 1.0, February 2005.

[169] M. I. Rahman, N. Marchetti, D. V. P. Figueiredo *et al.*, "Multi-antenna techniques in multi-user OFDM systems," Aalborg University, Denmark, JADE project Deliverable, D3.2[1], September 2004.

[170] H. Yaghoobi, "Scalable OFDMA physical layer in IEEE 802.16 wirelessMAN," *Intel Technical Journal*, August 2004.

[171] D. V. P. Figueiredo, M. I. Rahman, N. Marchetti *et al.*, "Transmit diversity Vs beamforming for multi-user OFDM systems," in *WPMC*, Abano Terme, Italy, September 2004.

[172] S. Sandhu, R. Nabar, D. Gore, and A. J. Paulraj, "Introduction to space-time codes," Smart Antenna Research Group, Stanford University, Technical Report, 2004.

[173] M. Okada and S. Komaki, "Pre-DFT combining space diversity assisted COFDM," *IEEE Transactions Vehicular Technology*, vol. 50, no. 2, March 2001.

[174] V. Tarokh, H. Jafarkhani, and A. R. Calderbank, "Space-time block coding for wireless communications: Performance results," *IEEE JSAC*, vol. 17, no. 3, March 1999.

[175] K. F. Lee and D. B. Williams, "A space-frequency transmitter diversity technique for OFDM systems," in *IEEE GLOBECOM*, vol. 3, pp. 1473–1477, November–December 2000.

[176] A. F. Molisch, M. Z. Win, and J. H. Winters, "Space-time-frequency (STF) coding for MIMO-OFDM systems," *IEEE Communications Letters*, vol. 6, no. 9, pp. 370–372, September 2002.

[177] J. C. Liberti and T. S. Rappaport, *Smart Antennas for Wireless Communications.* Prentice-Hall, April 1999.

[178] M. I. Rahman *et al.*, "Performance comparison between MRC receiver diversity and cyclic delay diversity in OFDM WLAN systems," in *Proceedings of 6th WPMC*, vol. 2, Yokosuka, Japan, pp. 198–202, October 2003.

[179] D. Schafhuber, *MIMO-OFDM Systems*, Seminar, Vienna University of Technology, Austria, June 2004.

[180] H. Bolcskei and A. J. Paulraj, "Performance analysis of space-time codes in correlated rayleigh fading environments," in *ACSSC*, October 2000.

[181] D. Gesbert *et al.*, "Outdoor MIMO wireless channels: Models and performance prediction," *IEEE Transactions on Communications*, July 2000.

[182] H. Bolcskei *et al.*, "Performance of spatial multiplexing in the presence of polarization diversity," in *ICASSP*, May 2001.

[183] B. Friedlander and S. Scherzer, "Beamforming versus transmit diversity in the downlink of a cellular communications system," *IEEE Transactions on Vehicular Technology*, vol. 53, no. 4, pp. 1023–1034, July 2004.

[184] H. Rohling *et al.*, "Comparison of multiple access schemes for an OFDM downlink system," in *Multi-Carrier Spread Spectrum*, K. Fazel and G. P. Fettewis, (eds.), Kluwer Academic Publishers, pp. 23–30, 1997.

[185] M. Schubert and H. Boche, "An efficient algorithm for optimum joint downlink beamforming and power control," in *Proceedings of VTC Spring*, pp. 1911–1915, May 2002.

[186] J. Medbo *et al.*, "Channel models for HIPERLAN/2 in different indoor scenarios," in *COST 259 TD(98)70*, (EURO-COST, ed.), Bradford, UK, April 1998.

[187] M. Patzold, "Mobile fading channels," John Wiley & Sons, June 2002.

[188] A. J. Paulraj and C. B. Papadias, "Space-time processing for wireless communications," *IEEE Signal Processing Magazine*, vol. 14, pp. 49–83, November 1997.

[189] H.-C. Kim, J.-H. Park, Y. Shin, and W.-C. Lee, "Transmit eigen-beamformer with space-time block code for MISO wireless communication systems," in *ITC-CSCC*, Phuket, Thailand, July 2002.

[190] B. Friedlander and S. Scherzer, "Beamforming vs. transmit diversity in the downlink of a cellular communications system," in *Proceedings of 35th Asilomar Conference on Signals, Systems and Computers*, vol. 2, pp. 1014–1018, November 2001.

[191] K. Witrisal, "OFDM air-interface design for multimedia communications," PhD dissertation, Delft University of Technology, the Netherlands, April 2002.

[192] K. F. Lee, and D. B. Williams, "A space-time coded transmitter diversity technique for frequency selective fading channels," in *IEEE Sensor Array and Multichannel Signal Processing Workshop*, Cambridge, USA, pp. 149–152, March 2000.

[193] Y. Li *et al.*, "Transmitter diversity for OFDM systems and its impact on high-rate data wireless networks," *IEEE JSAC*, vol. 17, no. 7, July 1999.

[194] A. Dammann and S. Kaiser, "Performance of low complex antenna diversity techniques for mobile OFDM systems," in *Proceedings of 3rd International Workshop on MC-SS*, Oberpfaffenhofen, Germany, September 2001.

[195] M. Bossert *et al.*, "On cyclic delay diversity in OFDM based transmission schemes," in *Proceedings of 7th International OFDM Workshop*, Hamburg, Germany, September 2002.

[196] M. I. Rahman *et al.*, "Optimum pre-DFT combining with cyclic delay diversity for OFDM based WLAN systems," in *Proceedings of VTC Spring*, Milan, Italy, May 2004.

[197] G. L. Stuber, *Principles of Mobile Communications*. minus Kluwer Academic Publisher, January 1996.

[198] S. B. Slimane, "A low complexity antenna diversity receiver for OFDM based systems," in *IEEE ICC*, vol. 4, Helsinki, Finland, pp. 1147–1151, June 2001.

[199] G. Bauch, "Differential modulation and cyclic delay diversity in orthogonal frequency-division multiplex," *IEEE Transactions on Communications*, vol. 54, no. 5, pp. 798–801, May 2006.

[200] G. Bauch and J. S. Malik, "Cyclic delay diversity with bit-intereaved coded modulation in orthogonal frequency-division multiple access," *IEEE Transactions on Wireless Communications*, vol. 5, no. 8, August 2006.

[201] Y. Li *et al.*, "MIMO-OFDM for wireless communications: Signal detection with enhanced channel estimation," *IEEE Transactions on Communications*, vol. 50, no. 9, September 2002.

[202] M. I. Rahman, N. Marchetti *et al.*, "Joint Quasi-orthogonal SFBC and spatial multiplexing in MIMO-OFDM systems," in *Proceedings of IEEE VTC Fall*, Montreal, Canada, September 2006.

[203] G. Bauch and J. S. Malik, "Differential modulation with cyclic delay diversity in OFDM," in *Proceedings of 9th International OFDM Workshop*, Dresden, Germany, pp. 123–127, September 2004.

[204] G. Auer, "Channel estimation for OFDM with cyclic delay diversity," in *Proceedings of 15th IEEE PIMRC'04*, Barcelona, Spain, 5–8 September 2004.

[205] G. Auer, "Channel estimation by set partitioning for OFDM with cyclic delay diversity," in *Proceedings of IEEE VTC Fall'04*, Los Angeles, USA, 26–29 September 2004.

[206] G. Auer, "Analysis of pilot-symbol aided channel estimation for OFDM systems with multiple transmit antennas," in *Proceedings of IEEE ICC'04*, Paris, France, 20–24 June 2004.

[207] H. Jafarkhani, "A quasi-orthogonal space-time block code," *IEEE Transactions on Communications*, vol. 49, no. 1, pp. 1–4, January 2001.

[208] X. Zhuang *et al.*, "Transmit diversity and spatial multiplexing in four-transmit-antenna OFDM," in *IEEE ICC*, vol. 4, pp. 2316–2320, May 2003.

[209] C.-B. Chae *et al.*, "Adaptive spatial modulation for MIMO-OFDM," in *Proceedings of WCNC*, vol. 1, pp. 87–92, March 2004.

[210] H. Skjevling, D. Gesbert, *et al.*, "Combining space-time codes and multiplexing in correlated MIMO channels: An antenna assignment strategy," in *Nordic Signal Processing Conference*, June 2003.

[211] M. H. Hairi *et al.*, "Adaptive MIMO-OFDM combining space-time block codes and spatial multiplexing," in *IEEE ISSSTA*, Sydney, Australia, August–September 2004.

[212] L. Zhao and V. K. Dubey, "Detection schemes for space-time block code and spatial multiplexing combined system," *IEEE Communications Letters*, vol. 9, no. 1, January 2005.

[213] H. Bolcskei and A. J. Paulraj, "Space-frequency coded broadband OFDM systems," in *IEEE WCNC*, vol. 1, Chicago, USA, pp. 1–6, September 2000.

[214] A. Stamoulis, Z. Liu and G. B. Giannakis, "Space-time block-coded OFDMA with linear precoding for multirate services," *IEEE Transactions on Signal Processing*, vol. 50, no. 1, pp. 19–129, January 2002.

[215] F. R. Farrokhi *et al.*, "Spectral efficiency of wireless systems with multiple transmit and receive antennas," in *IEEE PIMRC*, vol. 1, London, UK, pp. 373–377, September 2000.

[216] J. Coon *et al.*, "Adaptive frequency-domain equalization for single-carrier MIMO systems," in *IEEE ICC*, June 2004.

[217] Y. Zhu and K. B. Letaief, "A hybrid time-frequency domain equalizer for single carrier broadband MIMO systems," in *IEEE ICC*, June 2006.

[218] R. Kalbasi *et al.*, "Hybrid time-frequency layered space-time receivers for severe time-dispersive channels," in *IEEE SPAWC Workshop*, July 2004.

[219] D. Hwang, A. Pandharipande, and H. Park, "An interleaved TCM scheme for single carrier multiple transmit antenna systems," *IEEE Communications Letters*, vol. 10, no. 6, June 2006.

[220] L. Guo and Y.-F. Huang, "A multi-user SC-FDE-MIMO system for frequency-selective channels," in *IEEE ACSSC*, October–November 2005.

[221] J. Coon *et al.*, "A comparison of MIMO-OFDM and MIMO-SCFDE in WLAN environments," in *IEEE Globecom*, vol. 1, pp. 87–92, December 2003.

[222] N. Al-Dhahir, "Single-carrier frequency-domain equalization for space-time block-coded transmissions over frequency-selective fading channels," *IEEE Communications Letters*, vol. 5, no. 7, July 2001.

[223] S. Reinhardt *et al.*, "MIMO extensions for SC/FDE systems," in *IEEE European Conference on Wireless Technology*, May 2005.

[224] M. I. Rahman, N. Marchetti *et al.*, "Combining orthogonal space-frequency block coding and spatial multiplexing in MIMO-OFDM system," in *Proceedings of International OFDM Workshop, InOWo'05*, Hamburg, Germany, 31 August–01 September 2005.

[225] H. Bolcskei *et al.*, "Space-frequency coded MIMO-OFDM with variable multiplexing-diversity tradeoff," in *IEEE ICC*, pp. 2837–2841, May 2003.

[226] G. D. Golden, G. J. Foschini, R. A. Valenzuela, and P. W. Wolniansky, "Detection algorithm and initial laboratory results using V-BLAST space-time communication architecture," *Eletronics Letters*, vol. 35, pp. 14–16, January 1999.

[227] H. J. V. Tarokh and A. Calderbank, "Space-time block codes from orthogonal designs," *IEEE JSAC*, vol. 45, pp. 1456–1467, July 1999.

[228] G. V. Rangaraj, D. Jalihal, and K. Giridhar, "Exploting multipath diversity using space-frequency linear dispersion codes in MIMO-OFDM systems," in *IEEE ICC*, vol. 4, pp. 2650–2654, May 2005.

[229] J. Wu and S. D. Blostein, "Linear dispersion over time and frequency," in *IEEE ICC*, vol. 1, pp. 254–258, June 2004.

[230] J. Wu and S. D. Blostein, "Linear dispersion for single-carrier communications in frequency selective channels," in *IEEE VTC-fall*, 2006.

[231] T.-W. Yune, C.-H. Choi, and G.-H. Im, "Single carrier frequency-domain equalization with transmit diversity over mobile multipath channels," *IEICE Journal of Transactions on Communications*, vol. 89-B, no. 7, pp. 2050–2060, July 2006.

[232] N. Marchetti, E. Cianca, and R. Prasad, "Low complexity transmit diversity scheme for SCFDE transmissions over time-selective channels," in *IEEE International Conference on Communications (ICC)*, 24–28 June 2007.

[233] "Feasibility Study for Enhanced Uplink for ITRA FDD," 3GPP Technical Report 25.896, February 2004.

[234] "Requirements for Evolved-UTRA (E-UTRA) and Evolved-UTRAN (E-UTRAN)," 3GPP Technical Report 25.913 V7.2.0, 2005–06.

[235] "3GPP RAN WG1 LTE, R1-050665: Throughput Evaluations Using MIMO Multiplexing in Evolved UTRA Uplink," NTT DoCoMo, June 2005.

[236] "3GPP TSG RAN WG1, R1-060437: Basic Schemes in Uplink MIMO Channel Transmissions," NTT DoCoMo, NEC, Sharp, Toshiba Corporation, February 2006.

[237] "3GPP RAN1 LTE, R1-061481: User Throughput and Spectrum Efficiency for E-UTRA," InterDigital, May 2006.

[238] "3GPP RAN1 LTE, R1-060365: Extension of Uplink MIMO SC-FDMA with Preliminary Simulation Results," InterDigital, February 2006.

[239] "3GPP TSG-RAN WG1 Meeting no. 46, R1-062162: Uplink SDMA-MU-MIMO using Precoding for E-UTRA," InterDigital, August–September 2006.

[240] D. Grieco *et al.*, "Uplink single-user MIMO for 3GPP LTE," in *IEEE PIMRC*, pp. 1–5, September 2007.

[241] A. Goldsmith and S. Wicker, "Design challenges for energy-constrained *ad hoc* wireless networks," *IEEE Wireless Communications*, pp. 8–27, August 2002.

[242] D. D. Clark and D. L. Tennenhouse, "Architectural considerations for a new generation of protocols," *ACM SIGCOMM*, vol. 20, no. 4, pp. 200–208, August 1990.

[243] Z. Haas, "Design methodologies for adaptive and multimedia networks," *IEEE Personal Communications (guest editorial)*, vol. 39, no. 11, pp. 106–107, November 2001.

[244] C. S. Wijting and R. Prasad, "Cross-layer optimisation in wireless personal area networks," *Wireless Personal Communications, Kluwer Academic Publishers*, 2004 (Accepted for publication).

[245] B. Raman, P. Bhagwat, and S. Seshan, "Arguments for cross-layer optimizations in blue-tooth scatternets," *Symposium on Applications and the Internet*, pp. 176–184, January 2001.

[246] G. Pau, D. Maniezzo, S. Das, Y. Lim, J. Pyon, H. Yu, and M. Gerla, "A cross-layer frame-work for wireless LAN QoS support," *IEEE International Conference on Information Technology Research and Education (ITRE)*, Newark, New Jersey, USA, pp. 331–334, August 2003.

[247] "Protocol dictionary," http://www.javvin.com/protocolRTP.html.

[248] F. Fitzek, P. Seeling, and M. Reisslein, *Wireless Internet*, March 2004, no. ISBN: 0849316316, Electrical Engineering and Applied Signal Processing Series, ch. Video Streaming in Wireless Internet.

[249] J. Klaue, J. Gross, H. Karl, and A. Wolisz, "Semantic-aware link layer scheduling of MPEG-4 video streams in wireless systems," in *Proceedings of Applications and Services in Wireless Networks (AWSN)*, Telecommunication Networks Group Technical University of Berlin Einsteinufer 25, 10587 Berlin, Germany, July 2003.

[250] J. Holger Karl and A. Wolisz, "Throughput optimization of dynamic OFDM-FDMA systems with inband signalling," in *Proceedings of 2nd WiOpt'04 (Modeling and Optimization in Mobile, Ad Hoc and Wireless Networks)*, March 04.

[251] J. Jang and K. Bok Lee, "Transmit power adaptation for multiuser OFDM systems," *IEEE Journal on Selected Areas in Communications*, vol. 21, no. 2, pp. 171–178, February 2003.

[252] J. Gross, H. Karl, and A. Wolisz, "On the effect of inband signaling and realistic channel knowledge on dynamic OFDM-FDMA systems," *Proceedings of European Wireless 2004, Barcelona, Spain*, February 2004.

[253] V. Kawadia and P. Kumar, "A cautionary perspective on cross layer design," *IEEE Wireless Communication Magazine*, July 2003.

[254] Q. Wang and M. A. Aby-Rgheff, "Cross-layer signalling for next-generation wireless systems," *IEEE Wireless Communications and Networking Conference*, pp. 1084–1089, March 2003.

[255] "Internet doc: https://www.ist-winner.org," https://www.ist-winner.org.

[256] M. C. X. Yu, G. Chen and X. Gao, "Toward beyond 3G: The FuTURE project in China," *IEEE Communications Magazine*, January 2005.

[257] S. Abeta, H. Atarashi, and M. Sawahashi, "Broadband packet wireless access incorporating high-speed IP packet transmission," in *VTC fall*, vol. 4, pp. 844–848, September 2002.

[258] J. Mitola III, "Cognitive radio: An integrated agent architecture for software defined radio," PhD Thesis; KTH Royal Institute of Technology, Technical Report, 2000.

[259] F. B. Frederiksen and R. Prasad, "Strategies for improvement of spectrum capacity for WiMAX cellular systems by Cognitive Radio Technology supported by Relay Stations," in *Proceedings of International OFDM Workshop*, 2007.

[260] F. K. Jondral, "Software-defined radio — Basics and evolution to cognitive radio," Eurasip Journal on Wireless Communication and Networking, 2005.

[261] S. Haykin, "Cognitive radio: Brain-empowered wireless communications," *IEEE JSAC*, February 2005.

[262] I. F. Akyildiz *et al.*, "NeXt generation/dynamic spectrum access/cognitive radio wireless networks: A survey," *Computer Networks*, vol. 12, no. 2, April 2006.

[263] H. Lee, "3g lte and imt-advanced service," http://mmlab.snu.ac.kr/links/hsn/workshop/hsn2006/document/2.24.Fri/8-2.pdf, February 2006.

[264] S. Kumar, G. Costa, S. Kant, F. B. Frederiksen, N. Marchetti and P. Mogensen, "Spectrum sharing for next generation wireless communication networks," in *Proceedings of 1st International Workshop on Cognitive Radio and Advanced Spectrum Management (CogART)*, February 2008.

[265] "Wireless Networks That Build Themselves," http://www.sciencedaily.com/releases/2008/03/080311200326.htm, March 2008.

[266] "Is ZigBee best for low power wireless networks?" http://www.electronicsweekly.com/Articles/2008/04/16/43539/is-zigbee-best-for-low-power-wireless-networks.htm, April 2008.

[267] K. A. Townsend *et al.*, "Recent advances and future trends in low power wireless systems for medical applications," in *International Database Engineering and Application Symposium (IDEAS)*, 2005.

[268] K. D. Wise *et al.*, "Wireless implantable microsystems: High-density electronic interfaces to the nervous system," *Proceedings of IEEE*, vol. 92, pp. 76–97, January 2004.

Index

$\Delta > 2$, 305
2G, 1, 8
3G, 1, 5, 8, 383
3GPP, 5
4G, 4, 383

additive noise, 217
additive white gaussian noise, 253, 279
Alamouti, 224, 228, 230, 267, 276, 306, 308, 331, 336, 355–357, 362, 363, 372, 375
all-IP-based network, 7
angle diversity, 255, 256
angle spread, 227, 229, 230, 256
angular diversity, 255
angular spread, 234, 255–257, 259, 260, 263, 264
antenna correlation, 231, 256
antenna separation, 211, 226
antenna spacing, 231, 325, 327
array factor, 235
array gain, 215, 216, 218, 225, 228, 231, 233–235, 239, 241, 242, 245, 252, 258, 263, 264, 268, 270
autocorrelation, 281
average spectral efficiency, 327, 345
average spectral efficiency loss, 327

beamforming, 232–234, 245, 249, 251, 252, 256, 257, 262, 263, 374
bit error probability, 249
block coding, 308
bluetooth, 4
broadband channel, 276

cancelation, 314, 315
capacity, 216, 219, 232, 233, 244, 245, 247, 275, 276, 300, 377, 378, 380, 386
cell throughput, 389
cellular, 1, 234, 263, 305, 328, 354, 377
channel capacity, 216, 256, 297, 302
channel coding, 14, 39, 229, 230, 234, 300, 333, 340, 345, 348, 349, 352, 353
channel correlation, 234
channel covariance matrix, 256
channel estimation, 14, 151, 287, 292, 302, 320, 345
channel frequency response, 282, 315
channel frequency-selectivity, 271
channel impulse response, 269, 278
channel information, 218, 225, 301, 303
channel inversion, 9, 385
channel matrix, 297
channel response matrix, 297
channel transfer function, 269
circulant channel, 336
circular convolution, 278
closed-loop transmit diversity, 262
co-channel interference, 208
coding gain, 213, 214
coherence bandwidth, 210, 229, 310, 371
coherence distance, 211, 226
coherence time, 210, 223, 225, 291
coherent detection, 151
companion matrix, 310
computational complexity, 227, 265, 271, 281, 288, 290, 356, 375

convolutional coding, 266, 286, 291, 319, 345
cooperative communication, 378
correlation matrix, 281, 290
COST231-Hata, 18
covariance matrix, 217, 237, 253, 256, 280, 281
coverage, 226, 305, 328, 388, 389
cross layer design, 378
cross-layer optimization, 377, 381, 386
cyclic convolution, 284, 294
cyclic delay, 273, 274, 280, 282, 287, 289, 294

DAPSK, 172
data rate, 208, 211, 228, 231, 233, 266, 292, 356, 373, 374
decision-directed channel estimation, 161
delay, 226, 228, 229, 267, 275, 384
delay diversity, 274
delay spread, 9, 11, 385
deterministic capacity, 244
differential detection, 162
directivity, 385
dispersion, 353, 355, 358, 360, 362
dispersion matrices, 353, 355, 358–361, 363–365, 373
diversity, 221, 227–229, 233, 239, 245, 266, 271, 272, 275, 276, 278, 281, 290, 292, 301, 305, 319, 325, 329, 353, 359, 385
diversity branch, 278, 280, 283
diversity combining, 268, 271, 277, 279, 280, 285
diversity gain, 207, 211, 213, 218, 220, 226–229, 231, 234, 239–242, 249, 252, 270, 271, 275, 321, 342, 345, 388
Doppler, 25
downlink, 233

eigen-beamforming, 374
eigen-value decomposition, 336
eigenvalue spread, 299
eigenvalues, 256, 280, 297, 298
eigenvectors, 256, 280, 281

equalization, 301, 337, 385
equalizer, 332
ergodic capacity, 244, 245, 276, 300
error rate, 213, 297
exponential model, 320

fading, 208, 213, 214, 216, 219, 221, 235, 240, 249, 257, 263
fading correlation, 237
fast fading, 333
fast varying channel, 338
FDE, 109
feedback, 217, 218, 225, 228, 262, 304, 374, 378
first generation, 1
flat fading, 212, 230, 266, 267, 271, 355
flat fading channel, 277
flexibility, 377, 384
frequency correlation, 302
frequency dispersion, 363
frequency diversity, 210, 214, 228–230, 234, 265, 266, 268, 271, 274, 275, 289, 340
frequency diversity gain, 352, 371
frequency interleaving, 266
frequency overlay, 389
frequency selective channel, 272, 338, 345
frequency selectivity, 272, 274, 277, 285, 333, 339, 340, 345, 346, 348, 349, 356, 369, 373, 375
frequency-selective channel, 301
frequency-selective fading, 205, 227, 266

Gaussian noise, 331, 337
guard interval, 269

handoff, 388, 389
handover, 264
Hassibi, 356, 364
Hata-Okumura, 18
Hermitian, 280, 297, 298
heterogeneity, 377, 386
HomeRF, 1
HPA, 173

ICI, 50, 129
IMT-Advanced, 386
instantaneous capacity, 276
inter-element distance, 235
interference, 233, 254, 377, 388
interference cancelation, 309, 386
interference reduction, 209
interleaving, 14, 229, 230, 234, 289
ISI, 50
iterative decoding, 7

Jafarkhani, 365, 375
Jakes, 25

latency, 14, 228, 240, 253, 329, 383
linear receiver, 306, 307
link adaptation, 9, 384
link reliability, 226, 231, 378, 388, 389

macro diversity, 263, 264
macrocell, 234, 257
maximum delay spread, 338, 340, 359,
    367, 368
maximum doppler shift, 291
maximum likelihood detection, 292,
    306
MC-CDMA, 93
microcell, 257, 327
mobility, 329, 333, 356, 384, 391
Moore-Penrose pseudo-inverse, 312,
    314, 316
multi-antenna, 205, 211, 225, 231,
    233, 329
multi-carrier, 4, 9, 33, 234, 386, 387
multimedia, 377
multimedia communication, 3
multipath, 21
multipath channel, 229
multipath diversity, 266, 356
multipath fading, 226
multiple access, 10, 234, 263, 386, 387
multiple antenna, 13, 375, 385
multiplexing gain, 208, 218, 231, 232,
    353, 359, 363, 385
multiplexing-diversity trade-off, 220
multi-user, 7, 10, 232, 233, 302, 328,
    385, 391
multi-user diversity, 263

noise power density, 287
nonlinear receiver, 307, 313, 319
normalized coherence bandwidth,
    338, 369
normalized Doppler frequency,
    338, 340, 369
nulling, 314, 316, 317

OFDM, 31, 40
OFDM–CDMA, 63
OFDM–TDMA, 61
OFDMA, 62, 65
OFDMA-FSCH, 78
OFDMA-SSCH, 87
open loop transmit diversity, 240, 242,
    244, 263
orthogonal block coding, 307
orthogonal coding, 241
orthogonal design, 357
outage capacity, 245, 248, 249, 251,
    258, 263, 276, 300
outage channel capacity, 322
outage efficiency, 325
outage spectral efficiency, 307, 322,
    323, 325

packet scheduling, 387, 389
PAPR, 173
path loss, 18
phase noise, 129
picocells, 234, 255
pilots, 263
polarized antennas, 231
post-DFT combining, 227, 271
post-IDFT beamforming, 253
power adaptation, 11
power consumption, 211, 226, 227,
    377, 378, 383, 388, 389
power delay profile, 320
pre-coding, 374
pre-DFT combining, 227, 268, 288
pre-DFT diversity combining, 289
pre-IDFT beamforming, 253
pre-whitening, 306
processing delay, 352

QAM, 26, 28
quasi-orthogonal codes, 315

radio resource management, 388, 389
Rayleigh, 208, 213
Rayleigh-distributed component, 326
Rayleigh fading, 219, 285, 326
receive correlation, 325
receive correlation matrix, 323
receive diversity, 207, 208, 210, 226–230, 241, 242, 244, 246, 281
receiver, 43
reconfigurability, 377
reliability, 213, 222, 260, 303, 322, 345
retransmission diversity, 263
reuse, 377, 382
Ricean, 219, 231
Ricean channel, 288
Ricean fading, 285

SC-FDMA, 116
scalability, 377
scattering, 208, 212, 226, 231, 292, 327
SCFDE, 112
scheduling, 9, 380, 384, 385
SDNR, 183
sectorization, 385
self-interference, 217
shadowing, 19
signaling overhead, 303
single-carrier, 109
single-tap equalization, 301
small scale fading, 19
software-defined radio, 389
space diversity, 211, 226, 228, 229
spatial coding, 354
spatial correlation, 234, 256, 257, 263, 275, 302, 307, 323, 325, 327, 348
spatial covariance matrix, 256
spatial diversity, 215, 224, 230, 233, 239, 265, 266, 274, 276, 299, 301, 303
spatial fading, 226, 230
spatial modes, 298
spatial multiplexing, 208, 211, 213, 374
spatial multiplexing gain, 220
spatial rate, 345, 365
spatial separation, 290, 323, 327

spectral efficiency, 11, 12, 221, 226, 244, 268, 292, 303–305, 325, 377, 378, 388, 389
specular component, 326, 327
sub-carrier bandwidth, 310
sub-carrier combining, 268, 270
sub-carrier hopping, 253
sub-carrier spacing, 310
synchronization, 127, 259, 269, 338, 345
system capacity, 213, 265, 276, 292, 297–300, 304, 373, 386
system throughput, 328

Tarokh, 258, 259
thermal noise, 212, 318
throughput, 213, 231, 292, 374, 383
time diversity, 209, 214, 226, 229, 276
time selective channel, 338, 340
time selectivity, 333, 340, 349, 353, 356, 369
total degradation plot, 194, 202
trade-off diversity-multiplexing, 333, 341, 345, 375
transmit correlation, 231, 325
transmit correlation matrix, 323
transmit diversity, 207, 210, 228–230, 233, 234, 239, 240, 242, 245, 247, 249, 251–253, 255–260, 262–264, 267, 276, 334, 338, 341, 355, 356, 374
transmitter, 40

uncorrelated channel matrix, 323
unitary matrix, 280
user mobility, 262
user throughput, 389

velocity, 291, 338, 340, 368, 369
Viterbi decoder, 330

waterfilling, 218
wideband channel, 266
wideband frequency selective channel, 221, 292, 304
wireless broadband access, 377
wireless channel, 4, 205, 207, 228, 233, 273, 295, 301, 307, 309